ADVANCES IN FREE-RADICAL CHEMISTRY

VOL. IV

ADVANCES IN FREE-RADICAL CHEMISTRY

VOLUME IV

Edited by

G. H. Williams

1972

ACADEMIC PRESS NEW YORK/LONDON

ACADEMIC PRESS INC.
111 Fifth Avenue
New York, New York 10003

Library of Congress Catalog Card Number: 65–27405

Printed in Great Britain

PREFACE

This volume, the fourth in the series, contains four articles, one of which is rather longer than has been customary in previous volumes. This article by Dr. J. A. Howard, is on Absolute Rate Constants for Reactions of Oxyl Radicals, and a substantial review was necessary to afford adequate coverage of this topic. It is hoped that readers of this series will find this article a valuable addition to the literature on this hitherto inadequately documented subject. It is also hoped that reactions of other types of radicals will be dealt with in articles in subsequent volumes. The other chapters in this volume deal with aspects of free-radical chemistry in respect of which reviews are now appropriate. Dr. G. Sosnovsky and Mr. M. J. Rawlinson have reviewed Free-Radical Reactions in the presence of Metal Ions, Dr. A. Nechvatal has written on Allylic Halogenation and Dr. K. C. Bass and Mr. P. Nababsing have reviewed Homolytic Substitution Reactions of Heteroaromatic Compounds in Solution. In all these articles the authors have attempted to review the literature as comprehensively as possible up to mid-1970.

I should like to thank all the authors for their patience and co-operation, and particularly for their ready acceptance of editorial suggestions. I am also greatly indebted to the staff of Logos Press and particularly to Dr. Judith Clare and Mrs. Diane Baird, who prepared the subject and author indexes respectively.

G. H. Williams.

Bedford College,
London.
May 1971.

CONTRIBUTORS

Dr. K. C. Bass and Dr. P. Nababsing, Department of Chemistry, The City University, London.

Dr. J. A. Howard, Division of Chemistry, National Research Council of Canada, Ottawa, Ontario, Canada.

Dr. A. Nechvatal, Department of Chemistry, University of Dundee, Dundee, Scotland.

Dr. G. Sosnovsky, Department of Chemistry, The University of Wisconsin-Milwaukee, Milwaukee, Wisconsin, U.S.A.

Professor D. J. Rawlinson, Department of Chemistry, Western Illinois University, Macomb, Illinois, U.S.A.

CONTENTS

1

HOMOLYTIC SUBSTITUTION REACTIONS OF HETEROAROMATIC COMPOUNDS IN SOLUTION

K. C. BASS *and* P. NABABSING

Department of Chemistry, The City University, London, E.C.1.

A. INTRODUCTION

This chapter is a survey of the investigations which have been made on homolytic substitution reactions of fundamental heterocyclic systems in the liquid phase. The reactions of the systems with aryl and alkyl radicals are discussed, and, where the information is available, the observed homolytic reactivities of individual heterocyclic systems are compared with the reactivities predicted from theoretical calculations of free valence numbers and atom localisation energies.

The homolytic substitution reactions of pyridine and its derivatives, particularly phenylation, have been widely studied, but other heterocyclic systems have received less attention. Recently the homolytic reactivity of the thiazole series has been studied in some detail (see Section E. 2).

The homolytic substitution reactions of heteroaromatic compounds in the liquid phase have been reviewed previously by Norman and Radda (1963), who made a comprehensive survey of the reactions of heterocycles with aryl, alkyl and hydroxyl radicals in solution. The effect of substituents on homolytic substitution in the pyridine series has been reviewed by Abramovitch and Saha (1966a).

B. Theoretical Considerations

Quantum-mechanical calculations (Coulson, 1955) indicate that the replacement of a carbon–hydrogen centre in an aromatic ring by a nitrogen atom should lead to an increase in the reactivity of the system. Some evidence for the correctness of this prediction can be obtained from the work of Szwarc and his colleagues (Szwarc and Binks, 1959) on the methyl radical affinities of aromatic hydrocarbons and nitrogen heterocycles. The results obtained for some aromatic hydrocarbons and their heterocyclic analogues are given in

Table 1

Methyl affinities of aromatic hydrocarbons and nitrogen heterocycles (Szwarc and Binks, 1959)

Compound	Methyl affinity
Benzene	1
Pyridine	3
Pyrazine	*ca.* 18
Naphthalene	22
Quinoline	29
Isoquinoline	36
Anthracene	820
Acridine	430
Phenazine	*ca.* 250

Table 1. It can be seen that the reactivity increases in both the monocyclic and bicyclic series when a carbon–hydrogen centre is

replaced by a nitrogen atom to give the corresponding heterocyclic compound. Also, pyrazine is more reactive than pyridine. However, in the tricyclic series, acridine is less reactive than anthracene because the highly reactive *meso*-carbon atom at position 9 has been replaced by a nitrogen atom. Phenazine is even less reactive as both *meso*-carbon atoms in anthracene have been replaced by nitrogen atoms. The reactivity of acridine is approximately half that of anthracene, and this observation suggests that the addition of methyl radicals takes place only on the carbon–hydrogen centre and not on the nitrogen atom. The comparatively low reactivity of phenazine is in agreement with this hypothesis. The presence of two nitrogen atoms in phenazine is thought to activate the remaining positions, and therefore the decrease in reactivity from acridine to phenazine is less than twofold (Szwarc and Binks, 1959).

Quantum-mechanical calculations have also been used to predict the reactivities of each position in aromatic and heteroaromatic molecules towards homolytic attack. The values most commonly used in this respect are the free valence numbers (Coulson, 1946, 1947; Pullman and Pullman, 1958; Williams, 1960) and the atom localisation energies (Wheland, 1942; Williams, 1960). The free valence concept is based on the theoretical treatment of the isolated molecule, whereas the atom localisation energy takes into account the transition state, i.e. the activated complex formed between the molecule and the free radical.

According to the molecular orbital theory, the mobile bond-order $p_{r,s}$ between the nuclear carbon atoms r and s is defined by

$$p_{r,s} = \Sigma c_{r,j} \cdot c_{s,j}$$

where $c_{r,j}$ and $c_{s,j}$ are the coefficients of the molecular orbitals of the π-electrons in the bonds r—j and s—j. The total bonding N_r exhibited by the atom r is then

$$N_r = \Sigma p_{r,s} + C_r$$

where C_r is the contribution from the σ-electrons. The free valence number F_r of atom r is then given by

$$F_r = N_{\max} - N_r$$

where $N_{\max}$ is the maximum bonding that a carbon atom of the type r can obtain. The value of $N_{\max}$ is usually taken as $3 + \sqrt{3}$. The free valence number F_r is a measure of the amount of additional bonding in which the atom r could partake, and hence provides an index for homolytic reactivity. The values of free valence numbers for some aromatic hydrocarbons and nitrogen heterocyclic compounds, as calculated by several authors, are given in Tables 2 and 3.

TABLE 2

Free valence numbers for benzene, pyrrole, pyridine and naphthalene

Compound	N_{max}	Free valence numbers (F_r) Position 1	2	3	4	Ref.
Benzene	4·732	0·398	—	—	—	1
Pyrrole	4·732	0·623	0·453	0·404	—	2
Pyrrole	4·732	0·409	0·505	0·481	—	2
Pyridine	4·414	0·110	0·091	0·079	0·086	3
Pyridine	4·414	0·102	0·092	0·083	0·081	3
Pyridine	4·732	—	0·399	0·398	0·404	4
Pyridine	4·732	—	0·515	0·488	0·438	4
Pyridine	4·732	—	0·387	0·345	0·360	5
Pyridine	4·732	—	0·443	0·338	0·402	6
Naphthalene	4·732	0·448	0·400	—	—	1

References: (1) Hey and Williams (1953). (2) Brown (1955). (3) Davies (1955). (4) Brown and Heffernan (1956). (5) Sandorfy and Yvan (1950). (6) Metzger and Pullman (1948).

TABLE 3

Free valence numbers for quinoline and isoquinoline

Position	Quinoline (Ref. 1)	Quinoline (Ref. 2)	Isoquinoline (Ref. 2)
1	—	—	0·51
2	0·344	0·48	—
3	0·290	0·34	0·45
4	0·356	0·47	0·39
5	0·338	0·40	0·40
6	0·290	0·35	0·39
7	0·296	0·36	0·35
8	0·331	0.38	0·41

References: (1) Sandorfy and Yvan (1950). (2) Buu-Hoï and Daudel (1949).

The variations in the values of F_r for a particular compound are mainly due to the different values assigned to N_{max}, or to the electronegativity parameter h, which is used to account for the difference in electronegativity when a nitrogen atom replaces a carbon atom in the aromatic system. The values also differ, depending on whether or not allowance is made for the overlapping of adjacent orbitals, e.g. the values obtained by Davies (1955) for pyridine (see Table 2).

The atom localisation energy A_r is a measure of the energy required to localise one electron of the π-electron system in an aromatic molecule at the point of homolytic attack (Wheland, 1942). The formation of the intermediate adduct in a homolytic aromatic substitution may be regarded as the sum of two processes: one, the localisation of an electron at the point of attack, and the other, the pairing of this electron with that of the attacking radical. The energy change in the latter step would be independent of the nature of the aromatic nucleus and constant for a particular attacking radical. The energy of formation of the intermediate is therefore related to the atom localisation energy. Coulson (1955) has shown that a plot of the logarithm of methyl affinity per reactive position, which is a measure of the rate of substitution at that position, against the relevant localisation energy for a series of unsubstituted aromatic hydrocarbons, is linear. The assumption was made that only the most reactive positions in the aromatic hydrocarbon take part in the reaction, e.g. only the α-positions in naphthalene were considered, so that the methyl affinity of naphthalene was divided by four.

Brown (1956) has used the results obtained by Hey and Williams (1953) on the phenylation of pyridine to calculate the atom localisation energies of the different positions in pyridine. The values are given in Table 4, together with the partial rate factors for each position.

TABLE 4

Atom localisation energies for pyridine (Brown, 1956)

Position	A_r	Partial rate factor
2	2·510	1·91
3	2·542	0·86
4	2·536	1·01

Sandorfy and Yvan (1950) have calculated the atom localisation energies of all of the positions in quinoline, and the results indicate that the order of reactivity for homolytic substitution should be $2>3>6>7>8>4>5$.

The different values assigned to the electronegativity parameter h lead to different values of A_r. Moreover, the protonated forms of nitrogen heterocyclic compounds have different reactivities from those of the neutral molecules. Both the total reactivity of the molecule and the positional reactivities are altered. Brown and his co-workers have calculated the radical reactivities of all of the positions in quinoline (Brown and Harcourt, 1959), isoquinoline (Brown and Harcourt, 1960) and pyridine (Brown and Heffernan, 1956), for both the neutral and protonated forms. Their calculations of the atom localisation energies indicate that the positional reactivities of quinoline and the quinolinium ion are in the order $5>4>8>2>7>6>3$ and $2>5>4>8>7>6>3$, respectively. For isoquinoline and the isoquinolinium ion the orders of reactivities are $1>8>5>4>3>6$ and $1>3>8>5>4>6>7$. The values of free valence and atom localisation energies for pyridine and the pyridinium ion, as calculated by Brown and Heffernan (1956), are given in Table 5.

TABLE 5

Free valence numbers and atom localisation energies for pyridine and the pyridinium ion (Brown and Heffernan, 1956)

	Pyridine ($h=0{\cdot}5$)		Pyridinium ion ($h=2{\cdot}0$)	
Position	A_r	F_r	A_r	F_r
2	2·5124	0·399	2·2838	0·515
3	2·5381	0·398	2·5604	0·488
4	2·5374	0·404	2·5418	0·438

A low value for the atom localisation energy corresponds to a high value for the free valence number and indicates a high positional reactivity. The figures given in Table 5 show that the orders of reactivities indicated by the values of F_r and A_r are not in complete agreement. Thus for pyridine the atom localisation energies indicate the positional reactivities to be in the order $2 > 4 > 3$, whereas the free valence numbers predict the order to be $4 > 2 > 3$, and for the ion the orders are $2 > 4 > 3$ and $2 > 3 > 4$, respectively.

C. Homolytic Substitution Reactions of Six-Membered Nitrogen Heterocycles

1. *Pyridine Series: Arylation*

The homolytic phenylation of pyridine has been extensively investigated, but other arylation reactions of pyridine have received less attention. The earlier work has been comprehensively reviewed by Norman and Radda (1963), who have discussed both the quantitative and preparative studies which have been made of the homolytic arylation of pyridine. The quantitative aspects have also been discussed by Abramovitch and Saha (1966a).

Norman and Radda (1963) have given a summary of the sources of radicals which have been used for the arylation of pyridine and other heterocyclic compounds. The most commonly used sources are the aryldiazonium salts and diaroyl peroxides. The first investigation of the phenylation of pyridine was made by Möhlau and Berger (1893), who used the thermal decomposition of benzenediazonium chloride in the presence of pyridine and aluminium trichloride to prepare 2- and 4-phenylpyridine. In 1929, Overhoff and Tilman reported the formation of the same products from the thermal decomposition of benzoyl peroxide in pyridine. Substitution at positions 2 and 4 was also observed in the arylation of pyridine, using aqueous solutions of *para*-substituted benzenediazonium salts in the presence of an organic base (the Gomberg-Hey reaction) to give *p*-chloro-, *p*-bromo-, *p*-ethoxy- and *p*-carboxy-phenyl radicals (Butterworth, Heilbron and Hey, 1940). With phenyl and *o*-, *m*- and *p*-nitrophenyl radicals, produced from the corresponding diazonium salts, 2-, 3- and 4-arylpyridines were isolated (Haworth, Heilbron and Hey, 1940).

In much of the earlier work, the failure to isolate the 3-substituted pyridine was probably due to the limitations of the analytical methods used for the isolation and identification of the products. Consequently many of the earlier quantitative determinations of isomer ratios of products are not valid. More reliable quantitative results have been obtained with the advent of more sensitive and accurate physical methods of analysis, such as infrared and ultraviolet spectrophotometry and gas chromatography. A selection of the available data on the homolytic phenylation of pyridine is given in Table 6.

TABLE 6

Isomer ratios and relative reactivities for the homolytic phenylation of pyridine

Radical source	Temp. (°)	Isomer ratio (%) 2-	3-	4-	Relative reactivity (benzene = 1)	Ref.
Benzoyl peroxide	105	54	32	14	—	1
Lead tetrabenzoate	105	52	32·5	15·5	—	1
Phenyl iodosobenzoate	105	58	28	14	—	1
N-Nitrosoacetanilide	105	46	43	11	—	1
Phenylazotriphenyl-methane[a]	105	53	31	16	—	1
Electrolysis of benzoic acid	15–20	56	35	9	—	2
Photolysis of triphenylbismuth[b]	100	48	31	21	1·18[c]	3
Gomberg-Hey reaction[d]	40	52·4	29·6	18	1·14	4, 5
Benzenediazonium tetrafluoroborate	40	51·8	32·4	15·7	—	6

[a] Corrected values (see ref. 1). [b] Mean values for isomer ratio. [c] Measured at 80°. [d] Averaged values (see ref. 5).

References: (1) Hey, Stirling and Williams (1955). (2) Bunyan and Hey (1960). (3) Hey, Shingleton and Williams (1963a). (4) Abramovitch and Saha (1963). (5) Abramovitch and Saha (1964). (6) Abramovitch and Saha (1965).

Allowing for the differences in temperature at which the reactions were conducted, it can be seen from the results given in Table 6 that, with the exception of *N*-nitrosoacetanilide, the isomer ratio is hardly dependent on the source from which the phenyl radicals are derived.

Recent results for the isomer distribution in the homolytic phenylation of pyridine (Hey, Shingleton and Williams, 1963a; Abramovitch and Saha, 1964) (see Table 6) indicate that the order of positional reactivities is $2 > 4 > 3$, when allowance is made for the two pairs of equivalent sites corresponding to positions 2 and 3. Hey and Williams (1953) have reported atom localisation energies and free valence numbers for the different positions in pyridine, and also the rates of homolytic phenylation predicted by these values.

The order of reactivities predicted by the free valence numbers is $2>4>3$, whereas according to the atom localisation energies it should be $2>3>4$. The calculated rate of phenylation (3·36) was much higher than that actually observed (1·04) by Augood, Hey and Williams (1952). Brown (1956) has shown that by choosing a certain value for the electronegativity parameter, h, values can be obtained for the atom localisation energies which give the predicted order of reactivities as $2>4>3$, and also account for the partial rate factors found by Hey and Williams. Previous calculations (Metzger and Pullman, 1948; Sandorfy and Yvan, 1950) also gave the order of positional reactivities as $2>4>3$.

Abramovitch and Saha (1964, 1966b) have determined the isomer ratios and relative reactivities for the arylation of pyridine by substituted phenyl radicals produced from the corresponding diazonium salts. Their results are given in Table 7.

TABLE 7

Isomer ratios and relative reactivities for the homolytic arylation of pyridine (40°) (*Gomberg-Hey reaction*)

Radical	Isomer ratio (%)			Relative reactivity (benzene = 1)	Ref.
	2-	3-	4-		
Phenyl	52·4	29·6	18·0	1·14	1
o-Tolyl	52·1	30·7	17·2	1·72	1
p-Tolyl	56·0	28·2	15·8	1·44	1
o-Nitrophenyl	42·6	50·0	7·5	0·47	1
p-Nitrophenyl	44·6	42·7	12·9	0·78	1
o-Methoxyphenyl	49·3	35·0	15·6	1·27	2
p-Methoxyphenyl	57·5	26·3	16·2	1·30	2
o-Bromophenyl	43·2	46·3	10·5	0·70	2
p-Bromophenyl	53·3	33·4	13·3	0·87	2

References: (1) Abramovitch and Saha (1964). (2) Abramovitch and Saha (1966b).

Substituents which increase the nucleophilic character of the phenyl radical would be expected to effect a corresponding increase in the rate of arylation. The *o*- and *p*-tolyl and *o*- and *p*-methoxyphenyl radicals, accordingly, gave higher values for the relative reactivity than that obtained for the phenyl radical. On the other hand, the *o*- and *p*-nitrophenyl and *o*- and *p*-bromophenyl radicals, which would be expected to be more electrophilic than phenyl radicals, gave lower values. The inductive effect of a substituent is more pronounced when it is in the *ortho-* than when it is in the *para-* position. Thus the *o*-tolyl radical gave a higher value for the relative reactivity than that for the *p*-tolyl radical, and the *o*-nitrophenyl radical gave a lower value than that for the *p*-nitrophenyl radical. The higher relative reactivity for the *p*-methoxyphenyl radical, compared with that for its *ortho*-isomer, was thought by Abramovitch and Saha (1966b) to be due to the mesomeric effect of the methoxy-group, which would be larger when the substituent is in the *para-* position. The overall decrease in reactivity due to the electrophilic character of a radical should be accompanied by an increase in the proportion of the 3-substituted isomer produced. This is evident for the *o*-nitrophenyl and *o*-bromophenyl radicals, which are the least reactive of all of the radicals in the series. For these radicals, position 3 in pyridine was the most reactive, as can be seen from the results in Table 7.

The homolytic phenylation of the picolines has been studied by two groups of workers. Bonnier *et al.* (Bonnier and Court, 1965; Bonnier, Court and Gelus, 1966, 1967; Bonnier, Court and Fay, 1967) have used the thermal decomposition of benzoyl peroxide at 70°, and Abramovitch and Saha (1966c) have used the decomposition of benzenediazonium salts at 40°, as their sources of phenyl radicals. Their results are given in Table 8. The corresponding results for pyridine (Abramovitch and Saha, 1964) are included for comparison.

From Table 8 it can be seen that the orders of positional reactivities obtained for the phenylation of 4-picoline by the two groups of workers are not in agreement. Bonnier, Court and Gelus (1966, 1967) have calculated the free valence number and the partial rate factor for each of the positions in the three picolines. The calculated partial rate factors are in agreement with the experimental results. The effect of the methyl substituent is to increase the reactivities of the carbon atoms which are *ortho-* or *para-* to it. Thus, position 3 is more reactive than position 5 in 2-picoline and position 2 is more reactive than position 6 in 3-picoline. Also, position 5 in 2-picoline, which is *para-* to the methyl group, is more reactive than position 4, whereas in pyridine the order is reversed.

The calculations mentioned above (Bonnier, Court and Gelus, 1966, 1967) also showed that in a methyl-substituted pyridine the partial rate factor of a particular position in the substituted molecule is

TABLE 8

Isomer ratios and relative reactivities for the homolytic phenylation of pyridine and 2-, 3- and 4-picoline

Compound	Isomer ratio (%)					Relative reactivity (benzene = 1)	Ref.
	2-	3-	4-	5-	6-		
Pyridine	26·2	14·8	18·0	14·8	26·2	1·14	1
2-Picoline	—	31	15	20	34	1·25	2, 3, 4
3-Picoline	41·5	—	26	9·5	23	1·80	3
3-Picoline	43·3	—	28·7	6·9	21·1	1·39	5
4-Picoline	22·4	27·6	—	27·6	22·4	1·39	5
4-Picoline	27·5	22·5	—	22·5	27·5	1·30	2, 3, 6

References: (1) Abramovitch and Saha (1964). (2) Bonnier, Court and Gelus (1966). (3) Bonnier, Court and Gelus (1967). (4) Bonnier, Court and Fay (1967). (5) Abramovitch and Saha (1966c). (6) Bonnier and Court (1965).

2·3, 1·0 or 1·3 times that of its value in pyridine, depending on whether the position is *ortho*-, *meta*- or *para*- to the methyl group. However, the substituent itself eliminates one of the reactive sites, which causes a decrease in the total reactivity. This effect should manifest itself most strongly in 2-picoline, since position 2 is the most reactive site in pyridine. The order of the total reactivities of the picolines would therefore be the reverse of the positional reactivities in pyridine, so that 3-picoline > 4-picoline > 2-picoline, as was found from the experimental results (see Table 8) (Bonnier, Court and Gelus, 1967).

The increase in the total reactivity of pyridine when it is phenylated with benzoyl peroxide in glacial acetic acid was first reported by Dou and Lynch (1965, 1966a). Molecular orbital calculations by Brown and Heffernan (1956) predict that the reactivity of the pyridinium ion towards homolytic attack should be about 100 times that of the unprotonated molecule. Dou and Lynch (1965) found the value to be very much lower (0·66), from the competitive phenylation of mixtures of pyridine and nitrobenzene with benzoyl peroxide in the absence and presence of acetic acid. Bonnier and Court (1967) have shown that protonation is not complete when the ratio of pyridine to acetic acid is that used by Dou and Lynch (1965, 1966a). Through a study of the conductance of solutions of pyridine in mixtures of acetic acid and concentrated hydrochloric acid at various dilutions, protonation was found to be complete when suitable concentrations of the acids were used (Bonnier and Court, 1967). The ratio of isomeric phenylpyridines obtained in the homolytic phenylation of pyridine did not alter with further addition of hydrochloric acid. The ratios of isomers obtained by Bonnier and Court (1967) for the phenylation of pyridine, picolines and 2,6-lutidine in their protonated and unprotonated forms, using benzoyl peroxide as the radical source, are given in Table 9.

TABLE 9

Isomer ratios for the homolytic phenylation of pyridine, 2-, 3- and 4-picoline and 2,6-lutidine in acidic and non-acidic solution (Bonnier and Court, 1967)

Compound	Medium	Isomer ratio (%)				
		2-	3-	4-	5-	6-
Pyridine	Non-acidic	26·8	14·6	17·1	14·6	26·8
Pyridine	Acidic	32·3	2·3	30·8	2·3	32·3
2-Picoline	Non-acidic	—	31	15	20	34
2-Picoline	Acidic	—	8	41	4·5	46
3-Picoline	Non-acidic	41·5	—	26	9·5	23
3-Picoline	Acidic	41·5	—	34·5	1·5	22·5
4-Picoline	Non-acidic	27	23	—	23	27
4-Picoline	Acidic	43	7	—	7	43
2,6-Lutidine	Non-acidic	—	43·5	13	43·5	—
2,6-Lutidine	Acidic	—	17·7	64·6	17·7	—

Competitive reactions, using mixtures of benzene and pyridine, showed that although the reactivities of positions 2 and 4 in pyridine towards phenyl radicals were enhanced in acidic solution, that of position 3 remained practically unchanged (Bonnier and Court, 1967). The effect of the change in positional reactivities on the isomer ratios is evident from the results given in Table 9. The ratios of 2- to 3- and 4- to 3- isomers obtained in the phenylation of pyridine in acidic solution are higher than those obtained in non-acidic solution.

The homolytic phenylation of the picolines and the lutidines in acidic and non-acidic solution has also been studied by Vernin *et al.* (Vernin, Dou and Metzger, 1967b; Vernin and Dou, 1967). The relative reactivity of the heterocyclic compound and the partial rate factors of all of its positions have been obtained in each case, and the atom localisation energies and free valence numbers for the protonated and unprotonated forms of the bases have been calculated to account for the isomer ratios obtained. The relative reactivities obtained by Vernin and Dou (1967) for the homolytic phenylation of pyridine and some of its methyl derivatives in acidic and non-acidic solution are given in Table 10.

TABLE 10

Relative reactivities for the homolytic phenylation of pyridine, 2-, 3- *and* 4-*picoline and* 2,6- *and* 3,5-*lutidine in acidic and non-acidic solution* (Vernin and Dou, 1967)

Compound	Relative reactivities (benzene = 1)	
	Non-acidic	Acidic
Pyridine	0·97	1·61
2-Picoline	0·9	1·9
3-Picoline	1·45	7·0
4-Picoline	1·25	2·95
2,6-Lutidine	2·0	1·5
3,5-Lutidine	2·2	10·0

The relative reactivities are increased in acidic solution, with one exception: in 2,6-lutidine the protonated nitrogen atom reduces the effectiveness with which the adjacent methyl groups increase the reactivities of positions 3 and 5. Therefore the reactivities of these positions are decreased when the reaction is carried out in acidic solution. The extent of this decrease is not compensated by the increase in the reactivity of position 4 due to protonation of the molecule, as was shown from the calculation of the partial rate factors of positions 3 and 4 by Vernin and Dou (1967). The net result is an increase in the proportion of the 4-isomer, but a decrease in the overall reactivity of the molecule (see Tables 9 and 10). The high reactivity of 3,5-lutidine is due to the effects of the methyl groups and of protonation in increasing the reactivities of all three unsubstituted positions, 2, 4, and 6.

The results obtained by Vernin and Dou (1968b) for the arylation of pyridine and its methyl derivatives by *p*-bromophenyl and *p*-nitrophenyl radicals show that protonation is more effective in increasing the proportion of the 2- or 6-arylpyridine formed when the substituting radical is *p*-nitrophenyl. Vernin and Dou (1968b) postulated that the higher stability (or lower reactivity) of the *p*-nitrophenyl radical would cause it to be more susceptible to changes in the free valence of pyridine arising from protonation.

The homolytic phenylation of pyridine-metal complexes dissolved in *N,N*-dimethylformamide, using *N*-nitroso-*sym*-diphenylurea at 25° as the radical source, has been studied by Gritter and Godfrey (1964), whose results are given in Table 11. In each case the proportion of 3-phenylpyridine formed from the complex was lower and that of 4-phenylpyridine was higher than the corresponding proportions from pyridine itself. In the majority of cases the proportion of 2-phenylpyridine was also increased. These changes are similar to those observed in the homolytic phenylation of pyridine in acidic solution (Bonnier and Court, 1967) (see Table 9). The results are explained on the basis of back donation of electrons to the pyridine ring by the metal ions (Gritter and Godfrey, 1964).

TABLE 11

Isomer ratios for the homolytic phenylation of pyridine-metal complexes (Gritter and Godfrey, 1964)

Compound	Isomer ratio (%)		
	2-	3-	4-
Pyridine (Py)	41·1	39·1	19·8
$Co(Py)_4Cl_3$	66·7	13·1	20·2
$Cr(Py)_3Cl_3$	45·0	19·0	36·0
$Zn(Py)_2(SCN)_2$	46·0	23·0	31·8
$Mn(Py)_4(SCN)_4$	34·9	32·8	32·9
$Ni(Py)_4(SCN)_2$	23·8	34·5	41·8
$Cd(Py)_2(SCN)_2$	43·3	33·6	23·6

Dyall and Pausacker (1961) have investigated the homolytic phenylation of pyridine-*N*-oxide, using diazoaminobenzene at 131° or 181° as the radical source. The ratio of isomers obtained indicated that the reactivities of the nuclear positions are in the order $2 > 4 > 3$, which is also that predicted from calculations of atom localisation energies for the *N*-oxide (Barnes, 1959). 2-Phenylpyridine-*N*-oxide was the major product and it accounted for 71–81% of the total phenylation products, whereas the 3-substituted isomer comprised only 5·6–9·6% of that total. The phenylpyridines were also found among the by-products of the reaction.

There are few reports of the arylation of pyridine and its derivatives by radicals other than phenyl or substituted phenyl. The arylation of pyridine by 2-naphthyl radicals, produced from either the corresponding peroxide (Hey and Walker, 1948) or the triazen (Elks and Hey, 1943), gave all three isomeric (2-naphthyl)-pyridines. Only one isomer was isolated from the arylation of pyridine by 1-naphthyl radicals, produced from di-1-naphthoyl peroxide (Hey and Walker, 1948). 3-Pyridyl radicals (produced by the Gomberg-Hey reaction) substituted in pyridine at all three positions and the yields of products indicated the order of reactivities to be $2 > 4 > 3$ (Frank and Crawford, 1958). With 4-picoline and 4-ethylpyridine, substitution by 3-pyridyl radicals gave both the 2- and the 3-substituted products (Frank and Crawford, 1958). Position 3 was more reactive than position 2, owing to the activation of the adjacent positions by the substituent at position 4.

2. *Pyridine Series: Alkylation*

The sources of radicals which have been used for the alkylation of pyridine and its derivatives have been summarised by Norman and Radda (1963). Methylation has been the most widely studied of the homolytic alkylation reactions of the pyridine series.

Rieger (1950) has used the thermal decomposition of lead tetraacetate and lead tetrapropionate as sources of methyl and ethyl radicals, respectively, for the alkylation of pyridine, 2-, 3-, and 4-picoline and 2,6-lutidine. The lead salts were added to heated mixtures of the appropriate heterocycle and an excess of acetic or propionic acid, or produced *in situ* by the addition of red lead to the heated mixtures. The products which were isolated indicated substitution at positions 2, 4 and 6, as would be expected since the reactions were carried out in the presence of an excess of acid. Thus, in the methylation reactions, pyridine gave 2- and 4-picoline, 3-picoline gave 2,3- and 2,5-lutidine, and 2,6-lutidine gave 2,4,6-collidine.

The thermal decomposition of diacyl peroxides and the electrolysis of aliphatic carboxylic acids have been used as sources of alkyl radicals for the alkylation of pyridine by Goldschmidt and Minsinger (1954). Their results are summarised in Table 12. The 2- and 4-alkylpyridines were isolated from the reactions, with the 2-isomer formed in larger amount. The figures for the isomer ratios given in Table 12 are not very reliable since the analysis of the products was carried out by fractional distillation; this probably accounts for the failure to detect any 3-alkylpyridines among the products.

TABLE 12

Homolytic alkylation of pyridine (Goldschmidt and Minsinger, 1954)

Radical	Diacyl peroxide		Electrolysis of carboxylic acid	
	Overall yield (%)	Ratio of 2- to 4-isomer	Overall yield (%)	Ratio of 2- to 4-isomer
Methyl	86	7·62	3·5	2·81
Ethyl	87	2·14	6·8–14·3	1·25–2·78
n-Propyl	84	2·4	4·38	5·07

The thermal decomposition of more complex diacyl peroxides of the type $[CH_3CO_2(CH_2)_nCH_2CO_2]_2$, where $n = 1$ and 3, in the presence of pyridine, has been studied by Goldschmidt and Beer (1961). The corresponding 2-, 3- and 4-substituted pyridines were formed, with substitution occurring preferentially at the 2-position.

The ratios of isomeric picolines formed in the methylation of pyridine in non-acidic solution, with the thermal decompositions of t-butyl peroxide, acetyl peroxide and lead tetra-acetate as sources of methyl radicals, have been determined by several authors (Schwetlick and Lungwitz, 1964; Abramovitch and Kenaschuk, 1967; Nababsing, 1969; Bass and Nababsing, 1970). The effect of *N*-protonation on the homolytic methylation of pyridine, using t-butyl peroxide or lead tetra-acetate in glacial acetic acid, has also been investigated (Nababsing, 1969; Bass and Nababsing, 1970). A mixture of glacial acetic acid, concentrated hydrochloric acid and acetic anhydride was also used as the solvent with t-butyl peroxide, the anhydride being added to effect dissolution of the peroxide. Under these conditions, and with the concentrations of reagents used, pyridine has been shown to be completely protonated (Bonnier and Court, 1967). The ratios of isomeric picolines obtained in all of these experiments are given in Table 13.

TABLE 13

Isomer ratios for the homolytic methylation of pyridine in acidic and non-acidic media

Radical source	Medium	Isomer ratio (%) 2-	3-	4-	Ref.
t-Butyl peroxide	Non-acidic	58·0	23·0	19·0	1
t-Butyl peroxide	Non acidic	62·0	22·9	15·1	2
t-Butyl peroxide	Acetic acid	77·9	2·7	19·4	2
t-Butyl peroxide	Acetic acid-hydrochloric acid	93·2	0	6·8	2
Acetyl peroxide	Non-acidic	62·7	20·3	17·0	3
Lead tetra-acetate	Non-acidic	62·1	20·5	17·4	3
Lead tetra-acetate	Non-acidic	62·7	21·7	15·6	2
Lead tetra-acetate	Acetic acid	76·4	2·9	20·7	2

References: (1) Schwetlick and Lungwitz (1964). (2) Bass and Nababsing (1970). (3) Abramovitch and Kenaschuk (1967).

Several points arise from the data in Table 13. First, the ratio of isomeric picolines formed in non-acidic solution appears to be independent of the radical source used. Secondly, the ratios of isomers obtained when t-butyl peroxide was used as the radical source, in the absence or presence of acetic acid, were approximately the same as those obtained in the corresponding reactions using lead tetra-acetate (Bass and Nababsing, 1970). Therefore, in the latter case it was the uncomplexed pyridine in its protonated or unprotonated form that was being attacked and not its lead salt. Thirdly, when allowance is made for the two pairs of equivalent sites corresponding to positions 2 and 3, the order of positional reactivities for the homolytic methylation of pyridine in non-acidic solution is $2>4>3$, which is in agreement with that predicted from the calculations of the atom localisation energies of these positions (Brown, 1956; Brown and Heffernan, 1956). Fourthly, the ratios of 2- to 3- and 4- to 3-substitution in the homolytic methylation of pyridine in non-acidic solution are higher than those obtained in the phenylation of pyridine under similar conditions, using various radical sources (see Table 6). For example, when the thermal decomposition of benzoyl peroxide was used as the radical source, the percentages of the isomeric phenylpyridines formed were: 2-, 54; 3-, 32 and 4-, 14 (Hey, Stirling and Williams, 1955). Since nucleophilic substitution in pyridine occurs at positions 2 and 4, these results indicate that methyl radicals are more nucleophilic than phenyl radicals, as has been suggested previously (Cowley, Norman and Waters, 1959).

The effect of *N*-protonation on the homolytic methylation of pyridine with t-butyl peroxide or lead tetra-acetate is evident from the isomer ratios given in Table 13. The reactivities of positions 2 and 4 in pyridine were enhanced when the reactions were carried out in acetic acid. When a mixture of acetic acid and hydrochloric acid was used as the solvent, with t-butyl peroxide as the radical source, the proportion of the 2-isomer formed was increased at the expense of the 4-isomer and substitution at position 3 was not observed. The competitive methylation of mixtures of pyridine and nitrobenzene with t-butyl peroxide in the presence and absence of acetic acid showed that the total reactivity of pyridine was increased by a factor of 13·2 in acidic solution, and the reactivities of positions 2, 3 and 4 were increased by factors of 16·1, 1·7 and 17·1, respectively (Nababsing, 1969).

The homolytic methylation of the picolines in acidic and non-acidic media has been studied by Nababsing (1969) and Abramovitch and Kenaschuk (1967). The thermal decompositions of t-butyl peroxide, acetyl peroxide and lead tetra-acetate were used as sources of methyl radicals in the investigations. The isomer ratios of substitution products obtained from the methylation of the picolines under various conditions are given in Table 14.

The suggestion that the nucleophilic character of the methyl radical is greater than that of the phenyl radical (Cowley, Norman and Waters, 1959) is supported by a comparison of the results obtained from the homolytic methylation and phenylation of 2-picoline in non-acidic solution. The average percentage isomer ratio obtained in the methylation of 2-picoline with t-butyl peroxide and lead tetra-acetate in the absence of acid was: 3-, 37·9; 4-, 11·0; 5-, 8·0 and 6-, 44·1 (see Table 14). In the phenylation of 2-picoline with benzoyl peroxide in non-acidic solution the corresponding percentages of isomers were found to be 31, 15, 20 and 34, respectively (Bonnier and Court, 1967). The ratio of 5- to 6-substitution for phenylation was higher than that for methylation because of the lower nucleophilic character of the phenyl radical. In both cases, however, the reactivity of position 3 was higher than that of position 4, indicating that the methyl substituent in 2-picoline had increased the proportion of substitution

TABLE 14

Isomer ratios for the homolytic methylation of 2-, 3- and 4-picoline in acidic and non-acidic media

Compound	Radical source	Medium	Isomer ratio (%)					Ref.
			2-	3-	4-	5-	6-	
2-Picoline	t-Butyl peroxide	Non-acidic	—	37·6	11·4	7·5	43·5	1
2-Picoline	t-Butyl peroxide	Acetic acid	—	1·8	18·0	1·1	79·1	1
2-Picoline	t-Butyl peroxide	Acetic acid-hydrochloric acid	—	0	12·7	0	87·3	1
2-Picoline	Lead tetra-acetate	Non-acidic	—	36·3	10·5	8·5	44·7	1
2-Picoline	Lead tetra-acetate	Acetic acid	—	1·4	16·2	1·5	80·9	1
3-Picoline	t-Butyl peroxide	Non-acidic	55·4	—	20·5	5·0	19·1	1
3-Picoline	t-Butyl peroxide	Acetic acid	66·1	—	14·8	0	19·1	1
3-Picoline	t-Butyl peroxide	Acetic acid-hydrochloric acid	67·0	—	12·9	0	20·1	1
3-Picoline	Acetyl peroxide	Non-acidic	55·5	—	20·5	4·6	19·4	2
3-Picoline	Lead tetra-acetate	Non-acidic	57·2	—	19·3	5·2	18·3	1
3-Picoline	Lead tetra-acetate	Acetic acid	65·2	—	16·3	0	18·5	1
4-Picoline	t-Butyl peroxide	Non-acidic	33·6	16·4	—	16·4	33·6	1
4-Picoline	t-Butyl peroxide	Acetic acid	50·0	0	—	0	50·0	1
4-Picoline	t-Butyl peroxide	Acetic acid-hydrochloric acid	50·0	0	—	0	50·0	1
4-Picoline	Acetyl peroxide	Non-acidic	32·2	17·8	—	17·8	32·2	2
4-Picoline	Lead tetra-acetate	Non-acidic	33·2	16·8	—	16·8	33·2	1
4-Picoline	Lead tetra-acetate	Acetic acid	50·0	0	—	0	50·0	1

References: (1) Nababsing (1969). (2) Abramovitch and Kenaschuk (1967).

at the adjacent carbon atom. This effect was more pronounced in methylation than in phenylation. Thus the decrease in the reactivity of position 3 in 2-picoline towards methylation, which is due to the greater nucleophilic character of the methyl radical, was offset by the increase in reactivity due to the adjacent methyl substituent.

In the homolytic methylation of 3-picoline in non-acidic solution the 3-methyl substituent increased the reactivities of the adjacent positions, so that position 2 was more reactive than position 6. The proportion of substitution at position 2 in the homolytic methylation of 4-picoline in non-acidic solution (average figure: 33 %, see Table 14) was higher than that for the phenylation of 4-picoline with benzoyl peroxide under similar conditions, which gave the following percentage isomer ratio: 2-, 27; 3-, 23; 5-, 23 and 6-, 27 (Bonnier and Court, 1967). These results are in agreement with the greater nucleophilicity of the methyl radical.

In acidic solution the picolines were methylated mainly at positions 2, 4 and 6 (see Table 14). The ratios of 2- to 4- or 6- to 4-isomer for the methylation of 2- and 3-picoline were higher in acidic media. These results show that protonation is more effective in increasing the reactivities of positions 2 and 6, which are adjacent to the nitrogen atom, than in increasing the reactivity of position 4. The isomer ratios obtained in the methylation of 3-picoline in acetic acid (see Table 14) were similar to that reported by Hardegger and Nikles (1957) for the methylation of 3-n-butylpyridine with lead tetraacetate in acetic acid. The percentage isomer ratio obtained in the latter reaction was: 2-, 60; 4-, *ca.* 20; 5-, *ca.* 2 and 6-, 20. The n-butyl substituent increased the reactivities of the adjacent positions, so that position 2 was more reactive than position 6, and the reactivity of position 4 was approximately the same as that of position 6.

The orientation of the substitution products formed in the homolytic methylation of pyridine and the picolines in acidic media (Nababsing, 1969) is similar to that observed by Claret and Williams (1969) for the *C*-substituted alkylpyridinium halides formed in the thermal rearrangement of *N*-alkylpyridinium halides and their methyl-substituted derivatives (the Ladenburg rearrangement). The similarity supports the proposal that the rearrangement involves a homolytic mechanism (Claret and Williams, 1969).

The reactions of pyridine and the picolines with benzyl radicals in acidic and non-acidic media have been investigated (Nababsing, 1969; Bass and Nababsing, 1969). The thermal and photochemical decomposition of dibenzylmercury and the thermal decomposition of lead tetraphenylacetate were used as the sources of benzyl radicals. Both the thermolysis and the photolysis of dibenzylmercury in pyridine in non-acidic solution gave bibenzyl (91% from the photolytic reaction) as the sole organic product, in agreement with the work of Cadogan, Hey and Sanderson (1960), who found that benzyl radicals generated from di-(9-benzyl-9-fluorenyl) peroxide did not react with pyridine. In contrast, the thermal decomposition of dibenzylmercury in a solution of pyridine in acetic acid gave 2- and 4-benzylpyridine (2 > 4). The benzylpyridines were also formed from the thermal decomposition of lead tetraphenylacetate in pyridine in the presence of an excess of phenylacetic acid (Bass and Nababsing, 1969). The results indicated the enhanced reactivity of the pyridinium ion towards homolytic attack, compared with that of the free base, as predicted by molecular orbital calculations of atom localisation energies for the pyridinium ion (Brown and Heffernan, 1956). The absence of any attack by benzyl radicals at position 3 in pyridine is in accordance with the work of Bonnier and Court (1967), who found that although the reactivities of positions 2 and 4 towards phenyl radicals were increased in acidic solution, that of position 3 remained practically unchanged.

The thermal decomposition of dibenzylmercury in 2-, 3- or 4-picoline in non-acidic solution gave toluene, bibenzyl, and the corresponding phenethylpyridine and 1,2-dipyridylethane (Nababsing, 1969). Nuclear substitution by benzyl radicals was not observed. The products were accounted for by hydrogen abstraction from the side-chain of 2-, 3- or 4-picoline by benzyl radicals to give toluene and 2-, 3- or 4-pyridylmethyl radicals, which either combined with benzyl radicals to give the phenethylpyridines or dimerised to 1,2-dipyridylethanes. The phenethylpyridines were formed in greater amounts than the 1,2-pyridylethanes since benzyl radicals were more abundant than pyridylmethyl radicals, as was shown by the high proportions of bibenzyl produced in the reactions. The increasing amounts of bibenzyl formed indicated that the order of reactivities of the picolines towards α-hydrogen abstraction by benzyl radicals was 2- > 3- > 4-. Johnston and Williams (1960) have shown that the reactivities of 3- and 4-picoline towards α-hydrogen abstraction by t-butoxy-radicals are in the ratio of 0·44 to 0·33 (relative to toluene = 1·00). In previous work the thermal decomposition of a mixture of di-(9-benzyl-9-fluorenyl) peroxide and t-butyl peroxide in 4-picoline was shown to give 4-phenethylpyridine and 1,2-di-4′-pyridylethane (Cadogan, Hey and Sanderson, 1960). Also, the photolysis of dibenzylmercury in 4-picoline at 80° has been reported to give bibenzyl (89%) as the only organic product (Hey, Shingleton and Williams, 1963b).

The reaction of 2-, 3- or 4-picoline with dibenzylmercury in acetic acid gave the benzylmethylpyridines whose yields are given in Table 15 (Nababsing, 1969). Nuclear substitution occurred at positions 2, 4 and 6 only. The benzylmethylpyridines were also formed from the thermal decomposition of lead tetraphenylacetate in 2-, 3- or 4-picoline in the presence of an excess of phenylacetic acid. The proportions of isomers formed in the benzylation of pyridine and 2- and 3-picoline with dibenzylmercury in acidic solution indicated a higher reactivity of positions 2 and 6 in the heterocyclic bases than that of position 4. In the reaction with 3-picoline, position 2 was more reactive than position 6, indicating that the methyl substituent at position 3 had increased the reactivity of the adjacent position. The results were similar to those obtained in the homolytic phenylation (Bonnier and Court, 1967; Dou and Lynch, 1965, 1966a) and methylation (Nababsing, 1969; Bass and Nababsing, 1970) of pyridine and the picolines in acidic media.

TABLE 15

Products of homolytic benzylation of 2-, 3- and 4-picoline in acetic acid (Nababsing, 1969)

Compound	Products	Yield (%)[a]
2-Picoline	4-Benzyl-2-methylpyridine	6·0
	6-Benzyl-2-methylpyridine	35·1
3-Picoline	2-Benzyl-3-methylpyridine	20·3
	2-Benzyl-5-methylpyridine	13·8
	4-Benzyl-3-methylpyridine	7·5
4-Picoline	2-Benzyl-4-methylpyridine	34·8

[a] Yields of products calculated as % benzyl from dibenzylmercury.

The substitution products formed in the homolytic benzylation of pyridine and the picolines in acidic media are similar to those formed from the thermal rearrangement of *N*-benzylpyridinium and *N*-benzylpicolinium chlorides (Von Braun, Nelles and May, 1937; Crook, 1948; Tsuda, Satoh and Saeki, 1953; Dou and Lynch, 1966b).

The rearrangements have been shown to give products which are formed by the migration of a benzyl group from the nitrogen atom to positions 2, 4 or 6 of the heterocyclic ring. The similarity supports the suggestion that the rearrangements probably occur through a homolytic mechanism involving the protonated molecule (Dou and Lynch, 1966b).

3. *Pyridazine and Pyrimidine*

The homolytic phenylation of pyridazine, with three different radical sources, namely benzoyl peroxide, *N*-nitrosoacetanilide and benzenediazonium hydroxide, has been shown by Atkinson and Sharpe (1959) to give 4-phenylpyridazine only. The first two methods gave better yields than the third. Calculations of free valence numbers for pyridazine indicate that position 4 should be the most favoured point of homolytic attack (Davies, 1955). More recently, Dou and Lynch (1966a, 1966c) have reported the formation of 2- and 4-phenylpyridazine ($2 > 4$) from the phenylation of pyridazine with benzoyl peroxide in acetic acid.

The homolytic methylation of imidazo[1,2-*b*]pyridazine (I) with acetyl peroxide has been studied by Pollak, Stanovnik and Tisler (1968). The products identified from the reaction were the 8-methyl-, 7-methyl- and 7,8-dimethyl-derivatives of (I), in the approximate ratio of 1:2:5. The results suggested that the initial attack on (I) by methyl radicals occurred at position 8, with subsequent formation of the 7,8-dimethyl-derivative. Evidently under the conditions employed there was no attack on the imidazole ring in (I). This

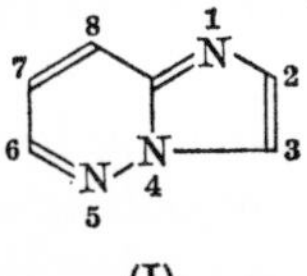

(I) 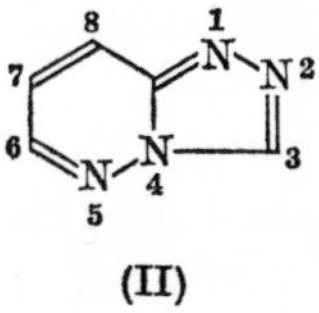 (II)

work has been extended by Japelj, Stanovnik and Tisler (1969) to include the homolytic methylation of *s*-triazolo[4,3-*b*]pyridazine (II) and its 7-methyl- and 7-chloro- analogues. The reaction of (II) with acetyl peroxide (in the molar ratio of 1 : 10) gave the 7,8-dimethyl- and 6,7,8-trimethyl-derivatives of (II) as the main products, together with several other monomethyl- and dimethyl-derivatives, including 3,7-dimethyl-*s*-triazolo[4,3-*b*]pyridazine.

The arylation of pyrimidine with *p*-nitrophenyl radicals, produced from the corresponding acylarylnitrosamine, gave the 2- and 4-substituted products (Lythgoe and Rayner, 1951). Position 2 in pyrimidine is *ortho*- to both of the nitrogen atoms, whereas position 4 is *ortho*- to one nitrogen atom and *para*- to the other. Therefore both positions are activated towards homolytic attack, with position 2 more reactive than position 4 since the activation of a position *ortho*- to a nitrogen atom in a heterocyclic system is greater than that of a position *para*- to it. However, since there are two equivalent sites corresponding to position 4 in pyrimidine, the yield of 4-(*p*-nitrophenyl)pyrimidine (58 %) produced in the reaction exceeded that of 2-(*p*-nitrophenyl)pyrimidine (42 %). The observed order of positional reactivities, $2 > 4$, is the same as that predicted from the free valence numbers of these positions in pyrimidine (Davies, 1955; Brown and Heffernan, 1956), although calculations of the atom localisation energies for pyrimidine indicate the reverse order (Brown and Heffernan, 1956).

4. *Quinoline and Isoquinoline*

The homolytic phenylation of quinoline with benzoyl peroxide was shown by Hey and Walker (1948) to give 4-phenylquinoline and a smaller amount of the 5-isomer. In a later investigation of this reaction by Pausacker (1958) all of the phenylquinolines were isolated. The percentage isomer ratio was: 2-, 6; 3-, 14; 4-, 20; 5-, *ca.* 12; 6-, 8; 7-, 8 and 8-, 30. The order of positional reactivities, $8 > 4 > 3,5 > 2,6,7$, is not in agreement with that predicted from either the free valence numbers (Sandorfy and Yvan, 1950) or the atom localisation energies (Sandorfy and Yvan, 1950; Brown and Harcourt, 1959) for quinoline.

Dou and Lynch (1966a, 1966c) have shown that the total reactivities of quinoline and isoquinoline towards phenylation with benzoyl peroxide are increased in acidic solution. No detectable change in the isomer ratios of products was observed, compared with those obtained in non-acidic solution, but the actual percentages were not reported.

Rieger (1950) has used the thermal decomposition of lead tetraacetate and lead tetrapropionate, in the presence of an excess of the corresponding carboxylic acid, as sources of methyl and ethyl radicals for the alkylation of quinoline and isoquinoline. Quinoline gave the 2- and 4-alkylated products and isoquinoline gave the 1-alkyl isomer only. Similarly, acetyl and methyl radicals, produced by the decomposition of t-butyl hydroperoxide in dilute sulphuric acid in the presence of acetaldehyde and ferrous sulphate, attacked quinoline at positions 2 and 4 (Caronna, Gardini and Minisci, 1969).

The homolytic methylation of quinoline and isoquinoline in acidic and non-acidic media, with t-butyl peroxide and lead tetra-acetate as the radical sources, has been studied (Nababsing, 1969; Bass and Nababsing, 1970). The isomer ratios obtained for the methylation of quinoline are given in Table 16. The separation of 3-, 6- and 7-

TABLE 16

Isomer ratios for the homolytic methylation of quinoline in acidic and non-acidic media (Bass and Nababsing, 1970)

		Isomer ratio (%)				
Radical source	Medium	2-	4-	5-	8-	3,6 & 7-
t-Butyl peroxide	Non-acidic	10·2	24·2	16·1	30·8	18·7
t-Butyl peroxide	Acetic acid	33·6	38·7	8·8	13·2	5·7
t-Butyl peroxide	Acetic acid-hydrochloric acid	49·0	50·0	←	~1·0	→
Lead tetra-acetate	Non-acidic	8·6	23·4	17·7	32·0	18·3
Lead tetra-acetate	Acetic acid	34·4	40·7	8·4	10·7	5·8

methylquinoline could not be achieved in the gas chromatographic analysis of the products. The order of positional reactivities for methylation in non-acidic solution, $8 > 4 > 5 > 2$, is the same as that found in the phenylation of quinoline (Pausacker, 1958). The reactivities of positions 2 and 4 towards methylation were enhanced in acidic media. When quinoline was methylated with t-butyl peroxide in a mixture of acetic acid and hydrochloric acid, 2- and 4-methylquinoline were formed almost exclusively, and in approximately equal amounts. The ratio of 2- to 4- isomer was highest in the mixed acid solution and lowest in non-acidic solution, and therefore the increase in reactivity due to *N*-protonation was greater at position 2 than at position 4, as predicted from the atom localisation energies for the quinolinium ion (Brown and Harcourt, 1959). The methylation of isoquinoline with t-butyl peroxide or lead tetra-acetate in both acidic and non-acidic media gave 1-methylisoquinoline only, in agreement with the atom localisation energies for isoquinoline and the isoquinolinium ion, which predict that position 1 should be the most reactive site towards homolytic attack (Brown and Harcourt, 1960). The yield of 1-methylisoquinoline was increased in acidic solution.

Competitive reactions with mixtures of quinoline or isoquinoline and naphthalene showed that the order of total reactivities towards methylation with t-butyl peroxide in non-acidic solution was isoquinoline > quinoline > naphthalene (Bass and Nababsing, 1970). The relative reactivities (1·52:1·25:1·0) were in reasonable agreement with the methyl affinities of the compounds (see Table 1) (Szwarc and Binks, 1959). Similar competitive reactions in the presence of acetic acid showed that the increase in the reactivities of quinoline and isoquinoline towards methylation in acidic solution was higher than that observed by Dou and Lynch (1966a, 1966c) in the competitive phenylation of mixtures of quinoline or isoquinoline and nitrobenzene with benzoyl peroxide in acetic acid.

Elkobaisi and Hickinbottom (1960) have shown that benzyl radicals, produced in the thermal rearrangement of benzyl phenyl ether or benzyl 2,4,6-trimethylphenyl ether, react with quinoline to give 2- and 4-benzylquinoline. 1-Benzylisoquinoline was isolated from the products of the rearrangement of benzyl phenyl ether in isoquinoline. The effect of *N*-protonation on the homolytic benzylation of quinoline and isoquinoline, with dibenzylmercury and lead tetraphenylacetate as thermal sources of benzyl radicals, has been investigated (Nababsing, 1969; Bass and Nababsing, 1969). The reaction of dibenzylmercury with quinoline in non-acidic solution gave small amounts of 2- and 4-benzylquinoline and 2,4-dibenzylquinoline. Under similar conditions isoquinoline gave small amounts of 1-, 3- and 4-benzylisoquinoline. Bibenzyl was the main product from both of the reactions. In acetic acid the yields of 2- and 4-benzylquinoline (2 > 4) and 1-benzylisoquinoline were increased considerably, and similar results were obtained from the thermal decomposition of lead tetraphenylacetate in quinoline and isoquinoline in the presence of an excess of phenylacetic acid. The same effect of *N*-protonation was observed in the homolytic methylation of quinoline and isoquinoline in acidic media (Nababsing, 1969; Bass and Nababsing, 1970). The competitive benzylation of mixtures of pyridine and quinoline or isoquinoline with dibenzylmercury in acetic acid showed that the ratios of relative reactivities of quinoline and isoquinoline to pyridine were 16·3:1 and 36·7:1, respectively (Nababsing, 1969).

5. *Quinoxaline, Phthalazine and Cinnoline*

The homolytic phenylation, methylation and benzylation of quinoxaline have been studied. The phenylation of quinoxaline with benzoyl peroxide, *N*-nitrosoacetanilide and benzenediazonium hydroxide gave 2-, 5- and 6-phenylquinoxaline ($2 > 5 > 6$) (Atkinson and Sharpe, 1959). The positions adjacent to the nitrogen atoms were the most reactive towards homolytic attack. The same order of positional reactivities was observed in the methylation of quinoxaline with t-butyl peroxide and lead tetra-acetate in the absence and presence of acetic acid (Nababsing, 1969). The average percentage isomer ratio of methylquinoxalines formed in the absence of acid was 2-, 71; 5-, 23·3 and 6-, 5·7. In acetic acid the average percentage isomer ratio was 2-, 88·9; 5-, 7·1 and 6-, 4·0. The increase in the reactivity of position 2 for methylation in acidic solution was greater than that observed by Dou and Lynch (1966a, 1966c) in the phenylation of quinoxaline with benzoyl peroxide in acetic acid. The benzylation of quinoxaline with dibenzylmercury or lead tetraphenylacetate in acidic and non-acidic solution gave 2-benzylquinoxaline only (Nababsing, 1969). The yield of 2-benzylquinoxaline was increased in acidic solution. Evidently the resonance-stabilised benzyl radicals were less reactive and more selective than methyl or phenyl radicals, and benzylation of quinoxaline occurred at only the most reactive position towards homolytic attack.

The homolytic phenylation of phthalazine and cinnoline with *N*-nitrosoacetanilide has been reported (Atkinson and Sharpe, 1959). Phthalazine gave a low yield of 5-phenylphthalazine, but the main product from the reaction with cinnoline was 4,4′-bicinnolyl, although a small amount of 4-phenylcinnoline was obtained.

6. *Acridine, Benzacridines and Phenazine*

Goldschmidt and Beer (1961) have shown that the thermal decomposition of diacyl peroxides of the type $[CH_3CO_2(CH_2)_nCH_2CO_2]_2$, where $n = 1$ and 3, in the presence of acridine gives the corresponding 9-substituted acridine derivatives.

The reactions of benzyl radicals with acridine, 9-phenylacridine, 1,2- and 3,4-benzacridine, and phenazine have been studied by Waters and Watson (1957, 1959a, 1959b). The decomposition of t-butyl peroxide in boiling toluene was used as the source of benzyl radicals in the investigations. Acridine gave 9-benzylacridine (17 %) and 9,10-dibenzylacridan (18 %), but no biacridan (Waters and Watson, 1957). The same products were formed from the thermal decomposition of dibenzylmercury in the presence of acridine (Taylor, 1969). In contrast, the products of the reaction of benzyl radicals with anthracene included 9,10-dibenzyl-9,10-dihydroanthracene and 10,10′-dibenzyl-9,9′,10,10′-tetrahydro-9,9′-bianthryl (Beckwith and Waters, 1957; Bass and Nababsing, 1965). These results indicated that the initial addition of a benzyl radical to acridine occurred at the *meso*-carbon atom and not at the nitrogen atom. A similar conclusion was reached from a comparison of the methyl radical affinities of acridine and anthracene, discussed in Section B (Szwarc and Binks, 1959), and it is further supported by the fact that the reaction of benzyl radicals with 9-phenylacridine proceeded less easily than their reaction with acridine (Waters and Watson, 1957). The products of the reaction with 9-phenylacridine were 9,10-dibenzyl-9-phenylacridan (2·5 %) and 4-benzyl-9-phenylacridine (10 %).

Benzyl radicals were shown to attack 3,4-benzacridine (IV) at the exposed *meso*-carbon atom to give 9-benzyl-3,4-benzacridine in 65 % yield (Waters and Watson, 1959a). From the sterically hindered 1,2-benzacridine (III), the addition product 9,10-dibenzyl-1,2-benzacridan was obtained, but only in 7·5 % yield (Waters and Watson, 1959a). Thus spatial access of the benzyl radical to the *meso*-carbon atom in both of the benzacridines is of importance in controlling the courses of reaction. The conclusion that the initial attack of benzyl radicals on acridine occurs at the *meso*-carbon atom is further supported by these results.

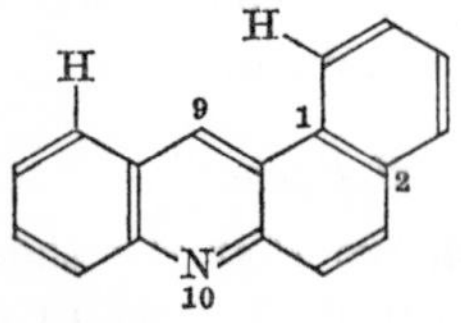

(III) 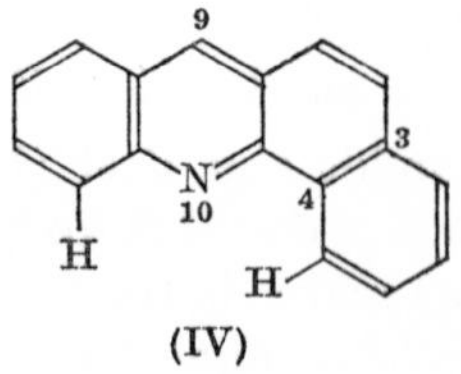 (IV)

The reaction of benzyl radicals, generated by the thermal decomposition of t-butyl peroxide in toluene, with phenazine has been reported to give 1-benzylphenazine and 5,10-dibenzyl-5,10-dihydrophenazine (V) in the approximate molar ratio of 3:1, and smaller amounts of more highly *C*-substituted phenazines, which were not identified (Waters and Watson, 1959b). The products were isolated by adsorption chromatography. Further study of this reaction, using the same radical source and high resolution gas chromatography for the identification of the products, has shown that a small amount of 2-benzylphenazine is also formed (Taylor, 1969). The reaction of benzyl radicals, generated by the thermal decomposition of dibenzylmercury, with phenazine gave 1- and 2-benzylphenazine (1 > 2), and a dibenzylphenazine, possibly the 1,6-isomer, but 5,10-dibenzyl-5,10-dihydrophenazine (V) was shown to be absent from the products of the reaction (Taylor, 1969).

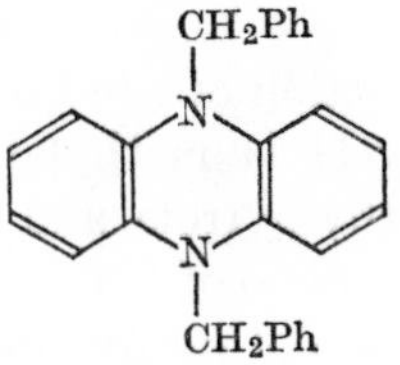

CH_2Ph
N
N
CH_2Ph

(V)

D. Homolytic Substitution Reactions of Five-Membered Heterocycles Containing One Heteroatom

1. *Pyrrole and Indole*

The products formed from the thermal decomposition of t-butyl peroxide in pyrrole and 1-methylpyrrole have been investigated by Gritter and Chriss (1964). When a mixture of t-butyl peroxide and pyrrole was heated at 150°, the t-butoxy-radicals generated from the peroxide abstracted hydrogen from position 2 in pyrrole to give 2-pyrrolyl radicals, which reacted with the diene system of pyrrole to give 2,2′-(1′-pyrrolinyl)pyrrole (VI). When the peroxide decomposed in 1-methylpyrrole, hydrogen abstraction from the methyl group by t-butoxy-radicals gave 1-pyrrolylmethyl radicals, which dimerised to give 1,2-di-(1′-pyrrolyl)ethane (VII) or substituted in 1-methylpyrrole at position 3 to give 1-methyl-3-(1′-pyrrolylmethyl)-pyrrole (VIII), after radical chain transfer. In an earlier investigation, Conant and Chow (1933) reported that triphenylmethyl radicals add to the diene system in pyrrole to give 2,5-*bis*(triphenylmethyl)-3-pyrroline.

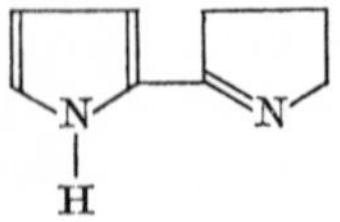

(VI)

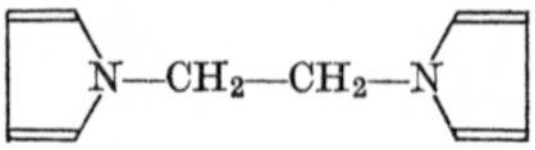

N—CH_2—CH_2—N

(VII)

CH_2—N

CH_3

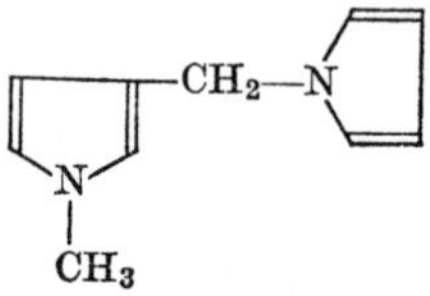

(VIII)

Further study of the reaction of t-butyl peroxide with pyrrole has shown that the heterocycle is methylated at positions 2 and 5 (Nababsing, 1969). 2-Methylpyrrole and 2,5-dimethylpyrrole were formed in the approximate ratio of 5 : 1.

The homolytic benzylation of pyrrole in non-acidic solution, using the thermal decomposition of dibenzylmercury as the radical source, gave a small amount of 2-benzylpyrrole (Nababsing, 1969). The yield of 2-benzylpyrrole was increased considerably when the reaction was carried out in acetic acid.

The substitution at position 2 in pyrrole by methyl and benzyl radicals is in agreement with the suggestion by Lynch and Chang (1964a) that, on the basis of simple resonance concepts, the reactivity in five-membered heterocyclic compounds should be high at positions adjacent to the hetero-atom. The free valence numbers (Brown, 1955; Minkin, Pozharskii and Ostroumov, 1966) and atom localisation energies (Del Re and Scarpati, 1964) for pyrrole also predict that position 2 should be more reactive than position 3 towards homolytic attack.

The products obtained from the autoxidation of alkylpyrroles have been accounted for by a reaction scheme involving attack on the pyrrole ring at position 2 by the peroxy-radicals derived from the alkylpyrroles (Smith and Jensen, 1967). Rinkes (1943) has shown that the homolytic phenylation of ethyl 1-pyrrolecarboxylate with *N*-nitrosoacetanilide gives the corresponding 2-phenyl derivative.

Hutton and Waters (1965) have studied the homolytic benzylation of indole, using the thermal decomposition of t-butyl peroxide in toluene as the radical source. The initial products of benzylation were mainly 1- and 3-benzylindole, together with a minor product, which was tentatively identified as 4-benzylindole. Further attack at the α-carbon atom of the benzyl group in 1-benzylindole gave α-(1-indolyl)bibenzyl (IX), while further attack on 3-benzylindole gave 1,3- and 2,3-dibenzylindole. 2-Benzylindole did not appear to be an initial reaction product. In contrast, the homolytic methylation of indole with t-butyl peroxide in non-acidic solution gave 2-methylindole only (Nababsing, 1969). This result was unexpected, since methyl radicals would be expected to be more reactive and less selective than benzyl radicals. The free valence numbers for indole predict that position 2 should be the favoured point of homolytic attack (Longuet-Higgins and Coulson, 1947).

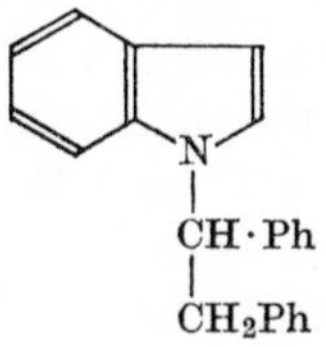

N

CH·Ph

CH_2Ph

(IX)

The difference between the reactivities of methyl and benzyl radicals towards indole is similar to that observed by Gritter and Chriss (1964) for the reactivities of 2-pyrrolyl and 1-pyrrolylmethyl radicals towards pyrrole and 1-methylpyrrole, respectively. 2-Pyrrolyl radicals, which have a similar reactivity to that of phenyl radicals, attacked pyrrole at position 2, whereas 1-pyrrolylmethyl radicals, which are resonance-stabilised like benzyl radicals, attacked 1-methylpyrrole at position 3. The reactions of methyl radicals would be expected to be similar to those of phenyl radicals, and therefore the methyl radical behaved like the 2-pyrrolyl radical and attacked indole at position 2 (Nababsing, 1969). The substitution in indole at position 3 by benzyl radicals (Hutton and Waters, 1965) is similar to the substitution in 1-methylpyrrole at position 3 by 1-pyrrolylmethyl radicals.

2. *Thiophen, Benzothiophen and Dibenzothiophen*

Synthetic and quantitative studies of the homolytic arylation of thiophen have been made by several groups of workers. The Gomberg reaction (Gomberg and Bachmann, 1924; Buu-Hoï and Hoàn, 1950; Smith and Boyer, 1951; Degani, Pallotti and Tundo, 1961) and the decomposition of *N*-nitrosoacetanilide and its derivatives (Degani, Pallotti and Tundo, 1961) have been used as radical sources for the synthesis of substituted 2-arylthiophens. The arylation of thiophen with *p*-chlorophenyl and *p*-tolyl radicals also gave small amounts of the 2,5-disubstituted derivatives (Buu-Hoï and Hoàn, 1950). In addition to 2-phenylthiophen, small amounts of the 3-isomer were produced in the phenylation of thiophen with *N*-nitrosoacetanilide or benzenediazonium salts (Degani, Pallotti and Tundo, 1961). Further substitution of 2- and 3-phenylthiophen gave 2,3-diphenyl-, 2,5-diphenyl-, 2-(*p*-biphenylyl)- and 3-(*p*-biphenylyl)-thiophen, together with 2-phenyl-5-(*p*-biphenylyl)thiophen, indicating homolytic attack on both the phenyl group and the thiophen nucleus in the phenylthiophens.

Quantitative investigations of the isomer ratios formed in the homolytic phenylation of thiophen have been made by Degani, Pallotti and Tundo (1961), who used the thermal decomposition of *N*-nitrosoacetanilide as their source of phenyl radicals, and Griffin and Martin (1965), who used three radical sources: the thermal decompositions of benzoyl peroxide and phenylazotriphenylmethane and the photolysis of iodobenzene in the presence of metallic silver. The isomer ratios obtained in these investigations are given in Table 17.

TABLE 17

Isomer ratios for the homolytic phenylation of thiophen

Radical source	Isomer ratio (%)		Ref.
	2-	3-	
N-Nitrosoacetanilide	95	5	1
Benzoyl peroxide	100	0	2
Phenylazotriphenylmethane in air	86·7	13·3	2
Phenylazotriphenylmethane in nitrogen	63	37	2
Photolysis of iodobenzene	91·8	8·2	2

References: (1) Degani, Pallotti and Tundo (1961). (2) Griffin and Martin (1965).

From the results given in Table 17, it can be seen that the order of positional reactivities for the homolytic phenylation of thiophen is $2>3$, which is in agreement with that predicted by molecular orbital calculations of the free valence numbers and atom localisation energies for thiophen (Metzger and Pullman, 1948). The phenylation of thiophen with benzoyl peroxide gave 2-phenylthiophen as the only phenylated product, together with 2,2′- and 2,3′-bithienyl (Griffin and Martin, 1965). This result contradicted a previous report by Ford and Mackay (1957) that the decomposition of benzoyl peroxide in thiophen gives 2-benzoyloxythiophen as the only substitution product. The bithienyls were formed by the combination of thienyl radicals or the substitution of thiophen by these radicals. Griffin and Martin (1965) have suggested that the thienyl radicals were formed *via* a one-electron oxidation of thiophen at the sulphur atom to give a radical cation (X), followed by an acid-base reaction involving the removal of a proton by a benzoate ion. The bithienyls

$$\text{C}_4\text{H}_4\ddot{\text{S}} + (\text{PhCO}_2)_2 \longrightarrow \text{C}_4\text{H}_4\text{S}^{\bullet+}\ (\text{X}) + \text{PhCO}_2{\bullet} + \text{PhCO}_2^- \qquad (1)$$

$$\text{C}_4\text{H}_4\text{S}^{\bullet+} + \text{PhCO}_2^- \longrightarrow \text{C}_4\text{H}_3\ddot{\text{S}}{\bullet} + \text{PhCO}_2\text{H} \qquad (2)$$

were not formed from the phenylation of thiophen with phenylazotriphenylmethane. This result was as expected since the radical source could not readily yield a stable nucleophile (e.g. a benzoate ion) as required in reactions (1) and (2) (Griffin and Martin, 1965). When the photolysis of iodobenzene was used as the radical source, the bithienyls were detected, but a control experiment showed that they were produced in this reaction by the direct photolysis of thiophen (Griffin and Martin, 1965).

The ratios of 2- to 3-phenylthiophen formed in the phenylation of thiophen with various radical sources (Table 17) show a wide variation, particularly for the reaction with phenylazotriphenylmethane, which gave a very low ratio of 2- to 3-phenylthiophen. Camaggi, Leardini, Tiecco and Tundo (1969) have made a detailed study of the latter reaction, and have attributed the unusually high proportion of the 3-isomer to the trapping of the σ-complex (XI) leading to 2-phenylthiophen by triphenylmethyl radicals, produced by the decomposition of phenylazotriphenylmethane, to give two stereoisomeric 2,5-dihydro-2-phenyl-5-triphenylmethylthiophens, (XIIa) and (XIIb).

$$\text{(XI)} + \cdot CPh_3 \longrightarrow \text{(XIIa)} + \text{(XIIb)}$$

A similar reaction has been observed in the phenylation of benzene with phenylazotriphenylmethane (Hey, Perkins and Williams, 1965); the intermediate phenylcyclohexadienyl radical, derived from the attack of the phenyl radical on benzene, was trapped by the stable triphenylmethyl radical to give a mixture of two isomeric dihydro-triphenylmethylbiphenyls.

The reactions of 2-thienyl and 3-thienyl radicals, generated by the thermal decomposition of the corresponding thenoyl peroxide, with thiophen have been reported (Ford and Mackay, 1957; Mackay, 1966). The reaction with 2-thienyl radicals gave 2,2′-bithienyl and 2-thienyl thenoate, with the bithienyl formed in larger amount (Ford and Mackay, 1957), and 3-thienyl radicals gave low yields of 2,3′- and 2,2′-bithienyl, and a trace of 3,3′-bithienyl (Mackay, 1966). The formation of 2,2′-bithienyl in the latter reaction was unexpected since it indicated that 2-thienyl radicals were formed in the reaction. The results showed that the 2-position in thiophen is more reactive towards homolytic attack than the 3-position, in agreement with previous work on the homolytic phenylation of thiophen (see Table 17).

Benzyl radicals, produced by the thermal decomposition of di-(9-benzyl-9-fluorenyl) peroxide reacted with thiophen to give 2-benzyl-thiophen (Cadogan, Hey and Sanderson, 1960). Similarly, the thermal decomposition of t-butyl peroxide in toluene in the presence of thiophen gave 2-benzylthiophen (Ford and Mackay, 1957).

The reactions of alkylthio-radicals with thiophen and 2-methyl-thiophen have been investigated by Gol'dfarb, Pokhil and Belen'kii (1966, 1967). The radicals were generated by the decomposition of hydrogen peroxide in alkylthiols in the presence of ferrous sulphate and sulphuric acid. Ethylthio- and n-butylthio-radicals reacted with thiophen to give the corresponding 2,5-disubstituted derivatives, and t-butylthio-radicals gave 2-(t-butylthio)thiophen. All three alkylthio-radicals attacked 2-methylthiophen at position 5.

The homolytic phenylation of benzothiophen with *N*-nitroso-acetanilide has been shown to give 2-phenylbenzothiophen only

(Gaertner, 1952). In contrast, the products obtained from the homolytic phenylation of dibenzothiophen (XIII) with benzoyl peroxide and benzenesulphonyl chloride showed that substitution occurred at all four of the available positions (McCall, Neale and Rawlings,

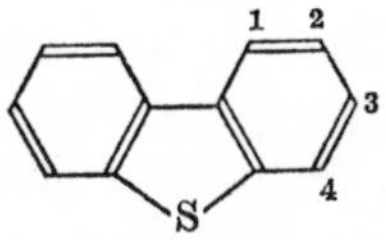

(XIII)

1962). The ratio of isomers obtained indicated the order of positional reactivities to be $1 \sim 4 > 3 > 2$. The high proportion of the sterically hindered 1- and 4-phenyl isomers was in contrast to the results obtained in the homolytic phenylation of phenanthrene with benzoyl peroxide and diazoaminobenzene (Beckwith and Thompson, 1961). In the latter reactions no substitution could be detected at position 4 (corresponding to position 1 in dibenzothiophen), and this was attributed to steric hindrance. However, the high reactivities of positions 1 and 4 in dibenzothiophen towards homolytic phenylation were in accordance with their free valence numbers and atom localisation energies (Berthier and Pullman, 1950).

3. *Furan and Dibenzofuran*

The Gomberg reaction has been widely used for the synthesis of 2-arylfurans. The homolytic arylation of furan with aryldiazonium salts has been used for the preparation of the following 2-aryl derivatives: phenyl, *p*-tolyl, *m*- and *p*-chlorophenyl, *p*-bromophenyl, *p*-nitrophenyl, *p*-methoxyphenyl and 1-naphthyl (Johnson, 1946; Ayres and Smith, 1968). A report by Johnson (1946) that the reaction with *p*-chlorophenyl radicals also gives a small amount of the 3-substituted isomer was not confirmed by Benati, La Barba, Tiecco and Tundo (1969), who used gas chromatography to confirm the absence of the 3-isomer. 2-Furoic acid has been arylated with the following substituted phenyl radicals, produced from the corresponding aryldiazonium chlorides: *p*-chloro-, *p*-bromo-, and *m*- and *p*-nitro- (Mathur and Mehra, 1961). Substitution at position 5 occurred in each case, since decarboxylation of the products gave 2-arylfurans. Freund (1952) has studied the reaction of 2-furylacrylic acid with diazonium salts. 5-Aryl-2-styrylfurans were obtained as the main products of the reactions, and 2-styrylfurans and β-(5-aryl-2-furyl)-acrylic acids were formed as by-products.

The reaction of furan with benzoyl peroxide at temperatures between 35° and 80° has been shown to proceed without loss of carbon dioxide to give in good yield the *cis*- and *trans*- forms of 2,5-dibenzoyloxy-2,5-dihydrofuran, indicating that the major reaction was one of addition of two benzoyloxy-radicals at the 2- and 5-positions in furan (Kolb and Black, 1969). Previously Kolb and Wilson (1966) had shown that acetyloxy-radicals add to furan to give 2,5-diacetoxy-2,5-dihydrofuran when a solution of acetic acid, sodium acetate and furan is electrolysed.

A systematic quantitative study of the homolytic phenylation of furan with several phenylating agents has been made by Benati, La Barba, Tiecco and Tundo (1969). The four radical sources used were the decomposition of *N*-nitrosoacetanilide at room temperature, the decomposition of phenylazotriphenylmethane at 75°, the aprotic diazotisation of aniline with pentyl nitrite and the Gomberg reaction, and in every case 2-phenylfuran was obtained as the only substitution product. The use of phenylazotriphenylmethane as the phenylating agent gave a low yield of 2-phenylfuran, and the main products of this reaction were two stereoisomeric 2,5-dihydro-2-phenyl-5-triphenylmethylfurans, (XVa) and (XVb). The formation of these compounds was accounted for by the interception of the σ-complex (XIV) leading to 2-phenylfuran by triphenylmethyl radicals, which are themselves products of the decomposition of phenylazotriphenyl-

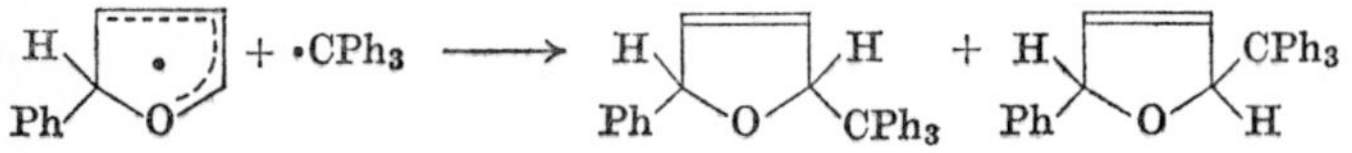

(XIV) + $\cdot CPh_3$ ⟶ (XVa) + (XVb)

methane. Similar reactions were observed in the phenylation of thiophen (Camaggi, Leardini, Tiecco and Tundo, 1969) and of benzene (Hey, Perkins and Williams, 1965) with phenylazotriphenylmethane. Competitive reactions, using mixtures of furan and benzene and of furan and thiophen showed that furan was 11·5 times more reactive than benzene and 4·4 times more reactive than thiophen towards substitution by phenyl radicals generated by the aprotic diazotisation of aniline with pentyl nitrite (Benati, La Barba, Tiecco and Tundo, 1969).

The reaction of dibenzofuran with carboxymethyl radicals, $\cdot CH_2CO_2H$, generated by the thermal decomposition of chloroacetylpolyglycollic acid, $ClCH_2CO_2(CH_2CO_2)_nCH_2COOH$, has been studied by Southwick, Munsell and Bartkus (1961). Substitution occurred at positions 1, 3 and 4, and the carboxymethyl products were obtained in a high yield (50–60 %). The percentage isomer distribution (1-, 55; 3-, 15 and 4-, 30) was essentially the same as that obtained when the carboxymethyl radicals were generated by the thermal decomposition of t-butyl peroxide in acetic acid (1-, 51; 3-, 15 and 4-, 34). The results were in qualitative agreement with the order of reactivities, $1 > 4 > 3 > 2$, predicted by the atom localisation energies for dibenzofuran, which were calculated by the simple Hückel method (Southwick, Munsell and Bartkus, 1961).

E. Homolytic Substitution Reactions of Five-Membered Heterocycles Containing Two Heteroatoms

1. *Pyrazole and Imidazole*

Lynch and Chang (1964a) have studied the homolytic phenylation of 1-methylpyrazole and 1-methylimidazole with benzoyl peroxide and *N*-nitrosoacetanilide. The results obtained for the two radical sources were similar and the average percentage isomer ratio of substituted products for the reaction with 1-methylpyrazole was 5-, 94; 3-, 5 and 4-, 1. 1-Methylpyrazole was found to be less reactive towards phenylation than benzene, having a relative reactivity of 0·62. 1-Methylimidazole gave 67 and 33 % of the 2- and 5-substituted phenyl derivatives, respectively, together with a trace amount of the 4-isomer. Thus both of the phenylation reactions showed that the positions adjacent to the *N*-methyl group were the most reactive sites towards homolytic attack in 1-methylpyrazole and 1-methylimidazole. 1-Methylimidazole has two such positions, and its relative reactivity (1·2) towards phenylation was found to be higher than that of benzene. When the thermal decomposition of benzenediazonium tetrafluoroborate was used as the radical source for the phenylation of 1-methylimidazole, the percentage distribution of isomeric phenylation products was found to be dependent on the concentration of the diazonium salt (Lynch and Chang, 1964b). Thus, when a molar ratio of benzenediazonium tetrafluoroborate to 1-methylimidazole of 1 : 250

was used, the percentage isomer ratio of substituted products was 2-, 61 and 4- plus 5-, 39, but when the molar ratio of reactants was increased to 1:3, the percentage isomer ratio of products was 2-, 81 and 4- plus 5-, 19. The abnormal isomer distribution at high concentration of benzenediazonium tetrafluoroborate has been attributed by Lynch and Chang (1964b) to radical substitution in the 3-phenylazo-1-methylimidazolium ion, which they regarded as an analogue of the methylimidazolium ion.

Lynch and Chang (1964a) have suggested, on the basis of simple resonance concepts, that the reactivity in five-membered heterocycles would be high at positions adjacent to the hetero-atoms. Thus the radical species (XVI), resulting from attack by a phenyl radical at the position adjacent to the 1-nitrogen atom in 1-methylpyrazole, is stabilised by conventional odd-electron delocalisation, while for the radical species (XVII), delocalisation necessarily involves a higher energy charge-separated structure (*b*). These conditions should hold for all five-membered heterocycles. In general, the results obtained

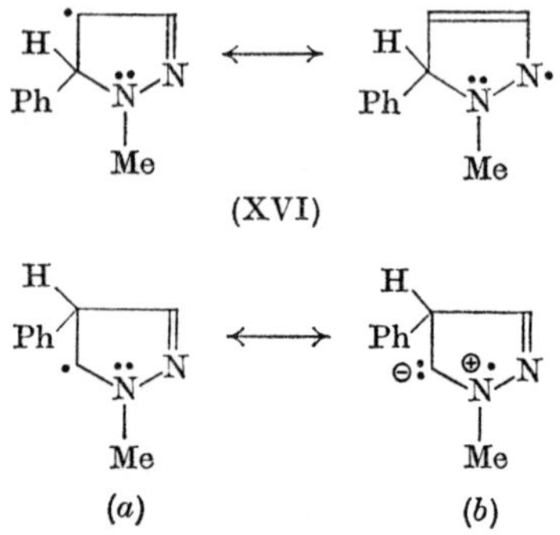

(XVI)

(*a*) (*b*)

(XVII)

in the homolytic reactions of five-membered heterocycles support this theory. However, in the case of 1-phenylpyrazole, phenylation with benzoyl peroxide gave substitution at position 3 (Lynch and Kahn, 1963), although the resonance theory would favour substitution at position 5. Calculations of free valence numbers for pyrazole predict the order of positional reactivities to be $3 > 4 > 5$ (Lynch and Kahn, 1963). The phenyl group in 1-phenylpyrazole was also substituted by the phenyl radical, giving all of the three possible isomers.

When the phenylation of 1-methylpyrazole was carried out with benzoyl peroxide in acetic acid the reactivity of the heterocyclic compound was enhanced and the proportion of the 3-substituted product was increased from 5 to 40 %, at the expense of the 5-isomer (Dou and Lynch, 1965, 1966a). 1-Methylimidazole was also more reactive towards phenylation with benzoyl peroxide in acetic acid, and the 2-substituted product only was obtained in this case (Dou and Lynch, 1965, 1966a).

2. *Thiazole Series, Benzothiazole and Isothiazole*

The homolytic phenylation of thiazole and its derivatives has been studied in some detail by Vernin, Dou and Metzger and their colleagues (Vernin and Metzger, 1963; Vitry-Raymond and Metzger, 1963; Dou and Lynch, 1965, 1966a; Dou and Metzger, 1966; Vernin, Dou and Metzger, 1966, 1967c, 1968; Dou, Vernin and Metzger, 1966, 1967; Vernin, 1967; Vernin and Dou, 1968a). The results obtained in the phenylation of thiazole and its alkyl derivatives with benzoyl peroxide in the absence and presence of acetic acid have been summarised by Vernin, Dou and Metzger (1967c), and are given in Table 18.

The free valence numbers calculated by Metzger and Pullman (1948) indicate the order of positional reactivities in the unprotonated thiazole molecule to be $2>4>5$, but the experimental results for homolytic phenylation in non-acidic solution give the order as $2>5>4$ (Dou, Vernin and Metzger, 1967). The atom localisation energies and free valence numbers for thiazole have been re-calculated, using modified parameters, and have been shown to be in agreement with the experimental order of reactivities (Vernin, Dou and Metzger, 1967c).

The phenylation of thiazole with benzoyl peroxide also gave 2,2′-, 2,5′- and 5,5′-bithiazole, thiazole benzoates, 2,5-diphenylthiazole and a trace of 2,4,5-triphenylthiazole, in addition to phenylthiazoles (Dou, Vernin and Metzger, 1967). The bithiazoles, together with phenylthiazoles (2-, 80 %; 4- plus 5-, 20 %) were also detected in the phenylation of thiazole with benzenediazonium chloride (Vernin and Metzger, 1963).

From the results given in Table 18, it can be seen that in non-acidic solution the methyl substituent in 4-methylthiazole increased the reactivity of the adjacent position, so that position 5 was more

TABLE 18

Isomer ratios for the homolytic phenylation of thiazole and its alkyl derivatives in acidic and non-acidic solution (Vernin, Dou and Metzger, 1967c)

Compound	Medium	Isomer ratio (%)		
		2-	4-	5-
Thiazole	Non-acidic	47	11·5	41·5
Thiazole	Acidic	83	4	13
2-Methylthiazole	Non-acidic	—	30	70
2-Methylthiazole	Acidic	—	30	70
4-Methylthiazole	Non-acidic	42·5	—	57·5
4-Methylthiazole	Acidic	75	—	25
5-Methylthiazole	Non-acidic	80	20	—
5-Methylthiazole	Acidic	95	5	—
4-Isopropylthiazole	Non-acidic	64·5	—	35·5
4-Isopropylthiazole	Acidic	93	—	7
4-t-Butylthiazole	Non-acidic	85	—	15
4-t-Butylthiazole	Acidic	100	—	0

reactive than position 2. In 4-t-butylthiazole, the steric hindrance due to the t-butyl group decreased the reactivity of position 5. In acidic solution the proportion of substitution at position 2 was increased in every case, with the exception of 2-methylthiazole, for which there was no change in the ratio of 4- to 5-substitution (see Table 18). Competitive reactions, using nitrobenzene as the reference compound, showed that the total reactivities of the alkylthiazoles towards homolytic phenylation were increased in acidic solution (Dou and Metzger, 1966).

Further evidence for the order of positional reactivities in thiazole towards homolytic phenylation ($2 > 5 > 4$) was obtained from the results of two series of experiments, (*a*) and (*b*), one of which was carried out in acetic acid (Vernin, Dou and Metzger, 1966) and the other in non-acidic solution (Vernin, 1967).

(*a*) The phenylation of 2,4-, 2,5- and 4,5-diphenylthiazole would give the same product, namely 2,4,5-triphenylthiazole. The decomposition of benzoyl peroxide in the presence of diphenylthiazoles was impractical because of their high melting-points, but in acetic acid phenylation with benzoyl peroxide under identical conditions showed that the following percentages of diphenylthiazoles were converted to 2,4,5-triphenylthiazole: 17% of 2,4-, 2–3% of 2,5- and 25% of 4,5-diphenylthiazole (Vernin, Dou and Metzger, 1966). 2,5-Diphenylthiazole was the least reactive compound because the site of substitution, position 4, is the least reactive position in thiazole. Conversely, 4,5-diphenylthiazole was the most reactive compound in the series because the site of substitution, position 2, is the most reactive position in thiazole.

(*b*) Since the presence of a substituent at a particular position in thiazole removes that site from substitution, there is a corresponding decrease in the reactivity of the molecule. For the methylthiazoles the decrease in reactivity should be largest in 2-methylthiazole and least in 4-methylthiazole, since position 2 is the most reactive and position 4 the least reactive of the positions in thiazole. Therefore, the total reactivities of the methylthiazoles towards homolytic phenylation should be in the reverse order to that of the positional reactivities in thiazole. Vernin (1967) found that the total reactivities of 2-, 4- and 5-methylthiazole towards homolytic phenylation, relative to that of benzene = 1, were 0·57, 1·2 and 0·8, respectively, so that the order of their reactivities was, as expected, 4-methyl > 5-methyl > 2-methyl.

The arylation of thiazole and its alkyl derivatives with *p*-bromophenyl and *p*-nitrophenyl radicals, produced from the corresponding diaroyl peroxides in the absence and presence of acetic acid, has been reported by Vernin, Dou and Metzger (1967a). The results were similar to those for phenylation. With the less reactive *p*-nitrophenyl radical there was a higher proportion of the 2-isomer among the products of arylation of thiazole.

Vernin and Dou (1968a) have studied the homolytic phenylation of thiazole and 4-methylthiazole with benzoyl peroxide in a series of organic acids of varying strengths. With an increase in the strength of the acid there were corresponding increases in the proportions of the 2-isomer produced and in the total reactivities of the heterocyclic compounds. The proportions of 2-phenylthiazole obtained from thiazole in the absence of acid, in acetic acid and in trifluoroacetic acid were 47, 83 and 96%, respectively, and the corresponding partial rate factors were 2·14, 6·2 and 30. Also, with increasing strength of the acid, there was a decrease in the amounts of secondary products formed, such as bithiazoles and thiazole benzoates.

The homolytic methylation of thiazole, methylthiazoles and dimethylthiazoles with lead tetra-acetate in acetic acid has been investigated by Dou (1966). The average proportions of 2- and 5-methylthiazole produced from the reaction with thiazole were 81 and 19% respectively. 4-Methylthiazole was not formed under the conditions of the reaction. Substitution in the methylthiazoles and dimethylthiazoles also occurred at positions 2 and 5 only. Thus there was no substitution with 2,5-dimethylthiazole, whereas both 2,4- and 4,5-dimethylthiazole gave 2,4,5-trimethylthiazole. Competitive methylation, using mixtures of thiazole and nitrobenzene in the absence of acid, showed that the methyl affinity of thiazole is in the range 1·5—2·0, which is between those of benzene (methyl affinity = 1) and pyridine (methyl affinity = 3) (see Table 1).

The homolytic phenylation of benzothiazole with benzoyl peroxide has been reported to give 2-phenylbenzothiazole (Nagasaka, Oda and Nukina, 1954). Dou and Lynch (1966c) found that the 2-isomer accounted for 94% of the total substitution products obtained from the phenylation of benzothiazole with benzoyl peroxide in non-acidic solution. The overall reactivity of benzothiazole was increased in acidic solution. The homolytic phenylation of 2-phenylbenzothiazole with benzoyl peroxide in non-acidic solution has been investigated by Vernin, Dou and Metzger (1969). Substitution occurred at all seven of the available positions in the fused benzene ring and the phenyl group, with positions 4 and 7 the most reactive.

Hübenett and Hofmann (1963) have shown that the decomposition of benzoyl peroxide in isothiazole at 100° gives a mixture of 3-, 4- and 5-phenylisothiazole. A quantitative study of the phenylation of isothiazole and its 3-, 4- and 5-methyl derivatives with benzoyl peroxide in non-acidic and acidic media has been made by Dou and his colleagues (Dou, Poite, Vernin and Metzger, 1969; Poite, Vernin, Loridan, Dou and Metzger, 1969). The percentage isomer ratio for the phenylation of isothiazole in non-acidic solution was: 3-, 47; 4-, 9 and 5-, 44, and the reactivity of position 3 was increased in acidic solution. Competitive reactions in non-acidic solution, using benzene (reactivity = 1) as the reference compound, showed that the relative reactivities of isothiazole, 3-methylisothiazole, 4-methylisothiazole and 5-methylisothiazole towards homolytic phenylation were 0·95, 0·65, 1·85 and 0·3, respectively. In acidic solution, the relative reactivities were 2, 0·8, 6·8 and 0·9. Thus substitution of the isothiazole ring by a methyl group gave the greatest increase in reactivity when the substituent occupied a weakly reactive position.

F. ADDENDUM

Since the preparation of this review, further work has been published, and a brief summary of this work is given.

Pyridine. Turkina and Gragerov (1969) have studied the reaction of phenyl radicals, generated by the photolysis of diphenylmercury, the thermal decomposition of benzoyl peroxide, and the reaction of iodobenzene with magnesium (at 60°), with heavy pyridine, C_5D_5N. The reactions gave phenyltetradeuteropyridines and monodeuterated benzene, which was consistent with the intermediate involvement of free phenyl radicals, according to the following scheme (for 4-substitution).

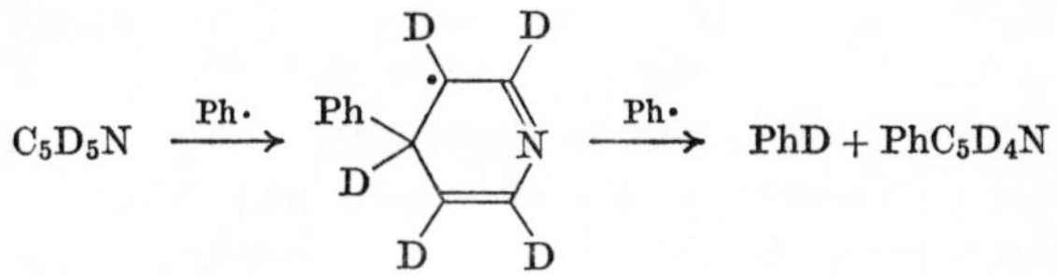

$$C_5D_5N \xrightarrow{Ph\cdot} [\text{Ph, D-substituted dihydropyridinyl radical}] \xrightarrow{Ph\cdot} PhD + PhC_5D_4N$$

The homolytic dissociation of nitrobenzene at 600° has been used by Fields and Meyerson (1970a) as a source of phenyl radicals for the phenylation of pyridine. Phenylpyridines and bipyridyls were the major products and the isomer distribution of the phenylpyridines was similar to that obtained from phenyl radical sources in the liquid phase.

Bonnier, Court and Gelus (1970) have extended their previous work on the homolytic phenylation of pyridine and its methyl derivatives in neutral and acid solution (see Tables 8 and 9) to include all of the lutidines.

The reactions of the 2-thiazolyl radical, produced by the decomposition of the diazonium salt of 2-aminothiazole, with pyridine, 4-picoline and 2,6-dimethylpyridine have been reported by Vernin, Barré, Dou and Metzger (1969). The results showed that the 2-thiazolyl radical is strongly electrophilic in character.

Travecedo and Stenberg (1970) have shown that pyridine is photoalkylated in hydrochloric acid-methanol solutions to give 2- and 4-picoline and the dimers, 1-(2′-pyridyl)-2-(4′-pyridyl)ethane and 1,2-di-4′-pyridylethane. The reaction was interpreted as proceeding *via* the protonated intermediates (XVIII) and (XIX).

(XVIII) (XIX)

Quinoxaline. The homolytic acylation of quinoxaline by means of acyl radicals obtained by the oxidation of aldehydes has been developed as a route to alkyl and aryl quinoxalin-2-yl ketones by Gardini and Minisci (1970). The aldehydes were oxidised by a redox system consisting of a mixture of t-butyl hydroperoxide, ferrous sulphate and dilute sulphuric acid.

Thiophen. The pyrolysis of nitrobenzene at 600° in the presence of thiophen has been shown by Fields and Meyerson (1970b) to give 2- and 3-phenylthiophen, in the approximate ratio of 3:1, and bithienyls. High product yields were obtained using a 1:20 mole ratio of nitrobenzene to thiophen. The reaction was also used to prepare phenyl-d_5-thiophens from nitrobenzene-d_5. In competitive experiments at 600°, with nitrobenzene as the radical source, the order of relative reactivities of benzene, pyridine and thiophen towards homolytic phenylation was found to be thiophen > pyridine > benzene.

Martelli, Spagnolo and Tiecco (1970) have reported that phenylethynyl radicals, generated by the photolysis of 1-iodo-2-phenylacetylene, react with thiophen to give 2- and 3-thienylphenylacetylene in the ratio of 3:1.

References

Abramovitch, R. A. and Kenaschuk, K. (1967) *Canad. J. Chem.*, **45**, 509.
Abramovitch, R. A. and Saha, J. G. (1963) *Tetrahedron Letters*, 301.
Abramovitch, R. A. and Saha, J. G. (1964) *J. Chem. Soc.*, 2175.
Abramovitch, R. A. and Saha, J. G. (1965) *Tetrahedron*, **21**, 3297.
Abramovitch, R. A. and Saha, J. G. (1966a) *Advances in Heterocyclic Chemistry*, Ed. Katritzky, A. R., Academic Press, New York, **6**, 229 (esp. p. 320 *et seq.*).
Abramovitch, R. A. and Saha, M. (1966b) *J. Chem. Soc.* (*B*), 733.
Abramovitch, R. A. and Saha, M. (1966c) *Canad. J. Chem.*, **44**, 1765.
Atkinson, C. M. and Sharpe, C. J. (1959) *J. Chem. Soc.*, 3040.
Augood, D. R., Hey, D. H. and Williams, G. H. (1952) *J. Chem. Soc.*, 2094.
Ayres, D. C. and Smith, J. R. (1968) *J. Chem. Soc.* (*C*), 2737.
Barnes, R. A. (1959) *J. Amer. Chem. Soc.*, **81**, 1935.

Bass, K. C. and Nababsing, P. (1965) *J. Chem. Soc.*, 4396.
Bass, K. C. and Nababsing, P. (1969) *J. Chem. Soc.* (*C*), 388.
Bass, K. C. and Nababsing, P. (1970) *J. Chem. Soc.* (*C*), 2169.
Beckwith, A. L. J. and Thompson, M. J. (1961) *J. Chem. Soc.*, 73.
Beckwith, A. L. J. and Waters, W. A. (1957) *J. Chem. Soc.*, 1001.
Benati, L., La Barba, N., Tiecco, M. and Tundo, A. (1969) *J. Chem. Soc.* (*B*), 1253.
Berthier, G. and Pullman, B. (1950) *Compt. rend.*, **231**, 774.
Bonnier, J. M. and Court, J. (1965) *Bull. Soc. chim. France*, 3310.
Bonnier, J. M. and Court, J. (1967) *Compt. rend.*, **265**, *C*, 133.
Bonnier, J. M., Court, J. and Fay, T. (1967) *Bull. Soc. chim. France*, 1204.
Bonnier, J. M., Court, J. and Gelus, M. (1966) *Compt. rend.*, **263**, *C*, 262.
Bonnier, J. M., Court, J. and Gelus, M. (1967) *Compt. rend.*, **264**, *C*, 1023.
Bonnier, J. M., Court, J. and Gelus, M. (1970) *Bull. Soc. chim. France*, 139.
Brown, R. D. (1955) *Austral. J. Chem.*, **8**, 100.
Brown, R. D. (1956) *J. Chem. Soc.*, 272.
Brown, R. D. and Harcourt, R. D. (1959) *J. Chem. Soc.*, 3451.
Brown, R. D. and Harcourt, R. D. (1960) *Tetrahedron*, **8**, 23.
Brown, R. D. and Heffernan, M. L. (1956) *Austral. J. Chem.*, **9**, 83.
Bunyan, P. J. and Hey, D. H. (1960) *J. Chem. Soc.*, 3787.
Butterworth, E. C., Heilbron, I. M. and Hey, D. H. (1940) *J. Chem. Soc.*, 355.
Buu-Hoï, Ng. Ph. and Daudel, R. (1949) *Bull. Soc. chim. France*, **16**, 801.
Buu-Hoï, Ng. Ph. and Hoàn, Ng. (1950) *Rec. Trav. chim.*, **69**, 1455.
Cadogan, J. I. G., Hey, D. H. and Sanderson, W. A. (1960) *J. Chem. Soc.*, 3203.
Camaggi, C. M., Leardini, R., Tiecco, M. and Tundo, A. (1969) *J. Chem. Soc.* (*B*), 1251.
Caronna, T., Gardini, G. P. and Minisci, F. (1969) *Chem. Comm.*, 201.
Claret, P. A. and Williams, G. H. (1969) *J. Chem. Soc.* (*C*), 146.
Conant, J. B. and Chow, B. F. (1933) *J. Amer. Chem. Soc.*, **55**, 3475.
Coulson, C. A. (1946) *Trans. Faraday Soc.*, **42**, 265.
Coulson, C. A. (1947) *Discuss. Faraday Soc.*, **2**, 9.
Coulson, C. A. (1955) *J. Chem. Soc.*, 1435.
Cowley, B. R., Norman, R. O. C. and Waters, W. A. (1959) *J. Chem. Soc.*, 1799.
Crook, K. E. (1948) *J. Amer. Chem. Soc.*, **70**, 416.
Davies, D. W. (1955) *Trans. Faraday Soc.*, **51**, 449.
Degani, J., Pallotti, M. and Tundo, A. (1961) *Ann. Chim.* (*Italy*), **51**, 434.
Del Re, G. and Scarpati, R. (1964) *Rend. Accad. Sci. fis. mat.* (*Napoli*), **31**, 88 (*Chem. Abs.*, **64**, 6425).
Dou, H. J. M. (1966) *Bull. Soc. chim. France*, 1678.
Dou, H. J. M. and Lynch, B. M. (1965) *Tetrahedron Letters*, 897.
Dou, H. J. M. and Lynch, B. M. (1966a) *Bull. Soc. chim. France*, 3815, 3820.
Dou, H. J. M. and Lynch, B. M. (1966b) *Compt. rend.*, **263**, *C*, 682.
Dou, H. J. M. and Lynch, B. M. (1966c) *Compt. rend.*, **262**, *C*, 1537.
Dou, H. J. M. and Metzger, J. (1966) *Compt. rend.*, **262**, *C*, 687.
Dou, H. J. M., Poite, J. C., Vernin, G. and Metzger, J. (1969) *Tetrahedron Letters*, 779.
Dou, H. J. M., Vernin, G. and Metzger, J. (1966) *Compt. rend.*, **263**, *C*, 1243.
Dou, H. J. M., Vernin, G. and Metzger, J. (1967) *Tetrahedron Letters*, 2223.

Dyall, L. K. and Pausacker, K. H. (1961) *J. Chem. Soc.*, 18.
Elks, J. and Hey, D. H. (1943) *J. Chem. Soc.*, 441.
Elkobaisi, F. M. and Hickinbottom, W. J. (1960) *J. Chem. Soc.*, 1286.
Fields, E. K. and Meyerson, S. (1970a) *J. Org. Chem.*, **35**, 62.
Fields, E. K. and Meyerson, S. (1970b) *J. Org. Chem.*, **35**, 67.
Ford, M. C. and Mackay, D. (1957) *J. Chem. Soc.*, 4620.
Frank, R. L. and Crawford, J. V. (1958) *Bull. Soc. chim. France*, 419.
Freund, W. (1952) *J. Chem. Soc.*, 3068.
Gaertner, R. (1952) *J. Amer. Chem. Soc.*, **74**, 4950.
Gardini, G. P. and Minisci, F. (1970) *J. Chem. Soc.* (*C*), 929.
Gol'dfarb, Ya. L., Pokhil, G. P. and Belen'kii, L. I. (1966) *Doklady Akad. Nauk S.S.S.R.* (*Chem. Sect.*), **167**, 823 (transl. p. 385).
Gol'dfarb, Ya. L., Pokhil, G. P. and Belen'kii, L. I. (1967) *Zhur. obshchei Khim.*, **37**, 2670 (transl. p. 2541).
Goldschmidt, S. and Beer, L. (1961) *Liebig's Ann.*, **641**, 40.
Goldschmidt, S. and Minsinger, M. (1954) *Chem. Ber.*, **87**, 956.
Gomberg, M. and Bachmann, W. E. (1924) *J. Amer. Chem. Soc.*, **46**, 2339.
Griffin, C. E. and Martin, K. R. (1965) *Chem. Comm.*, 154; *also* Martin, K. R. (1965) Ph.D. Thesis, University of Pittsburgh (*Diss. Abs.*, 1966, **27**, *B*, 761).
Gritter, R. J. and Chriss, R. J. (1964) *J. Org. Chem.*, **29**, 1163.
Gritter, R. J. and Godfrey, A. W. (1964) *J. Amer. Chem. Soc.*, **86**, 4724.
Hardegger, E. and Nikles, E. (1957) *Helv. Chim. Acta*, **40**, 2421.
Haworth, J. W., Heilbron, I. M. and Hey, D. H. (1940) *J. Chem. Soc.*, 349.
Hey, D. H., Perkins, J. M. and Williams, G. H. (1965) *J. Chem. Soc.*, 110.
Hey, D. H., Shingleton, D. A. and Williams, G. H. (1963a) *J. Chem. Soc.*, 5612.
Hey, D. H., Shingleton, D. A. and Williams, G. H. (1963b) *J. Chem. Soc.*, 1958.
Hey, D. H., Stirling, C. J. M. and Williams, G. H. (1955) *J. Chem. Soc.*, 3963.
Hey, D. H. and Walker, E. W. (1948) *J. Chem. Soc.*, 2213.
Hey, D. H. and Williams, G. H. (1953) *Discuss. Faraday Soc.*, **14**, 216.
Hübenett, F. and Hofmann, H. (1963) *Angew. Chem. Internat. Edn.*, **2**, 325.
Hutton, J. and Waters, W. A. (1965) *J. Chem. Soc.*, 4253.
Japelj, M., Stanovik, B. and Tisler, M. (1969) *Monatsh.*, **100**, 671.
Johnson, A. W. (1946) *J. Chem. Soc.*, 895.
Johnston, K. M. and Williams, G. H. (1960) *J. Chem. Soc.*, 1446.
Kolb, K. E. and Black, W. A. (1969) *Chem. Comm.*, 1119.
Kolb, K. E. and Wilson, C. L. (1966) *Chem. Comm.*, 272.
Longuet-Higgins, H. C. and Coulson, C. A. (1947) *Trans. Faraday Soc.*, **43**, 87.
Lynch, B. M. and Chang, H. S. (1964a) *Tetrahedron Letters*, 617.
Lynch, B. M. and Chang, H. S. (1964b) *Tetrahedron Letters*, 2965.
Lynch, B. M. and Kahn, M. A. (1963) *Canad. J. Chem.*, **41**, 2086.
Lythgoe, B. and Rayner, L. S. (1951) *J. Chem. Soc.*, 2323.
Mackay, D. (1966) *Canad. J. Chem.*, **44**, 2881.
Martelli, G., Spagnolo, P. and Tiecco, M. (1970) *J. Chem. Soc.* (*B*), 1413.
Mathur, K. B. L. and Mehra, H. S. (1961) *J. Chem. Soc.*, 2576.
McCall, E. B., Neale, A. J. and Rawlings, T. J. (1962) *J. Chem. Soc.*, 5288.
Metzger, J. and Pullman, A. (1948) *Compt. rend.*, **226**, 1613.
Minkin, V. I., Pozharskii, S. F. and Ostroumov, Yu. A. (1966) *Khim. geterotsikl. Soedinenii*, **2**, 551 (transl. p. 413).
Möhlau, R. and Berger, R. (1893) *Ber.*, **26**, 1994.

Nababsing, P. (1969) Ph.D. Thesis, City University, London.
Nagasaka, A., Oda, R. and Nukina, S. (1954) *J. Chem. Soc. Japan, Ind. Chem. Sect.*, **57**, 227 (*Chem. Abs.*, **49**, 11626).
Norman, R. O. C. and Radda, G. K. (1963) *Advances in Heterocyclic Chemistry*, Ed. Katritzky, A. R., Academic Press, New York, **2**, 131.
Overhoff, J. and Tilman, G. (1929) *Rec. Trav. chim.*, **79**, 2941.
Pausacker, K. H. (1958) *Austral. J. Chem.*, **11**, 200.
Poite, J. C., Vernin, G., Loridan, G., Dou, H. J. M. and Metzger, J. (1969) *Bull. Soc. chim. France*, 3912.
Pollak, A., Stanovnik, B. and Tisler, M. (1968) *Tetrahedron*, **24**, 2623.
Pullman, B. and Pullman, A. (1958) *Progress in Organic Chemistry*, Ed. Cook, J. W., Butterworth, London, **4**, 31.
Rieger, W. H. (1950) U.S. Patent 2,502,174 (*Chem. Abs.*, **44**, 5396).
Rinkes, I. J. (1943) *Rec. Trav. chim.*, **62**, 116.
Sandorfy, C. and Yvan, P. (1950) *Bull. Soc. chim. France*, **17**, 131.
Schwetlick, K. and Lungwitz, R. (1964) *Z. Chem.*, **4**, 458.
Smith, E. B. and Jensen, H. B. (1967) *J. Org. Chem.*, **32**, 3330.
Smith, P. A. S. and Boyer, J. H. (1951) *J. Amer. Chem. Soc.*, **73**, 2626.
Southwick, P. L., Munsell, M. W. and Bartkus, E. A. (1961) *J. Amer. Chem. Soc.*, **83**, 1358.
Szwarc, M. and Binks, J. H. (1959) *Theoretical Organic Chemistry. The Kekulé Symposium, London, 1958*, Butterworth, London, p. 262.
Taylor, G. M. (1969) Ph.D. Thesis, University of London.
Travecedo, E. F. and Stenberg, V. I. (1970) *Chem. Comm.*, 609.
Tsuda, K., Satoh, Y. and Saeki, S. (1953) *Pharm. Bull.* (*Japan*), **1**, 307.
Turkina, M. Ya. and Gragerov, I. P. (1969) *Zhur. org. Khim.*, **5**, 585 (transl. p. 575).
Vernin, G. (1967) *Compt. rend.*, **265**, *C*, 744.
Vernin, G., Barré, B., Dou, H. J. M. and Metzger, J. (1969) *Compt. rend.*, **268**, *C*, 2025.
Vernin, G. and Dou, H. J. M. (1967) *Compt. rend.*, **265**, *C*, 828.
Vernin, G. and Dou, H. J. M. (1968a) *Compt. rend.*, **266**, *C*, 822.
Vernin, G. and Dou, H. J. M. (1968b) *Compt. rend.*, **266**, *C*, 924.
Vernin, G., Dou, H. J. M. and Metzger, J. (1966) *Compt. rend.*, **263**, *C*, 1310.
Vernin, G., Dou, H. J. M. and Metzger, J. (1967a) *Compt. rend.*, **264**, *C*, 336.
Vernin, G., Dou, H. J. M. and Metzger, J. (1967b) *Compt. rend.*, **264**, *C*, 1762.
Vernin, G., Dou, H. J. M. and Metzger, J. (1967c) *Bull. Soc. chim. France*, 4514, 4521, 4523.
Vernin, G., Dou, H. J. M. and Metzger, J. (1968) *Bull. Soc. chim. France*, 3280.
Vernin, G., Dou, H. J. M. and Metzger, J. (1969) *Compt. rend.*, **268**, *C*, 977.
Vernin, G. and Metzger, J. (1963) *Bull. Soc. chim. France*, 2504.
Vitry-Raymond, J. and Metzger, J. (1963) *Bull. Soc. chim. France*, 1784.
Von Braun, J., Nelles, J. and May, A. (1937) *Ber.*, **70B**, 1767.
Waters, W. A. and Watson, D. H. (1957) *J. Chem. Soc.*, 253.
Waters, W. A. and Watson, D. H. (1959a) *J. Chem. Soc.*, 2082.
Waters, W. A. and Watson, D. H. (1959b) *J. Chem. Soc.*, 2085.
Wheland, G. W. (1942) *J. Amer. Chem. Soc.*, **64**, 900.
Williams, G. H. (1960) *Homolytic Aromatic Substitution*, Pergamon, London, Chapter 2 (and refs. therein).

2

ABSOLUTE RATE CONSTANTS FOR REACTIONS OF OXYL RADICALS[1]

J. A. Howard

Division of Chemistry,
National Research Council of Canada,
Ottawa, Ontario, Canada

A. Introduction

Oxyl radicals are transient species with an unpaired electron located mainly on an oxygen atom, and they are involved in a number of free radical processes. For example, t-butoxyl radicals take part in free radical chain chlorinations with t-butyl hypochlorite, and peroxyl radicals are the chain carrying species in the reactions of many organic compounds with molecular oxygen at moderate temperatures (< 100°). Although the kinetics and mechanisms of reactions

[1] Issued as NRCC No. 11537.

involving oxyl radicals have been known for some years it is only recently that accurate absolute rate constants for reactions of these radicals have been measured. Absolute rate constants have generally been determined either from investigations of chain reactions involving oxyl radicals or from direct measurement of radical concentrations by electron spin resonance spectroscopy or ultraviolet spectroscopy.

There are three elementary reactions associated with chain reactions, namely initiation, propagation and termination, and to evaluate rate constants for both propagation and termination reactions it is necessary to carry out three experimental determinations. These determinations are (*a*) the overall rate of reaction, (*b*) the rate of chain initiation and (*c*) the average lifetime of the radical chain or absolute concentration of the chain carrying species.

The overall rate of reaction is relatively easily measured by following the disappearance of one of the reactants, the appearance of one of the products, or by following the change in some physical property which is proportional to the rate of reaction.

It is comparatively easy to initiate a chain reaction by adding to the reaction mixture a thermally or photochemically unstable compound which decomposes to give free radicals. However, it is desirable that the rate of initiation should be constant for the duration of the experiment and that it can be measured accurately.

The overall rate of reaction, R, and the rate of chain initiation, R_i, enable the ratio of rate constants, $k_p/(2k_t)^{1/2}$, to be calculated, since R often obeys the kinetic expression,

$$R = \frac{k_p[\mathrm{RH}]\, R_i^{1/2}}{(2k_t)^{1/2}}$$

where k_p and $2k_t$ are the rate controlling propagation and termination rate constants. These rate constants can be separated by measuring the average lifetime, τ, of the chain carrying free radical since,

$$\tau = 1/(2k_t\, R_i)^{1/2}.$$

The finite lifetime of chain carrying intermediates in liquid-phase reactions was first demonstrated by Briers, Chapman and Walters (1926) using the method of intermittent illumination. These workers investigated the photochemically initiated reaction between iodine and potassium oxalate and studied the dependence of the rate of reaction on the intensity of the initiating light. Variations in light intensity were obtained by rapidly rotating an opaque disc, from which sectors had been cut, in the light beam. The rate was found to be proportional to the square-root of the light intensity and it was therefore concluded that reaction continued some time into the dark periods. This means that the reaction must involve a transient species with an appreciable lifetime.

The fact that the steady state rate of a chain reaction is not reached immediately the initiating light is switched on, or that reaction does not stop immediately the light is switched off, is the basis of several non-stationary state methods of measuring average chain lifetimes.

The rotating sector technique is the most versatile and widely used non-stationary state method and in principle can be used to study any photochemical reaction in which the overall rate is dependent on any power of the light intensity other than the first. It is however most accurate when the rate is proportional to the square root of the light intensity. In these circumstances, the rate in intermittent light depends on the frequency of interruption and on the radical lifetime. At sufficiently low speeds, no radicals produced during one period of illumination survive until the next light period. Each light period therefore promotes an independent reaction and the total reaction observed is the sum of these. The other limit occurs when periods of darkness are so short that no appreciable decay of radical concentration occurs; the effect of interrupting the light is then the same as that of reducing the intensity to a value equal to its average per cycle. For example, in the particular case when the ratio of dark to light periods is 3:1, the rate at high frequencies is twice that at low frequencies. It is clear that the critical frequency range over which the average reaction rate changes will be determined by the lifetime of the radicals and can be used to determine this quantity. The mathematical theory behind this method is discussed in many text-books on free radical reactions and is the subject of a review by Burnett and Melville (1963). This technique was first used to measure absolute rate constants for peroxyl radicals by Bateman and Gee (1948). These workers separated the propagation and termination rate constants for autoxidation of cyclohexene, 1-methylcyclohexene, dihydromyrcene, and ethyl linoleate, and this method has since been used to determine termination rate constants for many oxyl and peroxyl radicals.

Bateman and Gee (1948) also recognised that, since peroxyl radicals are relatively long lived compared with other free radicals, the induction period, after a photochemical autoxidation had been started or stopped, could be followed directly to give termination rate constants. Several techniques have subsequently been developed to follow non-stationary state autoxidations directly.

Direct measurement of the concentration of oxyl or peroxyl radicals involved in a steady state chain reaction by electron spin resonance spectroscopy is in theory the simplest method of measuring absolute termination rate constants for these radicals. During a steady state reaction, the rate of chain initiation is equal to that of chain termination,

$$R_i = 2k_t[\mathrm{R}\cdot]_s^2$$

where $[\mathrm{R}\cdot]_s$ is the stationary state concentration of free radicals. Measurement of $[\mathrm{R}\cdot]_s$ at a known R_i gives $2k_t$ directly.

An alternative method of measuring termination rate constants spectroscopically is to generate large concentrations of free radicals and monitor their decay after production has ceased. If radical decay is second-order, rate constants can be obtained from the expression,

$$\mathrm{d}[\mathrm{R}\cdot]/\mathrm{d}t = 2k_t[\mathrm{R}\cdot]^2.$$

Many absolute rate constants for reactions involving oxyl radicals have been determined by the techniques described above and the present review deals respectively with reactions of hydroxyl radicals, alkoxyl radicals, ketones in their triplet state, peroxyl radicals, aroxyl radicals, and nitroxide radicals. A comprehensive coverage of the literature has been attempted only for peroxyl radicals.

B. The Hydroxyl Radical

In 1967, Anbar and Neta published *A Compilation of Specific Bimolecular Rate Constants for the Reactions of Hydrated Electrons, Hydrogen Atoms and Hydroxyl Radicals with Inorganic and Organic Compounds in Aqueous Solution*. The data on the hydroxyl radical cover over 40 inorganic species including, for example, H•, HO•, HOO•, H_2, H_2O_2, Br^-, Cl^-, I^-, NO_2^-, Fe^{2+} and $Fe(CN)_6^{4-}$. The data on organic substrates includes over 200 compounds such as acids, alcohols, amines, aromatics, esters, ketones and phenols. The rate constants were obtained at temperatures from 15–25° both by pulse radiolysis and by competitive techniques. Virtually all the rate constants lie in the range from 1×10^8 M^{-1} sec.$^{-1}$ to 1×10^{10} M^{-1} sec.$^{-1}$. This means that these reactions occur at rates which are relatively close to the diffusion controlled limit. Most of the small differences in the rate constants between different substrates probably have no simple interpretation. Many of them must, in fact, be the result of experimental error since different workers report quite different values from the same substrate. For example, the reported rate constants for reaction with Br^- vary from $3{\cdot}6 \times 10^{10}$ M^{-1} sec.$^{-1}$ at a pH of 0·5 to 2·0, to a value of 6×10^8 M^{-1} sec.$^{-1}$ at pH 7, and even at pH 7 values up to $1{\cdot}6 \times 10^9$ M^{-1} sec.$^{-1}$ are listed. Similarly, with phenol at pH 7 the reported rate constants vary from a low of $2{\cdot}1 \times 10^9$ M^{-1} sec.$^{-1}$ to a high of $1{\cdot}06 \times 10^{10}$ M^{-1} sec.$^{-1}$. The data listed in Table 1 cover a few

TABLE 1

Bimolecular rate constants for hydroxyl radical reactions in aqueous solution (15–25°) (Anbar and Neta, 1967)

Reactant	pH	Rate Constant (M^{-1} sec.$^{-1}$) $\times 10^{-8}$
$\cdot OH$[a]	7	40–50
$\cdot OOH$[b]	0·4–3	150
I^-	7	70–100
Fe^{2+}	1	0·16–3·2
H_2	7	0·35–0·6
H_2O_2	7	0·23–0·45
Ethyl alcohol	2–10·5	7·2–12
n-Butyl alcohol	7–9	22
t-Butyl alcohol	2–9	2·5–4·2
Acetone	7	0·58–0·77
Benzene	0·7–10·5	30–47
Benzoic acid[c]	3	21–43
Benzoate ion[d]	7–10·7	25–38
Phenylacetate ion	9	26
Anisole	9	36
Toluene	3	30
Phenol	7–9	42–120
Hydroquinone	7	120

Notes: [a] More recent values are 6×10^9 M^{-1} sec.$^{-1}$ (Sehested, Rasmussen and Fricke, 1968) and $1{\cdot}04 \times 10^{10}$ M^{-1} sec.$^{-1}$ (Pagsberg, Christensen, Rabani, Nilsson, Fenger and Nielson, 1969). [b] A more recent value is $7{\cdot}1 \times 10^9$ M^{-1} sec.$^{-1}$ (Sehested, Rasmussen and Fricke, 1968). [c] A more recent value is $4{\cdot}3 \times 10^9$ M^{-1} sec.$^{-1}$ (Wander, Neta and Dorfman, 1968). [d] A more recent value is 6×10^9 M^{-1} sec.$^{-1}$ (Wander, Neta and Dorfman, 1968).

representative compounds which have, in part, been chosen because different workers have obtained similar values for the rate constants for the reactions with hydroxyl radicals.

The high rate constants for radical–radical addition reactions,

$$HO\cdot + HO\cdot \longrightarrow H_2O_2$$

$$HO\cdot + HOO\cdot \longrightarrow H_2O_3 \text{ or } H_2O + O_2$$

are to be expected. The high rate constants for the electron transfer processes,

$$HO\cdot + I^- \longrightarrow HO^- + I\cdot$$

$$HO\cdot + Fe^{2+} \longrightarrow HO^- + Fe^{3+}$$

can be attributed to the high electron affinity of the $\cdot OH$ radical and the high rate constants for hydrogen atom abstraction to the strength of the HO—H bond (119 kcal. $mole^{-1}$; Kerr, 1966).

$$HO\cdot + H_2O_2 \longrightarrow H_2O + HOO\cdot$$

$$HO\cdot + CH_3CH_2OH \longrightarrow H_2O + CH_3\dot{C}HOH$$

The reactions of the hydroxyl radical with aromatic compounds involve, without exception, an initial addition to the aromatic ring. This is true not only of benzene where no other fast reaction is likely to occur [D(Ph—H) = 112 kcal. $mole^{-1}$] (Rodgers, Golden and Benson, 1967), but is also true of substrates such as benzoic acid and the benzoate ion (Wander, Neta and Dorfman, 1968), phenylacetic acid (Norman and Pritchett, 1967), anisole (Jefcoate and Norman, 1968), toluene (Cercek, 1968) and even phenol and hydroquinone (Adams, Michael and Land, 1966; Adams and Michael, 1967; Land and Ebert, 1967). The hydroxyl radical is therefore unlike other oxyl radicals which do not react appreciably with benzene, benzoic acid and the benzoate ion. Moreover, when other oxyl radicals do react with phenylacetic acid, toluene, anisole, phenol and hydroquinone, they do so by abstracting the benzylic, ethereal or phenolic hydrogen atom.

The cyclohexadienyl radicals formed by the addition of the hydroxyl radical to the aromatic ring may enter into bimolecular radical–radical reactions. Alternatively, if they have appropriate structural features, they may undergo an acid catalysed heterolytic bond rupture in which the hydroxide ion is eliminated from the ring and a proton, or a proton and carbon dioxide or formaldehyde, is lost from the side chain. Typical examples of this behaviour are illustrated below.

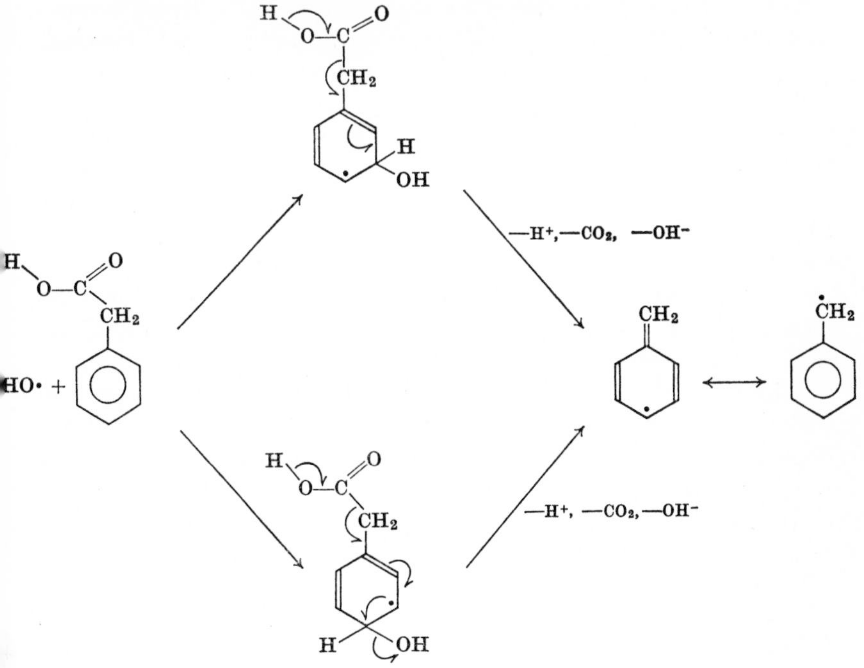

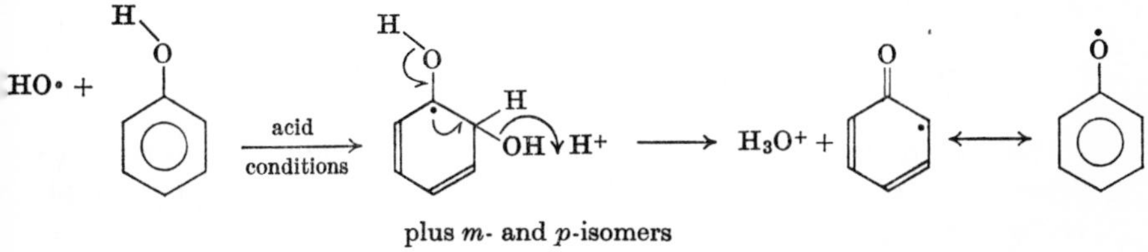

$HO\cdot$ + $C_6H_5CH_2COOH$

$-H^+, -CO_2, -OH^-$

$-H^+, -CO_2, -OH^-$

$HO\cdot$ + C_6H_5OH $\xrightarrow[\text{conditions}]{\text{acid}}$ $\longrightarrow$ H_3O^+ + $\longleftrightarrow$

plus *m*- and *p*-isomers

There are indications of different rates of catalysis for the different isomers. At pH 7 phenol yields benzosemiquinone radicals, presumably by the following reaction.

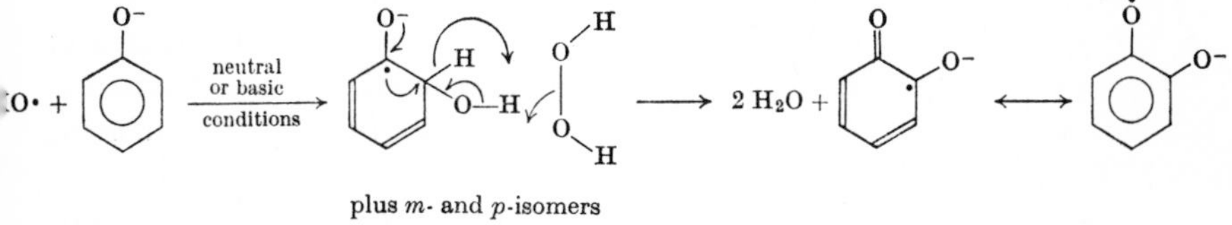

$HO\cdot$ + $C_6H_5O^-$ $\xrightarrow[\text{conditions}]{\text{neutral or basic}}$ $\longrightarrow$ 2 H_2O + $\longleftrightarrow$

plus *m*- and *p*-isomers

The driving force for all these reactions comes from the rearomatisation of the cyclohexadienyl radical.

The relative rate constants for the addition of hydroxyl radicals to substituted benzenes and to *para*-substituted benzoate ions can both be roughly correlated with the sigma constants of the substituents by means of the Hammett equation with $\rho = -0{\cdot}41$ in both cases (Anbar, Meyerstein and Neta, 1966a). The similarity in the ρ values means that the σ values in disubstituted benzenes are additive in this reaction. The negative sign of ρ supports the view that the mechanism of attack of hydroxyl radicals on aromatic compounds is analogous to an electrophilic substitution. The relative rates of hydrogen atom abstraction from a series of substituted methanes and a series of substituted acetate ions by the hydroxyl radical could also be roughly correlated by the Hammett sigma *para* constants of the substituents with $\rho = -0{\cdot}96$ (Anbar, Meyerstein and Neta, 1966b). This means that the reactivities of these substrates depend both on the inductive effect of the substituents and on the resonance stabilisation of the resultant substituted alkyl radical.

The addition of hydroxyl radicals to simple olefins is also a very fast reaction. J. K. Thomas (1967) has reported rate constants of $4{\cdot}1 \times 10^9$ M^{-1} sec.$^{-1}$ for ethylene, $6{\cdot}5 \times 10^9$ M^{-1} sec.$^{-1}$ for propylene, but-1-ene and butadiene and $5{\cdot}0 \times 10^9$ M^{-1} sec.$^{-1}$ for isobutylene. Cullis, Francis, Raef and Swallow (1967) found a value of $1{\cdot}0 \times 10^9$ M^{-1} sec.$^{-1}$ for ethylene.

$$\cdot\mathrm{OH} + \rangle\mathrm{C{=}C}\langle \longrightarrow \mathrm{HO{-}\overset{|}{\underset{|}{C}}{-}\overset{|}{\underset{|}{C}}}\cdot$$

The pK of the hydroxyl radical is $11{\cdot}8 \pm 0{\cdot}2$ (Rabani and Matheson, 1966; Weeks and Rabani, 1966) so that at pH's $> \sim 13$ the hydroxyl radical is completely converted to its basic form, O^-. This species has been found to have a reactivity towards hydrogen atom donors which is very similar to that of the neutral radical. For example, the rate constant for attack of O^- on ethanol is in the range $(8{\cdot}4\text{–}11{\cdot}3) \times 10^8$ M^{-1} sec.$^{-1}$ (Gall and Dorfman, 1969; Wander, Gall and Dorfman, 1970) while for attack of $\cdot$OH the rate constant is in the range $(7{\cdot}2\text{–}12) \times 10^8$ M^{-1} sec.$^{-1}$ (Table 1). However, O^- is much less reactive than $\cdot$OH towards the benzoate ion, the rate constant for the reaction of O^- being less than 6×10^6 M^{-1} sec.$^{-1}$ (Gall and Dorfman, 1969).

Franklin (1967) reviewed the mechanism and kinetics of hydrocarbon combustion in the gas phase and collected rate constants for many of the elementary reactions involved in these processes. Some of these rate constants for reactions involving the hydroxyl radical are listed in Table 2. The hydrogen atom abstraction reactions (except for H_2O_2) appear to be somewhat slower in the gas phase than in solution. A higher value of the rate constant for the attack of $\cdot OH$ on H_2O_2 in the gas phase has been reported by Greiner (1968a). He found that in the temperature range 300–458°K the rate constant for this reaction could be represented by

$$\log(k/\text{M}^{-1}\ \text{sec.}^{-1}) = 8{\cdot}39 \pm 0{\cdot}05 + \tfrac{1}{2}\log T - \frac{1200}{2{\cdot}303RT}.$$

This expression gives $k_{300°} = 5{\cdot}6 \times 10^8\ \text{M}^{-1}\ \text{sec.}^{-1}$; $A_{300°} = 4{\cdot}3 \times 10^9\ \text{M}^{-1}\ \text{sec.}^{-1}$; and $E = 1{\cdot}2$ kcal. mole^{-1}. Greiner (1968a) also reports a much higher rate constant than has been found by other workers for the bimolecular reaction between two hydroxyl radicals, viz., $3{\cdot}9 \pm 1{\cdot}5 \times 10^{10}\ \text{M}^{-1}\ \text{sec.}^{-1}$.

TABLE 2

Bimolecular rate constants for hydroxyl radical reactions in the gas phase (25–27°) (Franklin, 1967)

Reaction	Rate constant (M^{-1} sec.$^{-1}$)	log (A/M^{-1} sec.$^{-1}$)	E (kcal. mole^{-1})
$\cdot OH + \cdot OH \rightarrow H_2O + \cdot O \cdot$	$1{\cdot}5 \times 10^9$	—	—
$\cdot OH + H_2 \rightarrow H_2O + H\cdot$	$4{\cdot}3 \times 10^6$	10·3	5·2
$\cdot OH + CO \rightarrow CO_2 + H\cdot$	$1{\cdot}2 \times 10^8$	—	—
$\cdot OH + H_2O_2 \rightarrow H_2O + HOO\cdot$	$3{\cdot}5 \times 10^7$	—	—
$\cdot OH + CH_4 \rightarrow H_2O + CH_3\cdot$	$8{\cdot}9 \times 10^6$	11·7	6·5
$\cdot OH + H_2CO \rightarrow H_2O + H\dot{C}O$	$1{\cdot}6 \times 10^3$	12·7	13

With regard to the self-reactions of the hydroxyl radical it is worth pointing out that Caldwell and Back (1965) combined their measured yields of hydrogen and oxygen from the flash photolysis of water vapour with the earlier data of Black and Porter (1962a, b) and Del Greco and Kaufman (1962) to evaluate absolute rate constants for the two self-reactions of hydroxyl radicals in the gas phase, viz.,

$$2\cdot OH + M \longrightarrow H_2O_2 + M$$

$$2\cdot OH \longrightarrow H_2O + O.$$

The rate constant for the latter reaction was estimated to be $1{\cdot}5 \times 10^9\ \text{M}^{-1}\ \text{sec.}^{-1}$. The rate constants for the former reaction depend very much on the nature of the gaseous diluent M which supports the view that this reaction proceeds through an M—OH complex. The estimated rate constants in M^{-2} sec.$^{-1}$ units for these termolecular reactions are: M = He, $0{\cdot}31 \times 10^{12}$; M = A, $0{\cdot}35 \times 10^{12}$; M = Xe, $0{\cdot}47 \times 10^{12}$; M = N_2, $1{\cdot}2 \times 10^{12}$; M = O_2, $1{\cdot}8 \times 10^{12}$; M = CO_2, $1{\cdot}5 \times 10^{12}$; and M = H_2O, $6{\cdot}5 \times 10^{12}$.

$$\cdot OH + M \rightleftharpoons H\dot{O}M \xrightarrow{\cdot OH} H_2O_2 + M$$

The reactivity of •OD radicals towards hydrogen donors is similar to that of •OH radicals (Greiner, 1968b).

C. Alkoxyl Radicals

These radicals are less reactive in hydrogen atom abstractions than hydroxyl radicals but are more reactive than peroxyl radicals. These reactivity differences can be related to differences in the O—H bond strengths, viz., D(HO—H) = 119 kcal. mole^{-1}, D(RO—H) = 104 kcal. mole^{-1} and D(ROO—H) = 90 kcal. mole^{-1} (Benson and Shaw, 1968).

Alkoxyl radicals can be cleanly generated by the thermolysis or photolysis of peroxides:

$$ROOR \longrightarrow 2RO\cdot,$$

hyponitrites:

$$RON{:}NOR \longrightarrow 2RO\cdot + N_2$$

and some per-esters, particularly per-oxalates:

$$ROO\overset{\overset{O}{\|}}{C}.\overset{\overset{O}{\|}}{C}OOR \longrightarrow 2RO\cdot + 2CO_2.$$

Other commonly used sources of alkoxyl radicals include hydroperoxides:

$$ROOH \longrightarrow RO\cdot + \cdot OH,$$

nitrites:

$$RONO \longrightarrow RO\cdot + NO,$$

nitrates:

$$RONO_2 \longrightarrow RO\cdot + NO_2$$

and hypochlorites and hypobromites:

$$ROX \longrightarrow RO\cdot + X\cdot.$$

These last four methods of generating alkoxyl radicals suffer from the disadvantage that a second reactive species is formed in the homolysis. However, with the hypohalites this disadvantage is outweighed by the fact that it is possible to achieve a radical chain halogenation of suitable organic substrates in which the alkoxyl radical functions as one of the chain carriers.

$$RO\cdot + R'H \longrightarrow ROH + R'\cdot$$
$$R'\cdot + ROX \longrightarrow R'X + RO\cdot$$

No other source of alkoxyl radicals enters into chain reactions with a facility approaching that of the hypohalites simply because the second step of this chain reaction has a rate constant comparable with the rate constant for attack of RO· on the hydrocarbon.

Other important reactions which can yield alkoxyl include the following.

$$ROO\cdot + PR_3' \longrightarrow RO\cdot + PR_3'O$$
$$ROO\cdot + Ar_2N\cdot \longrightarrow RO\cdot + Ar_2NO\cdot$$
$$R\cdot + N_2O \longrightarrow RO\cdot + N_2$$
$$R\cdot + NO_2 \longrightarrow RO\cdot + NO$$

Alkoxyl radicals enter into all the usual types of radical reactions. That is, they react with radicals (combination and disproportionation with themselves and with other radicals), they react with molecules (intermolecular hydrogen abstraction, addition to unsaturated systems, and substitution at multivalent atoms, e.g.

$$RO\cdot + BR_3' \longrightarrow ROBR_2' + R'\cdot),$$

and they undergo unimolecular reactions (intramolecular hydrogen abstractions, additions, etc., and β-scission, i.e., $R_3CO\cdot \rightarrow R_2CO + R\cdot$).

Gray, Shaw and Thynne (1967) have written a comprehensive review entitled *The Rate Constants of Alkoxyl Radical Reactions* which covers the literature on this subject through 1965, with primary emphasis on reactions in the gas phase. Perhaps the most striking feature of this article, from the present point of view, was the absence of any directly measured *absolute* rate constants and Arrhenius parameters. The non-stationary state measurements which are a necessity if absolute rate constants are to be *measured* had not up to that time been made on any alkoxyl radical reaction. The measured rate constants were therefore relative, activation energies were differences or were unknown, and pre-exponential factors were obtained by theoretical calculation. Thus, for example, the authors pointed out that "no direct experimental data have been obtained for the velocity constants or Arrhenius parameters of (alkoxyl) dimerisation. The chief reasons for this are that by contrast with the alkyl series, it is far less easy to generate RO• radicals under conditions in which the peroxide dimer ROOR is stable, and far less easy to isolate and estimate quantitatively peroxides than alkanes."

Fortunately the thermochemistry of many alkoxyl reactions was fairly accurately known and so Gray, Shaw and Thynne were able to calculate indirect values for rate constants. For many of the reactions it was even possible to calculate a rate constant by more than one independent route. Thus, three methods were used to calculate the *A*-factor for the autocombination of methoxyl.

$$CH_3O\cdot + CH_3O\cdot \longrightarrow CH_3OOCH_3$$

These methods were: (*i*) analogy with experimental data for the corresponding reaction of two ethyl radicals (which gave $\log(A/\text{M}^{-1}\ \text{sec.}^{-1}) = 10$—$11$); (*ii*) transition state theory (which gave $\log(A/\text{M}^{-1}\ \text{sec.}^{-1} = 7{\cdot}3)$; (*iii*) combination of the known *A*-factor for the reverse reaction with the estimated overall entropy change (which gave $\log(A/\text{M}^{-1}\ \text{sec.}^{-1} = 9{\cdot}2)$. The last method was believed to give the most reliable results. Provided the dimerisation of alkoxyl does not involve any activation energy, the *A*-factors indicated that the rate constants for the dimerisation of methoxyl, n-propoxyl and t-butoxyl were all about $10^{9{\cdot}3}\ \text{M}^{-1}\,\text{sec.}^{-1}$. Theoretical rate constants obtained by these methods were subsequently combined with experimental rate constant ratios to derive absolute rate constants for many other reactions of alkoxyl radicals. A few of these estimated 'absolute' rate constants for various representative reactions of alkoxyl radicals in the gas phase have been taken from Gray, Shaw and Thynne's review and are listed in Table 3.

Although there is no particular reason to question the validity of the rate constants calculated by Gray, Shaw and Thynne, it is obviously desirable to have some experimentally measured absolute rate constants. Since their review was published there have been two direct experimental determinations of the rate constant for the association of t-butoxyl radicals in solution.

$$(CH_3)_3CO\cdot + (CH_3)_3CO\cdot \longrightarrow (CH_3)_3COOC(CH_3)_3$$

The kinetics of the t-butyl hypochlorite radical chain halogenation

TABLE 3

Estimated activation parameters for some reactions of alkoxyl radicals in the gas phase (Gray, Shaw and Thynne, 1967)

Reaction	$\log(A/\text{M}^{-1}\ \text{sec.}^{-1})$	E (kcal. mole^{-1})
$CH_3O\cdot + CH_3O\cdot \rightarrow CH_3OOCH_3$	9·2	0
$CH_3O\cdot + CH_3O\cdot \rightarrow CH_3OH + CH_2O$	10–11	0
$CH_3CH_2O\cdot + NO \rightarrow CH_3CH_2ONO$	7·5	0
$(CH_3)_3CO\cdot \rightarrow CH_3COCH_3 + CH_3\cdot$	10·5 ± 0·7[a]	13 ± 2
$CH_3O\cdot + CH_4 \rightarrow CH_3OH + CH_3\cdot$	8·8	11
$CH_3O\cdot$ + Propane $\rightarrow CH_3OH + Pr\cdot$	8·2	5·2
$CH_3O\cdot$ + Butane $\rightarrow CH_3OH + Bu\cdot$	7·4	2·9
$CH_3O\cdot + CH_3CO.OCH_3 \rightarrow CH_3OH + \cdot CH_2CO.OCH_3$	7·6	6·6
$CH_3O\cdot + H_2CO \rightarrow CH_3OH + H\dot{C}O$	8	3·0

[a] Sec.$^{-1}$.

of certain hydrocarbons in carbon tetrachloride indicates that chain termination can involve the autocombination of t-butoxyl radicals. The rotating sector method has been used to measure the rate constant for this process (Carlsson, Howard and Ingold, 1966; Carlsson and Ingold, 1967a, b). The rate constant has also been directly measured by an e.s.r. technique (Weiner and Hammond, 1969). The results of these experiments are described in some detail below.

In an impressive series of papers, Walling and co-workers have described the radical chain halogenation of hydrocarbons and other organic substrates in solution by t-alkyl hypohalites—principally t-butyl hypochlorite. In the first paper (Walling and Jacknow, 1960) the reaction was shown to be a free radical chain with the following propagating steps.

$$Bu^tO\cdot + RH \longrightarrow Bu^tOH + R\cdot \qquad (1)$$

$$R\cdot + Bu^tOCl \longrightarrow RCl + Bu^tO\cdot \qquad (2)$$

In this paper, and in subsequent papers in the series, the results of many competitive experiments (both inter- and intra-molecular) are reported, so there is now available a very large list of relative rate constants for t-butoxyl attack on organic substrates. Most of Walling's measurements were made at 40° but there were also measurements at temperatures as low as −78° and as high as 100°. The relative reactivities of many organic substrates in the liquid phase towards t-butoxyl at still higher temperatures have been determined by other workers (Williams, Oberright and Brooks, 1956; Brook, 1957; Johnston and Williams, 1958, 1960; Brook and Glazebrook, 1960; Patmore and Gritter, 1962; Wallace and Gritter, 1963; Ingold, 1963; Schwetlick, Karl and Jentzsch, 1963).

Kinetic studies of the radical chain halogenation of toluene in carbon tetrachloride by t-butyl hypochlorite were reported in preliminary form simultaneously by Carlsson, Howard and Ingold (1966) and Walling and Kurkov (1966), the full papers appearing the following year (Carlsson and Ingold, 1967a, b; Walling and Kurkov, 1967). The first group of workers initiated the reaction photochemically at room temperature.

$$\mathrm{Bu^tOCl} \xrightarrow{h\nu} \mathrm{Bu^tO\cdot + Cl\cdot}\ (\text{Rate} = R_i)$$

The overall rate of reaction was found to be proportional to the toluene concentration and to the square root of the rate of chain initiation, R_i. This implies that chain termination occurs exclusively by the self-reaction of t-butoxyl radicals.

$$\mathrm{Bu^tO\cdot + Bu^tO\cdot} \longrightarrow \mathrm{Bu^tOOBu^t} \tag{3}$$

The overall rate of reaction could be represented by the following expression:

$$\frac{-\mathrm{d[RH]}}{\mathrm{d}t} = k_1[\mathrm{RH}]\left(\frac{R_i}{2k_3}\right)^{1/2}.$$

Similar kinetics were obtained with p-xylene, t-butylbenzene, triphenylmethane and cyclohexane in solvents such as carbon tetrachloride, $CF_2Cl.CFCl_2$, and benzene. The rotating sector technique was applied to all these reactions and, in almost every case, gave $2k_3 \sim 2 \times 10^8\ \mathrm{M}^{-1}\ \mathrm{sec.}^{-1}$. The chain propagation rate constants reported by Carlsson and Ingold were based on an overall heat of reaction of -51 kcal. mole^{-1} for the process.

$$\mathrm{Bu^tOCl + PhCH_3} \longrightarrow \mathrm{Bu^tOH + PhCH_2Cl}$$

The overall heat of reaction for the analogous process with cyclohexane has now been accurately determined as $-42{\cdot}4$ kcal. mole^{-1} by Walling and Papaioannou (1968). Since $D(\mathrm{R—H}) - D(\mathrm{R—Cl})$ for toluene and cyclohexane must be quite similar, Carlsson and Ingold's propagation rate constants should be increased by a factor of 51/42·4. The revised values of the chain propagation rate constants for toluene in carbon tetrachloride from 10–55° can be represented by,

$$(k_1)_{PhCH_3} = 8{\cdot}0 \times 10^7 \exp(-5600/RT)\,\text{M}^{-1}\,\text{sec.}^{-1}$$
$$= 6 \times 10^3\,\text{M}^{-1}\,\text{sec.}^{-1} \text{ at } 24°.$$

The revised value for cyclohexane at 24° is $(k_1)_{C_6H_{12}} = 3{\cdot}1 \times 10^4$ $\text{M}^{-1}\,\text{sec.}^{-1}$.

In Walling and Kurkov's (1966, 1967) kinetic studies of the t-butyl hypochlorite chlorination of toluene and cyclohexane in carbon tetrachloride, the chains were initiated thermally with α,α′-azobisisobutyronitrile (AIBN). With toluene as the substrate the kinetics indicated that termination occurred by both reaction (3) and by the cross-termination reaction (4).

$$\text{Bu}^t\text{O}\cdot + \text{R}\cdot \longrightarrow \text{Inactive products} \qquad (4)$$

The relative importance of these two termination processes depends on the ratio of the concentrations of hypochlorite and toluene. With cyclohexane the kinetics indicated that termination occurred by reaction (4) and by the self-reaction of cyclohexyl radicals reaction (5).

$$\text{R}\cdot + \text{R}\cdot \longrightarrow \text{Inactive products} \qquad (5)$$

Walling and Kurkov concluded that, even with toluene, reaction (3) becomes dominant only at low hydrocarbon:hypochlorite ratios and it is here that β-scission and reaction of t-butoxyl radicals with the hypochlorite and its decomposition products become significant complications (particularly at the higher temperatures). This implies serious obstacles to non-steady state measurements. However, from the results on toluene, by using an ingenious analogy with the cage effect for the photo-production of iodine atoms from iodine (models of approximately the same mass, size and shape as t-butoxyl), they were able to estimate $2k_3 = 1{\cdot}4 \times 10^9\,\text{M}^{-1}\,\text{sec.}^{-1}$. This rate constant is close to the value of $2 \times 10^9\,\text{M}^{-1}\,\text{sec.}^{-1}$ which is generally found for the self-reaction of alkyl radicals (e.g. methyl, trichloromethyl, *n*-hexyl, cyclohexyl, t-butyl, etc.) in solvents of similar viscosity to cyclohexane (Carlsson and Ingold, 1968; Carlsson, Ingold and Bray, 1969 and references cited). The measured reaction rates (which were 3–4 times as large as those reported by Carlsson and Ingold) were combined with the estimated value of $2k_3$ to obtain $(k_1)_{PhCH_3} =$ $8 \times 10^4\,\text{M}^{-1}\,\text{sec.}^{-1}$ at 30°, and hence $(k_1)_{C_6H_{12}} = 4{\cdot}8 \times 10^5\,\text{M}^{-1}\,\text{sec.}^{-1}$.

At the time of these kinetic studies it was already known that chlorination of compounds containing benzylic hydrogen exhibited several curious features (Wagner and Walling, 1965). In 1967, Sakurai and Hosomi reported that the relative reactivities of ring substituted toluene derivatives towards t-butoxyl generated by the thermal decomposition of di-t-butyl peroxyoxalate at 45° could be correlated by means of the Hammett equation (σ^+ constants) with ρ^+ values in the range $-0{\cdot}32$ to $-0{\cdot}39$ in $CF_2Cl.CFCl_2$, chlorobenzene and acetonitrile. These ρ values were only about half as large as the values reported by earlier workers, who all used t-butyl hypochlorite as their t-butoxyl source. This led Sakurai and Hosomi to suggest that a chlorine atom chain might be involved in the t-butyl hypochlorite—toluene reaction.

$$R\cdot + Cl_2 \longrightarrow RCl + Cl\cdot \qquad (6)$$

$$Cl\cdot + RH \longrightarrow HCl + R\cdot$$

$$Bu^tOCl + HCl \longrightarrow Bu^tOH + Cl_2$$

Their suggestion has been amply confirmed by Walling and McGuinness (1969), who showed that the chlorine atom chain is normally important only for compounds containing benzylic hydrogen. (Many earlier tests for a chlorine chain had given negative results because substrates such as cyclohexane were used). The chlorine atom chain is successfully propagated by benzylic radicals but not by alkyl radicals because k_6/k_2 is very large for the former radicals (perhaps $>10^4$), but is probably near unity for the latter radicals. Walling and McGuinness concluded that in the earlier kinetic studies chlorine atom chains may have been involved. However, since chlorine atom reactions are very fast, and kinetics depend chiefly on slow steps, the reported rate constants are probably essentially correct. On the whole, it seems likely that the true propagation rate constants are reasonably close to the mean of the values estimated by Carlsson and Ingold and by Walling and Kurkov, viz., $\sim 2 \times 10^4$ M^{-1} sec.$^{-1}$ for toluene and $\sim 1 \times 10^5$ M^{-1} sec.$^{-1}$ for cyclohexane at ambient temperatures. These values may be compared with the values which can be calculated from the data in Table 3 for the analogous gas phase hydrogen atom abstraction by methoxyl radicals from propane ($k_{24°} = 2{\cdot}5 \times 10^4$ M^{-1} sec.$^{-1}$) and from n-butane ($k_{24°} = 2{\cdot}0 \times 10^5$ M^{-1} sec.$^{-1}$). They also allow the large quantity of information on relative rates obtained by Walling and by other workers to be put on a reasonably firm 'absolute' basis. Some representative results are given in Table 4.

TABLE 4

Some estimated rate constants for t-butoxyl radical attack on organic substrates in non-bonding solvents (CCl_4 *commonly*) *at ambient temperatures* (24–40°)

Substrate	Rate constant (M^{-1} sec.$^{-1}$)	Ref.
Toluene	$^{a}2 \times 10^4$	See text
Cyclohexane	$^{a}1 \times 10^5$	See text
t-Butylbenzene	6×10^3	Walling and Jacknow (1960)
Ethylbenzene	5×10^4	Walling and Jacknow (1960)
Cumene	6×10^4	Walling and Jacknow (1960)
2,3-Dimethylbutane	6×10^4	Walling and Jacknow (1960)
Cyclohexene	8×10^5	Walling and Thaler (1961)
Cyclopentene	5×10^5	Wagner and Walling (1965)
Diethyl ether (0°)	$^{b}2 \times 10^5$	Walling and Mintz (1967)
Tetrahydrofuran (0°)	$^{b}2 \times 10^5$	Walling and Mintz (1967)
Tri-n-butylborane	$^{c}7 \times 10^6$	Davies, Griller, Roberts and Tudor (1970)
Cumene hydroperoxide	$^{d}1 \times 10^7$	Howard and Ingold (1969a)

Notes: [a]Approximate values (uncertain by a factor of at least 3) which have been used to calculate the other rate constants. [b]Taking the toluene rate constant as 1×10^4 M^{-1} sec.$^{-1}$ at 0°. [c]This is a homolytic substitution at boron, not a hydrogen atom abstraction. $Bu^tO\cdot + Bu_3^nB \rightarrow Bu^tOBBu_2^n + Bu^n\cdot$. [d]Attack by cumyloxyl radical.

Although fairly stable, the t-butoxyl radical can undergo β-scission to yield acetone and a methyl radical.

$$Bu^tO\cdot \longrightarrow CH_3CO.CH_3 + CH_3\cdot \qquad (7)$$

This reaction has frequently been utilised to compare the relative rates of attack of t-butoxyl radicals on different substrates to avoid doing direct competition experiments. That is, it is possible to measure the ratio of t-butyl alcohol to acetone under standard conditions for each of two hydrocarbons R′H and R″H,

$$[Bu^tOH]'/[Me_2CO]' = k_1'[R'H]/k_7$$

$$[Bu^tOH]''/[Me_2CO]'' = k_1''[R''H]/k_7.$$

From these ratios it is easy to calculate k_1'/k_1''. Reaction (7) is subject to very pronounced solvent effects, k_7 increasing with increasing solvent polarity (Walling and Wagner, 1963, 1964). The rate constant for reaction (7) in two weakly bonding solvents (CCl_4 and $CF_2Cl.CFCl_2$) was estimated by Carlsson and Ingold (1967b). The revised rate constant is given by,

$$k_7 = 3{\cdot}4 \times 10^{12} \exp(-13{,}900/RT)\ \text{sec.}^{-1}.$$

Gray, Shaw and Thynne (1967) gave $k_7 = 10^{10{\cdot}5 \pm 0{\cdot}7} \exp[(-13{,}000 \pm 2000)/RT]\ \text{sec.}^{-1}$. Most other alkoxyl radicals undergo β-scission more rapidly than t-butoxyl. A greater rate of formation of carbonyl product may be useful for comparing the rates of alkoxyl attack on highly reactive substrates in which t-butoxyl would not undergo detectable β-scission.

Weiner and Hammond (1969) have tackled the problem of an absolute rate constant for an alkoxyl radical reaction in the most direct manner possible. Di-t-butyl peroxide was photolysed in the cavity of an e.s.r. spectrometer and a one-line spectrum at $g = 2{\cdot}004$ was assigned to the t-butoxyl radical. The rate of decay of this signal when the light was cut-off gave $2k_3 = 1{\cdot}3 \times 10^9\ \text{M}^{-1}\ \text{sec.}^{-1}$, in excellent agreement with Walling and Kurkov's (1967) estimate, and about six times as large as Carlsson and Ingold's (1967) measured value.

The assignment of the $g = 2{\cdot}004$ signal to t-butoxyl has been questioned by Symons (1969) on the grounds that in this radical the unpaired electron would be in an orbitally degenerate system, the ground state of which, being $^2\pi 3/2$, would have a g tensor ranging from 0 to 4. That is, one should not be able to see an e.s.r. signal from t-butoxyl. Symons (1969) suggests that the radical is actually a trioxide, possibly formed by the following reaction.

$$\text{Bu}^t\text{O}\cdot + \text{O}_2 \rightleftharpoons \text{Bu}^t\text{OOO}\cdot$$

To sum up, there are no accurately determined absolute rate constants for any alkoxyl radical reactions. It is to be hoped that this situation will not long continue.

D. Ketones in their Triplet States

The weak absorption band of ketones and aldehydes which occurs in the 2800–3600 Å region arises from the excitation of a nonbonding $2p$ electron from one of the unshared pairs of oxygen to the antibonding state of the carbon—oxygen π bond. The excitation corresponds to an n-π^* state which has two electrons in the bonding π

orbital, one in the antibonding π^* orbital and the remaining $2p$ electrons still largely localised on oxygen. The initial singlet state, $[n,\pi^*]^1$ undergoes rapid intersystem crossing to the triplet state, $[n,\pi^*]^3$ with high efficiency so that much of the photochemistry of carbonyl compounds arises from their triplet states.

$$R_2C{=}O \xrightarrow{h\nu} R_2C{=}O^{*(1)} \xrightarrow[\text{crossing}]{\text{intersystem}} R_2C{=}O^{*(3)}$$

In valence bond terms the ground state structure can be represented as:

$$\rangle C{=}\ddot{O}: \text{ (ground state)},$$

while the n,π^* state has an electron deficiency at the carbonyl oxygen which is produced by promotion of a non-bonding (n) electron to an antibonding (π^*) orbital:

$$\rangle \underset{\delta-}{\dot{C}}{=}\underset{\delta+}{\dot{O}}: \text{ } (n,\ \pi^* \text{ excited state}).$$

This state, with an unpaired $2p$ electron on oxygen, resembles an alkoxyl radical (which has seven electrons on singly bonded oxygen). This resemblance was first pointed out by Padwa (1964) and by Walling and Gibian (1965).

Aromatic carbonyl compounds form triplets with very high efficiency, but in some cases (e.g. naphthyl ketones) the lowest excited triplet corresponds to a π,π^* state, in which the excitation is redistributed throughout the entire conjugated π system (Wagner and Hammond, 1968). Since the carbonyl oxygen retains its two non-bonding $2p$ electrons, these triplets do not have appreciable alkoxyl character.

The chemical behaviour of carbonyl compounds in the n,π^* triplet state is remarkably similar to that of alkoxyl radicals. Thus, acyclic aliphatic carbonyl compounds can cleave into acyl and alkyl radicals.

$$RR'C{=}O^{*(3)} \longrightarrow R\cdot + R'\dot{C}{=}O \longrightarrow R'\cdot + CO$$

This is known as a Norrish type I process and is analogous to the β-scission of alkoxyls. [Actually, type I cleavage occurs from both singlet and triplet excited states, as has been shown by the fact that the photolysis of acetone is only incompletely quenched by high pressures of biacetyl or but-2-ene (Wagner and Hammond, 1968).

The extent to which dissociation can be quenched depends not only on the stability of the alkyl radical, R•, produced but also on temperature and on the wavelength of the exciting radiation.]

Carbonyl compounds that possess γ-hydrogen atoms can give olefins and smaller carbonyl compounds. The reaction proceeds by a shift of a γ-hydrogen to oxygen and cleavage to an olefin and an enol (Wagner and Hammond, 1968), e.g.,

$$CH_3.CD_2.CH_2.CH_2.\overset{\overset{O}{\|}}{C}.CH_3 \xrightarrow{h\nu} CH_3.CD{=}CH_2 + CH_2{=}\overset{\overset{DO}{|}}{C}{-}CH_3 \longrightarrow CH_2D{-}\overset{\overset{O}{\|}}{C}{-}CH_3.$$

Analogous intramolecular abstractions occur with alkoxyl radicals (Walling and Padwa, 1963). Intramolecular hydrogen transfer is also involved in the photolytic conversion of cyclopentanone and cyclohexanone to the corresponding ω-alkenals in solution, e.g.,

$$\text{cyclohexanone} \xrightarrow{h\nu} CH_2{=}CH.CH_2.CH_2.CH_2.CHO \left(+ CO + \text{cyclopentane} \right).$$

The commonest reaction of triplet ketones in solution is intramolecular hydrogen abstraction to yield an α-hydroxyalkyl radical, $RR'\dot{C}OH$. The photoreduction of benzophenone in propan-2-ol leads to benzopinacol and acetone by the reaction sequence,

$$Ph_2C{=}O \xrightarrow{h\nu} Ph_2C{=}O^{*(1)} \longrightarrow Ph_2C{=}O^{*(3)}$$

$$Ph_2C{=}O^{*(3)} + (CH_3)_2CHOH \longrightarrow Ph_2\dot{C}OH + (CH_3)_2\dot{C}OH$$

$$(CH_3)_2\dot{C}OH + Ph_2C{=}O \longrightarrow CH_3.CO.CH_3 + Ph_2\dot{C}OH$$

$$Ph_2\dot{C}OH + Ph_2\dot{C}OH \longrightarrow \underset{OH}{Ph_2C}{-}\underset{OH}{CPh_2}.$$

The quantum efficiency of this process approaches 2 at high benzophenone concentrations since the $(CH_3)_2\dot{C}OH$ radical reacts with benzophenone to form a second α-hydroxydiphenylmethyl radical.

The similarity between the benzophenone triplet and alkoxyl radicals even extends to the energetics of hydrogen atom abstractions.

Walling and Gibian (1965) calculated $D(\mathrm{Ph_2\dot{C}O{-}H}) = 104 \pm 3$ kcal. mole^{-1} which is the same as the O—H bond strength in alcohols, i.e., $D(\mathrm{RO{-}H}) = 104$ kcal. mole^{-1}. It is therefore not surprising to find that triplet ketones and alkoxyl radicals exhibit similar relative reactivities in hydrogen atom abstraction from organic substrates and, moreoever, that the absolute rate constants for hydrogen abstraction by the two species are similar.

Both Padwa (1964) and Walling and Gibian (1965) have carried out competitive experiments in which mixtures of two or more hydrocarbons and a ketone were photolysed, the relative consumptions of the hydrocarbons yielding the relative rate constants for attack by the triplet ketone. Benzophenone, some ring substituted benzophenones, acetophenone and propiophenone all exhibit almost identical selectivities. The ketone triplets appear to be somewhat more selective than t-butoxyl, particularly towards aliphatic compounds with relatively strong C—H bonds. Thus the reactivity of cyclohexane relative to toluene towards the benzophenone triplet is given as 4·08 by Padwa (1964) and as 2·2 by Walling and Gibian (1965). Towards t-butoxyl the reactivity ratio is about 6 (Walling and Jacknow, 1960; Walling and McGuinness, 1969). The primary hydrogens of 2,3-dimethylbutane are less than 1/300 as reactive as the tertiary hydrogens towards triplet benzophenone, but are only ~1/40 as reactive as the tertiary towards t-butoxyl (Walling and Gibian, 1965). The reactivities of ring substituted toluenes towards triplet benzophenone can be correlated by the Hammett equation using σ^+ constants with $\rho = -1{\cdot}16$. This value is about 3 times as large as that found for t-butoxyl (Sakurai and Hosomi, 1967), in agreement with our expectation that the triplet ketone is the more electrophilic species.

Relative rate constants for hydrogen atom abstractions by aromatic triplet ketones, particularly the benzophenone triplet, have been determined by several workers. These rate constants can be put on an absolute basis since they can be related, either directly or indirectly, to the rate constant, k_q, for the quenching of the triplet by various 'diffusion-controlled' quenchers such as naphthalene, azoisopropane and certain paramagnetic metal chelates.

$$\mathrm{Ph_2C{=}O^{*(3)} + RH \xrightarrow{k_a} Ph_2\dot{C}OH + R\cdot}$$

$$\mathrm{Ph_2C{=}O^{*(3)} + Q \xrightarrow{k_q} Ph_2C{=}O + Q^{*(3)}}$$

The average maximum rate constant for what are believed to be diffusion-controlled triplet energy transfers in benzene at 25°, k_q (diff), has been found to be $\sim 5 \times 10^9$ M^{-1} sec.$^{-1}$ (Sandros and Bäckström, 1962; Sandros, 1964; Herkstroeter and Hammond, 1966; Fry, Liu and Hammond, 1966). The average value of k_q (diff)/k_a for the attack of the benzophenone triplet on benzhydrol:

$$Ph_2C{=}O^{*(3)} + Ph_2CHOH \longrightarrow 2Ph_2\dot{C}OH,$$

(one of the most studied photo-reductions) is approximately 600 (Hammond and Leermakers, 1962; Hammond and Foss, 1964), giving a k_a for this abstraction of 9×10^6 M^{-1} sec.$^{-1}$. Bell and Linschitz (1963) measured this same rate constant directly by a flash spectroscopic technique and obtained a value of 2×10^6 M^{-1} sec.$^{-1}$. Wagner (1969) has pointed out that the extensive measurements of the relative reactivities of various molecules toward triplet benzophenone make it important to decide which is the correct rate constant. He favours the upper value for the rate constant on the grounds that there is no reason to question the accuracy of the competitive technique, whereas the flash equipment was being pressed to its limit in Bell and Linschitz's experiments. Recent flash-spectroscopic studies by Clark, Litt and Steel (1969a, b) support the high value of 5×10^9 M^{-1} sec.$^{-1}$ for k_q (diff) for triplet benzophenone, and hence indirectly support the high value of 9×10^6 M^{-1} sec.$^{-1}$ for hydrogen abstraction from benzhydrol by this triplet.

Several groups of workers (Beckett and Porter, 1963; Cohen and Sherman, 1963; Bell and Linschitz, 1963; Cohen and Stein, 1969) have reported a rate constant of ~ 1–2×10^6 M^{-1} sec.$^{-1}$ for abstraction by triplet benzophenone from propan-2-ol in concentrated propan-2-ol or water. However, Walling and Gibian (1965) estimated a value of only $8{\cdot}7 \times 10^4$ M^{-1} sec.$^{-1}$ for dilute benzene solutions of propan-2-ol. These workers combined Hammond, Baker and Moore's (1961) value of $k_q/k_a \sim 10^5$ for toluene [with ferric dipivaloylmethide as quencher with $k_q = 9 \times 10^8$ M^{-1} sec.$^{-1}$ (Bell and Linschitz, 1963; Wagner, 1969)] to obtain a value of k_a for toluene of 9×10^3 M^{-1} sec.$^{-1}$. Their competitive hydrogen atom abstraction experiments then gave the above mentioned value for propan-2-ol. Wagner and Hammond (1968) pointed out that this discrepancy might be due to a solvent effect. However, recent results by Cohen and Litt (1970) indicate that the rate constant for reaction of the benzophenone triplet with propan-2-ol is virtually the same in benzene, in alcoholic solution and in water. This is in line with the view that solvent effects on oxyl radical reactions are small for bimolecular hydrogen abstraction reactions, though they can be large for unimolecular cleavage reactions (Walling and Wagner, 1963, 1964). Taking Cohen and Litt's (1970) rate constant of $1{\cdot}8 \times 10^6$ M^{-1} sec.$^{-1}$ for abstraction from propan-2-ol rather than Walling and Gibian's estimated value of $8{\cdot}7 \times 10^4$ M^{-1} sec.$^{-1}$ gives a value of k_a for toluene of $1{\cdot}9 \times 10^5$ M^{-1} sec.$^{-1}$. This figure has been used to convert Walling and Gibian's relative reactivities to an absolute base. These rate constants are listed in Table 5.

TABLE 5

Rate constants for hydrogen atom abstraction by triplet benzophenone at ambient temperatures

Substrate	Rate constant $\times 10^{-5}$ (M^{-1} sec.$^{-1}$)
Toluene	1·9[a]
Ethylbenzene	5·8[b]
Cumene	6·3[a]
t-Butylbenzene	0·5[b]
m-Xylene	5·5[a]
p-Xylene	8·2[a]
Mesitylene	10[a]
p-Fluorotoluene	2·1[a]
p-Chlorotoluene	1·8[a]
p-Methoxytoluene	21[a]
Anisole	1·0[a]
Cyclohexane	4·2[a]
2,3-Dimethylbutane	3·8[a]
Propan-2-ol	18[a]
Octan-2-ol	20[a]

[a] Walling and Gibian (1965). [b] Padwa (1964).

A rate constant of $1{\cdot}9 \times 10^5$ M^{-1} sec.$^{-1}$ for the reaction of triplet benzophenone with toluene is in much better agreement than the value of 9×10^3 M^{-1} sec.$^{-1}$ with Beckett and Porter's (1963) estimate of $1{\cdot}07 \times 10^6$ M^{-1} sec.$^{-1}$ for this rate constant. The Beckett and Porter value was obtained from quenching experiments with naphthalene, and k_q was taken to be $1{\cdot}3 \times 10^{10}$ M^{-1} sec.$^{-1}$. Since this estimated quenching rate constant is probably too large by a factor of 2 or 3, Beckett and Porter's rate constant should be reduced to $3\text{–}5 \times 10^5$ M^{-1} sec.$^{-1}$. These results imply that Hammond, Baker and Moore's (1961) value of $k_q/k_a \sim 10^5$ for toluene is too large by at least an order of magnitude.

The rate constants for triplet benzophenone abstraction from hydrocarbons which are listed in Table 5 are of comparable magnitude to those estimated for attack of t-butoxyl on the same hydrocarbons (cf. Table 4). As has already been mentioned, the absolute, though not the relative, t-butoxyl rate constants are uncertain by at least a factor of 3. The rate constants in Table 5 are probably accurate to within a factor of 2. Since relative reactivities towards triplet aromatic ketones are virtually independent of the nature of the ketone (Padwa, 1964; Walling and Gibian, 1965) it is reasonable to assume that the absolute rate constants should also be rather similar. That is, the rate constants in Table 5 are probably applicable to ring substituted benzophenones and to alkyl aromatic ketones. Some support for this view comes from experiment. For example, Clark, Litt and Steel (1969a, b) found approximately the same lifetime of ~ 2 μsec for the benzophenone triplet and for the acetophenone triplet in iso-octane. Since the reciprocal of the lifetime is equal to k_a [solv.], they calculated $k_a \sim 7 \times 10^4$ M^{-1} sec.$^{-1}$ for both triplets. Similarly, Cohen, Laufer and Sherman (1964) reported $k_q/k_a = 2700$ for the naphthalene quenched photoreduction of acetophenone in propan-2-ol from which, by taking $k_q \sim 5 \times 10^9$ M^{-1} sec.$^{-1}$, we obtain $k_a \sim 18{\cdot}6 \times 10^5$ M^{-1} sec.$^{-1}$ which is the same, within experimental error, as the value found for the benzophenone triplet.

Aliphatic ketones also undergo photoreduction, as is evidenced by the easy reduction of acetone in cyclohexane and hexane (Wagner and Hammond, 1968). However, type I cleavage often competes with photoreduction so that, for example, photolysis of diethyl ketone in cumene apparently yields ethyl radicals to the exclusion of reduction products (Jarvie and Laufer, 1964). Thus fragmentation is much faster than hydrogen abstraction. Simonaites, Cowell and Pitts (1967) have reported that k_a for reaction of the cyclohexanone triplet with propan-2-ol has a value of $\sim 2 \times 10^6$ M^{-1} sec.$^{-1}$ and Chien (1967) has concluded that the triplet states of diethyl ketone and benzophenone abstract hydrogen with comparable reactivities. It seems likely that, when able to abstract, most aliphatic ketones in their triplet states will abstract from hydrogen donors with rate constants comparable to those found for triplet state phenyl and phenyl alkyl ketones. There are however some exceptions to this general rule. Thus, triplet cyclopentanone is about six times as reactive as triplet cyclohexanone towards propan-2-ol (Simonaites, Cowell and Pitts, 1967). In addition, dialkyl ketone triplets appear to be more reactive than phenyl alkyl ketone triplets in type II cleavage (intramolecular hydrogen atom abstraction from the γ-position). Thus, the rate constants for pentan-2-one and hexan-2-one are 2×10^8 sec.$^{-1}$ and 1×10^9 sec.$^{-1}$, respectively, while the corresponding rates for triplet butyrophenone and valerophenone are 8×10^6 and $1{\cdot}4 \times 10^8$ sec.$^{-1}$, respectively (Wagner and Hammond, 1968).

Bäckström and Sandros (1960) and Turro and Engel (1969) have shown that the biacetyl triplet is much less reactive than phenyl ketone triplets, presumably because of its much lower triplet energy. Thus, in benzene the rate constants for attack of the biacetyl triplet on propan-2-ol and on benzhydrol are only $3{\cdot}3 \times 10^3$ and $7{\cdot}0 \times 10^4$ M^{-1} sec.$^{-1}$, respectively (Turro and Engel, 1969). More reactive hydrogen donors such as tributyltin hydride and phenols, which react with most triplet ketones at, or near, the diffusion-controlled limit, also react relatively slowly with triplet biacetyl (Turro and Engel, 1969).

It has been stated that although triplet benzophenone has a reactivity comparable to t-butoxyl when the substrate is a hydrocarbon, it is much more reactive than t-butoxyl towards alcohols. We have not found any concrete evidence that this is really the case. There do not appear to be any data available on the reactivities of secondary alcohols towards attack by t-butoxyl radicals. The absence of such data is, in part, due to the fact that secondary alcohols react with t-butyl hypochlorite by an ionic mechanism. However, Walling and Mintz (1967) have found that primary alcohols react with the hypochlorite by a free radical chain process, and have obtained relative reactivities towards t-butoxyl for ethyl alcohol : propyl alcohol : benzyl alcohol : cyclohexane of 0·94 : 0·84 : 1·82 : 1·0. Unhindered secondary alcohols are about five times as reactive as cyclohexane towards the benzophenone triplet (Table 5). Ketones can be fairly easily photoreduced in primary alcohols, although the process is not normally as rapid as in secondary alcohols. For example, Bäckström and Sandros (1958) have reported rate constants for the reaction of triplet biacetyl with methyl alcohol, isopropyl alcohol and benzyl alcohol of $2{\cdot}6 \times 10^2$, $2{\cdot}7 \times 10^3$ (cf. Turro and Engel, 1969) and $6{\cdot}9 \times 10^3$ M^{-1} sec.$^{-1}$, respectively. It would appear that the hydrogen donating ability of most primary alcohols is comparable to that of cyclohexane towards both t-butoxyl and triplet ketones.

Sterically hindered secondary alcohols are less reactive towards triplet ketones than unhindered alcohols (see Table 6). A similar steric effect has been observed in the photoreduction of benzophenone by aliphatic amines (Cohen and Baumgarten, 1965). The latter reaction occurs in a stepwise manner, a very rapid charge transfer process being followed either by charge destruction and quenching or by hydrogen transfer and the formation of radicals (Cohen and Chao, 1968; Cohen and Litt, 1970).

$$Ar_2C{=}O^{*(3)} + RCH_2NR'_2 \xrightarrow{k_{ir}} [Ar_2\dot{C}{-}\bar{O} \quad RCH_2\overset{+\cdot}{N}R'_2] \longrightarrow Ar_2\dot{C}OH + R\dot{C}HNR'_2$$

Rate constants for the initial interaction, k_{ir}, have been determined for a variety of alkyl amines and alkyl anilines (Cohen and Stein, 1969; Cohen and Litt, 1970). The majority of these rate constants

TABLE 6

Rate constants for hydrogen atom abstraction from alcohols by triplet benzophenone at ambient temperatures

Substrate	Rate constant $\times 10^{-5}$ (M^{-1} sec.$^{-1}$)	Ref.
Benzhydrol	90	1
Benzhydrol	75–140	2
Cyclohexanol	28[a]	3
Cyclopentanol	23[a]	3
Heptan-2-ol	18[a]	3
Heptan-3-ol	12[a]	3
Methylisobutyl-carbinol	7·0[a]	3
Methylneopentyl-carbinol	3·2[a]	3
Di-isobutylcarbinol	1·3[a]	3
Isoborneol	80	4
Isoborneol	100	5

[a] Based on a value of 18×10^5 M^{-1} sec.$^{-1}$ for propan-2-ol.

References: (1) Wagner (1969). (2) Cohen and Litt (1970). (3) Pearson and Moss (1967). (4) Clark, Litt and Steel (1969a). (5) Guttenplan and Cohen (1969).

are close to the diffusion-controlled limit. However, a few amines are less reactive, e.g., $k_{ir} = 7 \times 10^7$ M^{-1} sec.$^{-1}$ for t-butylamine in benzene and for 2-aminobutane $k_{ir} = 2{\cdot}5 \times 10^8$ M^{-1} sec.$^{-1}$ in benzene and $6{\cdot}3 \times 10^7$ M^{-1} sec.$^{-1}$ in water. The actual rates of photoreduction are much lower, 2-aminobutane being about as effective a photoreducing agent as propan-2-ol (Cohen and Baumgarten, 1965). The reactions of triplet ketones with other types of compounds may also involve an initial transfer of charge (see for example, Guttenplan and Cohen, 1969).

E. Peroxyl Radicals

1. *Mechanisms of Autoxidation*

Peroxyl radicals take part in a variety of reactions with organic compounds, free radicals and metal ions, and these reactions have been documented in a recent review by Ingold (1969a). Although this Section is concerned with absolute rate constants for the reactions of peroxyl radicals and the experimental methods used to determine them, it is worth illustrating the importance of peroxyl radicals by considering the reactions and mechanisms involved in the autoxidation of organic compounds at moderate temperatures ($< 100°$). The following free radical chain process is usually used to describe autoxidations.

Initiation: Production of free radicals (8)

Propagation: $$\mathrm{R\cdot + O_2 \xrightarrow{k_9} ROO\cdot} \qquad (9)$$

$$\mathrm{ROO\cdot + RH \xrightarrow{k_{abs}} ROOH + R\cdot} \qquad (10)$$

$$\mathrm{ROO\cdot + RH \xrightarrow{k_{add}} ROO\dot{R}H} \qquad (11)$$

Termination:
$$\left.\begin{array}{ll} \mathrm{ROO\cdot + ROO\cdot \xrightarrow{2k_{12}}} & (12) \\ \mathrm{R\cdot + ROO\cdot \xrightarrow{4k_{13}}} & (13) \\ \mathrm{R\cdot + R\cdot \xrightarrow{2k_{14}}} & (14) \end{array}\right\} \text{Non-radical products}$$

In the above scheme, R· is the alkyl radical and ROO· is the alkylperoxyl radical derived from the organic substrate RH.

Autoxidations are usually initiated by addition of low concentrations ($< 10^{-2}$ M) of α,α'-azobisisobutyronitrile, α,α'-azobiscyclohexane carbonitrile, t-butyl hyponitrite or tetraphenylbutane. The rate of chain initiation, R_i, is given by,

$$R_i = 2ek_i[\mathrm{In}_2]$$

where e is the efficiency of radical production and k_i is the unimolecular rate constant for decomposition of the initiator In_2.

Alkyl radicals are formed either by free radical abstraction of the weakest hydrogen [reaction (10)] or by addition of a free radical to a double bond (11). Addition of oxygen to an alkyl radical (9) is extremely fast and in many cases is probably diffusion controlled, i.e., $k_9 \sim 10^9$ M^{-1} sec.$^{-1}$ (Ingold, 1969a). This means that the rate of oxidation above about 100 torr is independent of the oxygen pressure, that the rate controlling propagation step involves reaction of a peroxyl radical with the substrate, and that the rate controlling termination step involves the reaction of two peroxyl radicals to give non-radical products. Under these conditions the rate of oxidation, R_s, is given by,

$$R_s = \frac{-\mathrm{d}[\mathrm{O}_2]}{\mathrm{d}t} = \frac{k_p[\mathrm{RH}]\,R_i^{1/2}}{(2k_t)^{1/2}}$$

where k_p and $2k_t$ are the rate controlling propagation and termination rate constants. For many organic compounds k_p is equivalent to k_{abs} or k_{add} while $2k_t$ is equivalent to $2k_{12}$ and a hydroperoxide, ROOH, or a polyperoxide, —(ROOROOR)—$_n$, is the principal reaction product. In certain cases, e.g., autoxidation of indene, k_p is made up of contributions from k_{abs} and k_{add} since this hydrocarbon gives a mixture of hydroperoxide and polyperoxide. Measurement of the steady state rate of oxidation at a known rate of chain initiation and substrate concentration enables the ratio of rate constants, $k_p/(2k_t)^{1/2}$, to be calculated.

Some hydrocarbons and alcohols undergo autoxidation at 30° to give hydrogen peroxide as the major hydroperoxidic product. For cyclohexa-1,4-diene this is apparently due to the reaction of the cyclohexadienyl radical with oxygen to give benzene and the hydroperoxyl radical.

$$\text{(cyclohexadienyl)}^{\bullet} + \mathrm{O}_2 \longrightarrow \text{(benzene)} + \mathrm{HO}_2^{\bullet}$$

This chain oxidation is therefore propagated and terminated by hydroperoxyl radicals:

$$HO_2\cdot + \text{(cyclohexadiene)} \xrightarrow{k_p} HOOH + \text{(cyclohexadienyl radical)}$$

$$HO_2\cdot + HO_2\cdot \xrightarrow{2k_t} \text{Non-radical products.}$$

Howard and Koreck (1970) have suggested that the chain carrying species for the autoxidation of benzyl alcohol and α-methylbenzyl alcohol at 30° are α-hydroxyalkylperoxyl radicals, and that the production of hydrogen peroxide results from the instability of the hydroperoxides:

$$PhC(OH)(R)OOH \longrightarrow PhC(R){=}O + H_2O_2$$

where R = H or CH_3.

Hydrocarbons with very labile hydrogens, e.g. triphenylmethane, are often oxidised relatively slowly and the rate of oxidation is dependent on the oxygen concentration at atmospheric pressure. This is apparently due to the reversibility of the reaction of the resonance stabilised alkyl radical with oxygen [reaction (9)]. The alkyl radicals can then take part in termination via reaction (13).

Autoxidations are generally terminated by reaction of two peroxyl radicals to give molecular products. This reaction proceeds through a tetroxide intermediate, ROOOOR, which may be either an activated complex or a molecule with an appreciable lifetime.

Primary and secondary tetroxides are believed to decompose rapidly to give molecular products *via* a cyclic transition state (Russell, 1957; Howard and Ingold, 1968c, d).

$$R_2C(H)\text{–O–O–O–O–}C(H)R_2 \text{ (cyclic transition state)} \longrightarrow R_2C{=}O + O_2 + R_2CHOH$$

The mutual reaction of tertiary peroxyl radicals has been studied much more thoroughly and can be illustrated by considering some of the reactions involved in the oxidation of cumene (Blanchard, 1959; Bartlett and Traylor, 1963; Traylor and Russell, 1965; Thomas, 1967; Howard, Schwalm and Ingold, 1968; Hendry, 1967). Cumyl tetroxide, formed from the combination of two cumylperoxyl radicals, can decompose irreversibly to give cumyloxyl radicals, RO·, confined in a solvent cage:

$$\mathrm{ROO\cdot + ROO\cdot} \overset{K}{\rightleftharpoons} \mathrm{ROOOOR} \xrightarrow{k_a} [\mathrm{RO\cdot + O_2 + \cdot OR}]_{\mathrm{cage}}$$

where $K = [\mathrm{RO_2\cdot}]^2/[\mathrm{ROOOOR}]$ and k_a is the rate constant for irreversible decomposition of the tetroxide. The caged alkoxyl radicals can either combine to give dicumyl peroxide:

$$[\mathrm{RO\cdot + O_2 + \cdot OR}]_{\mathrm{cage}} \xrightarrow{k_b} \mathrm{ROOR + O_2},$$

or they can diffuse out of the cage:

$$[\mathrm{RO\cdot + O_2 + \cdot OR}]_{\mathrm{cage}} \xrightarrow{k_c} \mathrm{2\,RO\cdot + O_2}.$$

The cumyloxyl radicals which escape from the cage can either react with cumene and propagate a chain:

$$\mathrm{RO\cdot + RH} \longrightarrow \mathrm{ROH + R\cdot}, \tag{15}$$

or they can undergo β-scission to give acetophenone and methyl radicals:

$$\mathrm{RO\cdot} \xrightarrow{k_d} \mathrm{Acetophenone + CH_3\cdot}.$$

Methyl radicals will add oxygen to give methylperoxyl radicals which can either propagate the reaction:

$$\mathrm{CH_3OO\cdot + RH} \longrightarrow \mathrm{CH_3OOH + R\cdot}, \tag{10a}$$

or react with a cumylperoxyl radical in a 'crossed' termination reaction:

$$\mathrm{CH_3OO\cdot + ROO\cdot} \xrightarrow{4k_{16}} \text{Non-radical products.} \tag{16}$$

Measurement of absolute rate constants for oxidation of pure cumene (and many other substrates with labile tertiary hydrogens) therefore yields an overall termination rate constant, $2k_t$, which is given by:

$$2k_t = \frac{2fk_a}{K} + \frac{2(1-f)\,k_a}{K}\left[\frac{k_d}{(k_d+k_{15}[\mathrm{RH}])}\cdot\frac{4k_{16}[\mathrm{ROO}\cdot]}{(4k_{16}[\mathrm{ROO}\cdot]+k_{10a}[\mathrm{RH}]}\right],$$

where $f = k_b/(k_b+k_c)$, the fraction of alkoxyl radicals that escape the solvent cage, and

$$\frac{2fk_a}{K} \equiv 2k_{12}.$$

It is clear that the termination rate constants for tertiary peroxyl radicals must depend on k_a, K, f, the fraction of alkoxyl radicals which undergo β-scission, the nature of the alkyl radical formed in the β-scission process, and the fraction of these alkyl radicals which are consumed in the termination reaction (16).

Addition of cumene hydroperoxide to cumene during oxidation increases the rate of oxidation until a limiting rate is reached, whereupon addition of more hydroperoxide has no effect on the rate (Thomas, 1967). Under these conditions methylperoxyl radicals are eliminated from termination reactions by chain transfer with the added hydroperoxide, and termination involves only cumylperoxyl radicals (Thomas, 1967; Howard, Schwalm and Ingold, 1968). Consequently, oxidation of a tertiary hydrocarbon in the presence of its hydroperoxide enables absolute values of the rate constant for the self-reaction of tertiary peroxyl radicals, $2k_{12}$, to be measured.

2. *Absolute Propagation and Termination Rate Constants*

Individual rate constants have been measured for the autoxidation of many organic compounds, and for convenience they are divided into four groups, namely those determined by (*i*) electron spin resonance spectroscopy, (*ii*) ultraviolet spectroscopy, (*iii*) the rotating sector technique, and (*iv*) pre- or after-effect methods.

(i) *Electron spin resonance spectroscopy*

Absolute concentrations of peroxyl radicals were first measured by electron spin resonance spectroscopy (e.s.r.) during the initiated oxidation of cumene at moderate temperatures by Lebedev, Tsepalov and Shlyapintokh (1961) and Thomas (1963). Absolute concentrations were found to be compatible with concentrations calculated from rates of chain initiation and the corrected termination rate constant for cumylperoxyl radicals.

Electron spin resonance spectroscopy has since been used to measure absolute concentrations of many peroxyl radicals and to follow their bimolecular decay. Peroxyl radicals for e.s.r. studies have been generated by the following methods.

(*a*) Autoxidation of organic compounds (Thomas, 1963; Guk, Tsepalov, Shuvalov and Shylapintokh, 1968; D. R. Bowman, private communication).

(*b*) Free radical or photochemically induced decomposition of hydroperoxides (Thomas, 1965; Thomas and Ingold, 1968; Howard, Adamic and Ingold, 1969).

$$\mathrm{ROOH + In\cdot \longrightarrow ROO\cdot + InH}$$

$$\mathrm{ROOH \xrightarrow{h\nu} RO\cdot + \cdot OH \xrightarrow{ROOH} ROO\cdot}$$

(*c*) Photolysis of an azo-compound in an oxygenated inert solvent (Adamic, Howard and Ingold, 1969).

$$\mathrm{RN{:}NR \xrightarrow{h\nu} R\cdot + N_2}$$

$$\mathrm{R\cdot + O_2 \longrightarrow ROO\cdot}$$

(*d*) Complete oxidation of a hydroperoxide with ceric ammonium nitrate (Thomas, 1965; Thomas and Ingold, 1968).

$$\mathrm{ROOH + Ce^{4+} \longrightarrow ROO\cdot + Ce^{3+}}$$

(*e*) Pulse radiolysis of oxygenated hydrocarbons (McCarthey and MacLachlan, 1961a, b; Smaller, Remko and Avery, 1968).

$$\mathrm{RH \rightsquigarrow R\cdot}$$

$$\mathrm{R\cdot + O_2 \longrightarrow ROO\cdot}$$

(*f*) Photolysis of t-butyl peroxide in oxygenated alkanes (Bennett, Brown and Mile, 1970a, b).

$$\mathrm{Bu^tOOBu^t \xrightarrow{h\nu} 2Bu^tO\cdot}$$

$$\mathrm{Bu^tO\cdot + RH \longrightarrow Bu^tOH + R\cdot}$$

$$\mathrm{R\cdot + O_2 \longrightarrow ROO\cdot}$$

Absolute concentrations of peroxyl radicals have usually been measured by integrating the e.s.r. signal produced by these radicals and comparing it with the integrated signal obtained from a standard solution of a stable free radical, e.g. α,α'-diphenyl-β-picrylhydrazyl, under identical conditions.

Under steady state conditions at a known rate of chain initiation, termination rate constants can be calculated from the expression,

$$R_i = 2k_t[\mathrm{ROO}\cdot]_s^2$$

where $[\mathrm{ROO}\cdot]_s$ is the measured steady state concentration of peroxyl radicals. Alternatively, the second order decay of a peroxyl radical signal can be monitored by e.s.r. spectroscopy and a termination rate constant calculated from

$$\frac{-\mathrm{d}[\mathrm{ROO}\cdot]}{\mathrm{d}t} = 2k_t[\mathrm{ROO}\cdot]^2.$$

All the e.s.r. methods described above yield $2k_t$ (i.e., $2k_{12}$) for substrates which form primary and secondary peroxyl radicals (e.g. n-butane and cyclohexane). However, substrates which form tertiary peroxyl radicals (e.g. cumene) are more complicated because of the occurrence of both terminating and non-terminating reactions of these peroxyl radicals. Autoxidation of an organic compound and photolysis of t-butyl peroxide in an oxygenated compound give an overall termination rate constant, $2k_t$; because in these two cases termination via reaction (16) can occur. Free radical induced decomposition or photolysis of a hydroperoxide give $2k_{12}$ because chain transfer with the excess of hydroperoxide eliminates termination via reaction (16). Ceric ion oxidation of a hydroperoxide and photolysis of an azo-compound produce peroxyl radicals in an inert solvent, and since alkoxyl radicals are difficult to detect by e.s.r. spectroscopy, these methods give the rate of constant for the total decay of peroxyl radicals, i.e. $2k_a/K$.

Rate constants and Arrhenius parameters for the mutual reaction of peroxyl radicals, as measured by e.s.r. spectroscopy, are summarised in Table 7.

The extensive data for t-butylperoxyl radicals in Table 7 imply that the e.s.r. method of determining rate constants is not very precise, since absolute values of $2k_{12}$ for this radical vary from 7×10^1 to 3×10^5 M^{-1} sec.$^{-1}$. However, Maguire and Pink's (1967) rate constants appear too high when they are compared with absolute values of $2k_{12}$ for t-butylperoxyl radicals obtained by other workers using e.s.r. spectroscopy, the rotating sector technique and pre- and after-effect methods. The 'best' value of $2k_{12}$ at 30° for t-butylperoxyl radicals appears to be in the range $1\text{–}2 \times 10^3$ M^{-1} sec.$^{-1}$. Absolute values of $2k_a/K$ for t-butylperoxyl radicals vary from

TABLE 7

Second order rate constants and Arrhenius parameters for mutual reactions of peroxyl radicals as determined by e.s.r. spectroscopy

Peroxyl radical	$2k_t \times 10^{-6}$ (M^{-1} sec.$^{-1}$)	$2k_{12} \times 10^{-6}$ (M^{-1} sec.$^{-1}$)	$10^{-6} \times 2k_a/K$ (M^{-1} sec.$^{-1}$)	Temp. (°C)	Solvent	E (kcal. mole^{-1})	log (A/M^{-1} sec.$^{-1}$)	Exptl. Method	Ref.
t-Butyl		0·00007		30	CF_2Cl_2	$5·1 \pm 0·2$	5·5	*c*	1
t-Butyl		0·0004		22	Benzene	$10·2 \pm 1$	10·15	*b*	2
t-Butyl		0·001		30	CCl_4	—	—	*b*	3
t-Butyl		0·0013		30	Benzene	$4·5 \pm 1$	6·4	*c(sf)*	4
t-Butyl		0·0023		30	Benzene	$5·4 \pm 2$	7·3	*c(s)*	4
t-Butyl		0·0042		22	t-Butyl hydro-peroxide	—	—	*c(sf)*	2
t-Butyl		0·005		30	Methanol	$6·3 \pm 1$	8·2	*c(sf)*	4
t-Butyl		0·47		30	Hexane	7·2	10·93	*c(s)*	5
t-Butyl		0·82		30	Cyclopentane	8·5	12·04	*c(s)*	5
t-Butyl		1·1		30	Methylcyclohexane	8·5	12·17	*c(s)*	5
t-Butyl		1·1		30	3-Methylpentane	8·4	12·09	*c(s)*	5
t-Butyl		0·36		30	t-Butyl peroxide	8·3	11·55	*c(s)*	5
t-Butyl		0·0028		65	t-Butyl hydro-peroxide	—	—	*b*	6
t-Butyl			0·0025	30	CF_2Cl_2	$8·7 \pm 3$	9·7	*g*	1
t-Butyl			0·0035	30	Methanol	$9·5 \pm 2$	10·4	*d*	4
t-Butyl			0·0082	22	Methanol	—	—	*d*	2
t-Butyl			0·011	22	Methanol	—	—	*d*	2
t-Butyl			0·025	22	H_2O	—	—	*d*	2
2-Ethyl-2-propyl		0·0002		30	CF_2Cl_2	$4·7 \pm 0·5$	5·7	*c*	1
2-Ethyl-2-propyl		0·00045		30	CCl_4	—	—	*b*	3
2-Isopropyl-2-propyl		0·001		30	CCl_4	—	—	*b*	3
2-n-Propyl-2-propyl	0·025			30	2-Methylpentane	$9·3 \pm 1$	$11·1 \pm 1·0$	*e*	7
2-t-Butyl-2-propyl	0·006			30	2,2,3-Trimethyl-butane	$7·5 \pm 1·0$	$9·2 \pm 1·0$	*e*	7
2-Neopentyl-2-propyl		0·0006		30	CF_2Cl_2	$4·0 \pm 0·2$	5·7	*c*	1
2-Neopentyl-2-propyl		0·001		30	CCl_4	—	—	*b*	3
2-Neopentyl-2-propyl		0·0015		30	Methanol	$7·0 \pm 2$	9·2	*c(sf)*	1
2-Neopentyl-2-propyl			0·0083	30	Methanol	$9·8 \pm 1$	11·0	*d*	4

1-Methylcyclopent-1-yl		0·002		30	CF_2Cl_2	5·7 ± 0·3	7·5	*c*	1
1-Methylcyclopent-1-yl		0·023		30	CCl_4	—	—	*b*	3
1-Methylcyclohex-1-yl		0·001		30	CCl_4	—	—	*b*	3
1-Methylcyclohex-1-yl		0·002		30	CF_2Cl_2	6·1 ± 5	7·7	*c*	1
2,5-Dimethyl-2-hydroperoxyhexyl-5-peroxyl		0·01		30	Benzene	12·5 ± 3	13·0	*c*(*s*)	4
2,5-Dimethyl-2-hydroperoxyhexyl-5-peroxyl		0·018		30	Benzene	11·0 ± 3	12·7	*c*(*sf*)	4
2,5-Dimethylhexyl-2,5-diperoxyl			0·028	30	Methanol	9 ± 2	10·9	*d*	4
Cumyl		0·008		30	CCl_4	—	—	*b*	3
Cumyl		0·032		30	Benzene	7·1 ± 2	9·6	*c*(*s*)	4
Cumyl		0·037		30	Benzene	5·8 ± 1	8·8	*c*(*sf*)	4
Cumyl		0·044		25	Cumene hydroperoxide	—	—	*c*	8
Cumyl		0·044		30	H_2O	—	—	*c*(*sf*)	1
Cumyl		0·25		25	H_2O	—	—	*c*	9
Cumyl		0·08		65	Cumene hydroperoxide	—	—	*b*	6
Cumyl	0·7			65	Cumene	—	—	*a*	6
Cumyl	0·072			74	Benzene	—	—	*a*	10
Cumyl			0·12	30	Methanol	7·8	10·7	*d*	4
Cumyl			0·24	30	CF_2Cl_2	7·3 ± 3	10·7	*g*	1
2-Benzyl-2-propyl		0·0017		30	CCl_4	—	—	*b*	3
2-Phenyl-2-butyl		0·025–0·1		30	CCl_4	—	—	*b*	3
2-Phenyl-2-butyl			0·29	30	Methanol	5·5	9·4	*d*	4
1,1-Diphenylethyl		0·64		30	CCl_4	—	—	*b*	3
1,1-Diphenylethyl		2·2		30	Benzene	2·5 ± 1	8·1	*c*(*s*)	4
1,1-Diphenylethyl	1·0			30	1,1-Diphenylethane	7·02	11·0	*e*	11
1,1-Diphenylethyl			~1·0	30	Methanol	—	—	*d*	4

TABLE 7—*continued*

Peroxyl radical	$2k_t \times 10^{-6}$ (M^{-1} sec.$^{-1}$)	$2k_{12} \times 10^{-6}$ (M^{-1} sec.$^{-1}$)	$10^{-6} \times 2k_a/K$ (M^{-1} sec.$^{-1}$)	Temp. (°C)	Solvent	E (kcal. mole^{-1})	log (A/M^{-1} sec.$^{-1}$)	Exptl. Method	Ref.
Cyclopentenyl	~15			30	Methanol	—	—	*d*	4
Cyclopentyl	17			30	Cyclopentane	—	—	*f*	12
2,3-Dimethyl-2-butenyl	8			65	2,3-Dimethylbut-2-ene	—	—	*a*	6
Hexyl	11			30	Hexane	—	—	*f*	12
Cyclohexenyl	43			65	Cyclohexene	—	—	*a*	6
Cyclohexenyl	2·8			30	Methanol	6·0 ± 3	10·8	*d*	4
Cyclohexyl	1·6			30	Cyclohexane	~0	—	*f*	13
Cyclohexyl	6·7			30	Cyclohexane	—	—	*f*	12
From methyl cyclohexane	3·7			30	Methylcyclohexane	—	—	*f*	12
Heptyl	2·2			30	Heptane	—	—	*f*	12
sec-Heptyl	2·0			30	Heptane	1·9 ± 3	7·7 ± 1	*e*	7
Cycloheptyl	8·6			30	Cycloheptane	—	—	*f*	12
Cyclooctyl	14			30	Cyclooctane	—	—	*f*	12
Octyl	7·6			30	Octane	—	—	*f*	12
2,5-Dimethylhex-3-enyl	0·9			65	2,5-Dimethylhex-3-ene	—	—	*a*	6
Nonyl	2·2			30	Nonane	—	—	*f*	12
Decyl	3·0			30	Decane	—	—	*f*	12
Tridecyl	1·6			30	Tridecane	—	—	*f*	12
α-Tetralyl	50			65	Tetralin	—	—	*a*	6
α-Tetralyl	5·2			30	Methanol	4·6 ± 1	10·0	*d*	4
n-Butyl	30			7	Methanol	—	—	*d*	4
Neopentyl	> 400			−20	Neopentane	—	—	*e*	7

Methods: (*a*) Hydrocarbon autoxidation. (*b*) Free radical hydroperoxide decomposition initiated by t-butyl hyponitrite or α,α′-azobisisobutyronitrile. (*c*) Photolysis of the hydroperoxide either using a stop flow technique (*sf*) or photolysis in the cavity (*s*). (*d*) Ceric oxidation of the hydroperoxide using a stop flow technique. (*e*) Photolysis of di-t-butyl peroxide in the oxygenated alkane. (*f*) Pulse radiolysis of oxygenated hydrocarbons. (*g*) Photolysis of the azo-compound in oxygenated CF_2Cl_2.

References: (1) Adamic, Howard and Ingold (1969). (2) Thomas (1965). (3) Howard, Adamic and Ingold (1969). (4) Thomas and Ingold (1968). (5) Maguire and Pink (1967). (6) D. R. Bowman (private communication). (7) Bennett, Brown and Mile (1970a). (8) Zwolenik (1967). (9) Piette, Bulow and Loeffler (1964). (10) Thomas (1963). (11) Guk, Tsepalov, Shuvalov and Shlyapintokh

$2{\cdot}5 \times 10^2$ to $2{\cdot}5 \times 10^4$ M^{-1} sec.$^{-1}$. This rate constant must be larger than $2k_{12}$ and the 'best' value appears to be about 10^4 M^{-1} sec.$^{-1}$.

Data for other t-peroxyl radicals are not as extensive as for t-butylperoxyl radicals. However, in most cases rate constants agree fairly well with rate constants obtained by other techniques.

Termination rate constants, $2k_{12}$, for tertiary peroxyl radicals apparently depend on the nature of the alkyl group attached to oxygen and increase by two orders of magnitude as R is increased from t-butyl to 1,1-diphenylethyl. This difference is apparently due to a decrease in the stability of the tetroxide, i.e., k_a is larger for 1,1-diphenylethyl tetroxide than for t-butyl tetroxide.

Secondary peroxyl radicals appear to have termination rate constants in the range $1\text{–}10 \times 10^6$ M^{-1} sec.$^{-1}$ while rate constants for primary peroxyl radicals are apparently higher.

Arrhenius parameters listed in Table 7 imply that pre-exponential factors for the mutual reaction of peroxyl radicals are in the range expected for a bimolecular reaction ($10^8\text{–}10^{12}$ M^{-1} sec.$^{-1}$) and that there is an appreciable activation energy for the mutual reaction of tertiary peroxyl radicals. Though little information is available for secondary peroxyl radicals, the large difference in termination rate constants between secondary and tertiary peroxyl radicals ($\sim 10^3$) appears to be due to differences in activation energy.

Chien and Boss (1967b) and Boss and Chien (1969) have used an e.s.r. spectroscopic method to estimate rate constants for autoxidation of polyethylene at 110° ($k_p = 1{\cdot}9$ M^{-1} sec.$^{-1}$ and $2k_t = 3 \times 10^6$ M^{-1} sec.$^{-1}$) and polypropylene at 140° ($k_p = 1{\cdot}0$ M^{-1} sec.$^{-1}$ and $2k_t = 1{\cdot}24 \times 10^8$ M^{-1} sec.$^{-1}$).

Peroxyl radical—tetroxide equilibria. The equilibrium between t-butylperoxyl radicals and t-butyl tetroxide was first demonstrated by Bartlett and Guaraldi (1967) for peroxyl radicals generated by irradiation of di-t-butyl peroxycarbonate in frozen methylene chloride at $-196°$ and by oxidation of t-butyl hydroperoxide with lead tetraacetate at $-90°$ in the same solvent. Further e.s.r. support for this equilibrium has been obtained by photolysis of t-butyl peroxide in oxygenated alkanes (Bennett, Brown and Mile, 1970b), by photolysis of azo-compounds in dichlorodifluoromethane in the presence of oxygen, and by photolysis of hydroperoxides (Adamic, Howard and Ingold, 1969). At temperatures below $-115°$ a reduction in temperature causes the concentration of tertiary peroxyl radicals to decrease to a new value which, once achieved, does not vary with time, warming up the solution to a temperature of not above $-115°$ restores the original concentration of peroxyl radicals. The variation of the equilibrium constant K with temperature can be expressed by an integrated form of the van't Hoff isochore:

$$\ln K = \frac{\Delta S°}{R} - \frac{\Delta H°}{RT},$$

where

$$K = \frac{[\mathrm{ROO}\cdot]^2}{\mathrm{ROOOOR}}$$

and $\Delta S°$ and $\Delta H°$ are changes in the standard entropy and enthalpy.

TABLE 8

Thermodynamic constants for the equilibrium between tertiary alkylperoxyl radicals and tetroxide molecules

Peroxyl radical	$K \times 10^5$ (M) at $-120°$	$\Delta H°$ (kcal. mole^{-1})	$\Delta S°$ (cal. deg.$^{-1}$ mole^{-1})	Ref.
t-Butyl[a]	—	8.8 ± 0.4	34 ± 1	1
t-Butyl[b]	—	8.0 ± 0.2	31 ± 1	1
t-Butyl[c]	—	8.4 ± 0.4	30	1
2-Ethyl-2-propyl[d]	1·45	8.9 ± 1.0	36 ± 8	2
2-Ethyl-2-propyl[b]	—	7.5 ± 0.3	29 ± 1	1
2-n-Propyl-2-propyl[d]	6·7	8.9 ± 0.7	39 ± 6	2
2-Isopropyl-2-propyl[c]	—	8.6 ± 0.4	33 ± 1	1
2-Ethyl-2-butyl[a]	5·5	9.7 ± 0.5	44 ± 4	2
2,3-Dimethylbutyl[d]	5·3	8.2 ± 0.5	34 ± 4	2
2-t-Butyl-2-propyl[d]	7·7	8.7 ± 1.0	38 ± 7	2
Cumyl[d]	0·31	11.2 ± 0.9	48 ± 7	2
Cumyl[a]	—	9.2 ± 0.4	32 ± 1	1
Cumyl[b]	—	10.6 ± 0.8	*e*	1
1-Methyl-1-cyclohexyl[b]	—	7.0 ± 0.4	35	1
2-Neopentyl-2-propyl[b]	—	7.8 ± 0.5	31	1
1-Methyl-1-cyclopentyl[b]	—	8.0 ± 0.3	30 ± 1	1
2-Phenyl-2-butyl[b]	—	11 ± 1	*e*	1

Notes: [a] From irradiation of RN:NR and oxygen in CF_2Cl_2. [b] From irradiation of ROOH in CF_2Cl_2. [c] From irradiation of ROOH in isopentane. [d] From irradiation of t-butyl peroxide in oxygenated alkanes. [e] $\Delta S°$ was not measured because irreversible decay started to occur at a temperature below that at which the tetroxide was fully dissociated into peroxyl radicals.

References: (1) Adamic, Howard and Ingold (1969). (2) Bennett, Brown and Mile (1970b).

These two equations may be combined to give:

$$2\ln[\mathrm{ROO}\cdot]-\ln[\mathrm{ROOOOR}] = \frac{\Delta S^\circ}{R}-\frac{\Delta H^\circ}{RT}.$$

Since no irreversible decay occurs below $-115°$, the maximum peroxyl radical concentration at complete dissociation, $[\mathrm{ROO}\cdot]_{\max}$, is given by

$$[\mathrm{ROO}\cdot]_{\max} = 2[\mathrm{ROOOOR}]+[\mathrm{ROO}\cdot],$$

and it is clear that:

$$2\ln[\mathrm{ROO}\cdot]-\ln([\mathrm{ROO}\cdot]_{\max}-[\mathrm{ROO}\cdot])-\ln 2 = \frac{\Delta S^\circ}{R}-\frac{\Delta H^\circ}{RT}.$$

Values of K, ΔS°, and ΔH° for several tertiary tetroxide–peroxyl radical equilibria are given in Table 8. It appears that changes in entropy and enthalpy show little dependence on the nature of the tertiary tetroxide.

There is e.s.r. evidence for the formation of tetroxides from secondary peroxyl radicals. (Bennett, Brown and Mile, 1970a; Adamic, Howard and Ingold, 1969). However, thermodynamic constants could not be measured for these tetroxides because of low concentrations of peroxyl radicals and irreversible decay of the radicals below $-115°$.

Arrhenius parameters for irreversible tetroxide decomposition. Thermodynamic constants for tertiary tetroxide–peroxyl radical equilibria can be combined with kinetic constants for the termination reaction to calculate Arrhenius parameters for irreversible tetroxide decay. For convenience the overall reaction scheme used to describe the mutual reaction of tertiary peroxyl radicals is repeated below.

$$2\,\mathrm{ROO}\cdot \underset{}{\overset{K}{\rightleftharpoons}} \mathrm{ROOOOR} \xrightarrow{k_a} [\mathrm{RO}\cdot + \mathrm{O_2} + \cdot\mathrm{OR}]_{\text{cage}} \xrightarrow{k_b} \mathrm{ROOR} + \mathrm{O_2}$$

$$[\mathrm{RO}\cdot + \mathrm{O_2} + \cdot\mathrm{OR}]_{\text{cage}} \xrightarrow{k_c} 2\,\mathrm{RO}\cdot + \mathrm{O_2}$$

$$2\,\mathrm{RO}\cdot \xrightarrow{\text{RH or ROOH}} 2\,\mathrm{ROO}\cdot$$

$$2\,\mathrm{RO}\cdot \xrightarrow{k_d} \mathrm{R}\cdot + \text{a molecule of ketone}$$

If β-scission of the alkoxyl radicals is unimportant the rate of termination is given by:

$$\frac{-\mathrm{d}[\mathrm{ROO}\cdot]}{\mathrm{d}t} = 2fk_a[\mathrm{ROOOOR}] = \frac{2fk_a}{K}[\mathrm{ROO}\cdot]^2.$$

The measured rate constant $k_{\mathrm{e.s.r.}}$ is equivalent to $2fk_a/K$, and

$$\ln k_{\mathrm{e.s.r.}} = \ln 2 + \ln f + \ln k_a - \ln K.$$

In addition,

$$\ln A - E/RT = \ln 2 + \ln f - \Delta S^\circ/R + \Delta H^\circ/RT + \ln A_a - E_a/RT.$$

Using this equation, A-factors and activation energies for tetroxide decomposition can be calculated, and examples are given in Table 9.

The calculated Arrhenius parameters suggest that the unimolecular decomposition of a tetroxide is characterised by a high A-factor and a significant activation energy.

TABLE 9

Arrhenius parameters for the irreversible decomposition of some tetroxides

Peroxyl radical	E_a(kcal. mole^{-1})	$\log(A/\mathrm{sec.}^{-1})$	Ref.
2-n-Propyl-2-propyl	$18{\cdot}2 \pm 1{\cdot}7$	$\leqslant 19{\cdot}6 \pm 2{\cdot}2$	1
2-t-Butyl-2-propyl	$16{\cdot}2 \pm 2{\cdot}0$	$\leqslant 17{\cdot}5 \pm 2{\cdot}4$	1
t-Butyl	17·5	16·6	2
Cumyl	16·5	17·1	2

References: (1) Calculated from Bennett, Brown and Mile's (1970a) results. (2) Calculated from Adamic, Howard and Ingold's (1969) results.

Crossed termination rate constants for peroxyl radicals from e.s.r. studies. Bennett, Brown and Mile (1970a) have estimated 'crossed' and 'uncrossed' termination rate constants for secondary and tertiary peroxyl radicals derived from 2-methylpentane. Very few primary peroxyl radicals are formed from this substrate and there are three main termination reactions.

$$\left.\begin{array}{l} \mathrm{R^sOO}\cdot + \mathrm{R^sOO}\cdot \xrightarrow{2k_t^{ss}} \\ \mathrm{R^sOO}\cdot + \mathrm{R^tOO}\cdot \xrightarrow{4k_t^{st}} \\ \mathrm{R^tOO}\cdot + \mathrm{R^tOO}\cdot \xrightarrow{2k_t^{tt}} \end{array}\right\} \text{Non-radical products}$$

Values of $2k_t^{ss}$, $4k_t^{st}$, and $2k_t^{tt}$ were chosen to give the closest fit between predicted and experimental curves for the build-up and decay of the concentration of peroxyl radicals upon irradiation of the oxygen-saturated alkane containing t-butyl peroxide at $-40°$. The best values of $2k_t^{ss}$, $4k_t^{st}$ and $2k_t^{tt}$ were $1{\cdot}1 \times 10^5$, $2{\cdot}3 \times 10^4$, and $2{\cdot}2 \times 10^2$ M^{-1} sec.$^{-1}$, respectively. These rate constants give a ϕ-factor of 4·7.

(ii) *Ultraviolet spectroscopy*

Large concentrations ($> 10^{-4}$ M) of peroxyl radicals have been generated by pulse radiolysis of oxygenated solvents, and in a few cases ultraviolet spectra of the peroxyl radicals have been recorded. Termination rate constants have been estimated from the second order decay of the maximum absorption at 2700–2900Å. Rate constants obtained by this technique are presented in Table 10.

TABLE 10

Rate constants for peroxyl radical recombinations determined by optical spectroscopy (25°)

Peroxyl radical from	$2k_t \times 10^{-6}$ (M^{-1} sec.$^{-1}$)	Ref.
Ethanol	120	1
Oct-1-ene	9·2	2
Cyclohexane	2·8	2
Cyclohexanol	18	2

References: (1) MacLachlan (1965). (2) MacCarthey and MacLachlan (1961a).

(iii) *Rotating sector technique*

The rotating sector method has been used to measure termination rate constants for many peroxyl radicals generated by autoxidation of organic compounds either in the absence or presence of their hydroperoxides. This method is best suited to autoxidations with chain lengths greater than 5 and with rates of oxidation which are linear over a reasonable period of time.

The chief disadvantages of the rotating sector technique are the length of time it takes to perform an experiment, and the rather limited temperature range over which it can be used accurately. In order to obtain a termination rate constant it is usual to make a number of rate measurements on a single sample, e.g. a steady rate, a rate at very high sector speed, a rate at a sector speed in the region where the rate is most sensitive to the length of the dark period, another steady rate, and a dark rate. Over the length of time it takes to measure these rates, products accumulate and may interfere with the rate of chain initiation. For example, hydroperoxides formed during autoxidation may be efficient thermal or photochemical initiators. On the other hand a product may be an efficient inhibitor or retarder of chain reactions. Under these conditions the method becomes less precise because of doubts about the exact rate of photo-initiation and/or contribution from a thermal reaction. It is possible to use a fresh solution (and a large reaction volume) for each rate measurement to minimise the effect of the products, although this makes the technique even more laborious, and since free radical reactions are notoriously subject to trace impurities it is difficult to repeat the rate of an identical reaction to better than 5–10%. It is usually best to use the rotating sector method at one optimum temperature (usually 30–40°), since at higher temperatures than this thermal initiation from either the photochemical initiator or the products becomes significant and drastically lowers the accuracy of the rate constant determinations.

Because of the large number of results available, rate constants determined by the rotating sector technique have been divided into five sections based on classes of compounds, namely those determined by autoxidation of activated hydrocarbons, alkanes, ethers, aldehydes, and α-monosubstituted toluenes.

Activated hydrocarbons. Tables 11, 12 and 13 contain rate constants for autoxidation of aralkanes in the absence and presence of their hydroperoxides, aralkenes and alkenes.

Hydrocarbons which give primary peroxyl radicals such as *ortho*- and *para*-xylene, allylbenzene and oct-1-ene have termination rate constants of about 3×10^8 M^{-1} sec.$^{-1}$. Hydrocarbons which give secondary peroxyl radicals exhibit a wide variation in termination rate constants which, with a few exceptions, can be classified according to whether the peroxyl radical is derived from a benzylic, allylic, or cyclic system. In the benzylic group are compounds such as ethylbenzene, n-butylbenzene, styrene, bibenzyl and diphenylmethane with termination rate constants of the order of $2\text{–}6 \times 10^7$ M^{-1} sec.$^{-1}$. Peroxyl radicals from alkenes undergo termination more slowly than benzylic peroxyl radicals. It has been assumed that this is at least

TABLE II

Absolute rate constants for autoxidation of aralkanes (30°)

Aralkane	$k_p/(2k_t)^{1/2} \times 10^3$ (M$^{-1/2}$ sec.$^{-1/2}$)	k_p (M^{-1} sec.$^{-1}$)	$2k_t \times 10^{-6}$ (M^{-1} sec.$^{-1}$)	$2k_{12} \times 10^{-6}$ (M^{-1} sec.$^{-1}$)	Ref.
Toluene	0·014	0·24	300	—	1
m-Xylene	0·028	0·48	300	—	1
o-Xylene	0·033	0·42	154	—	1
p-Xylene	0·049	0·84	300	—	1
n-Butylbenzene	0·081	0·56	50	—	1
Bibenzyl	0·13	0·28	20	—	1
Ethylbenzene	0·21	1·3	40	—	1
Ethylbenzene	0·16	0·71	19	—	2
Diphenylmethane	0·38	(4·8)	(160)	—	1
Diphenylmethane[a]	0·40	2·1	28	—	3
Indan	1·70	4·8	8·2	—	1
Tetralin	2·30	6·4	7·6	—	1
Tetralin[a]	2·35	6·3	7·2	—	3
Tetralin	2·85	18·3	41	—	4
9,10-Dihydrophenanthrene	7·2	56	60	—	1
9,10-Dihydroanthracene	79	350	20	—	1
9,10-Dihydroanthracene[a]	69·3	310	20	—	3
sec.-Amylbenzene	0·11	0·07	0·44	—	1
Phenylcyclohexane	0·15	0·06	0·16	—	1
sec.-Butylbenzene	0·18	0·08	0·18	0·032[a]	1, 5
1,1-Diphenylethane	1·10	0·34	0·135	0·064[a]	1, 5
Cumene	1·50	0·18	0·015	0·006[a]	1, 5
Cumene	2·6	0·41	0·024	—	6

[a] In the presence of the hydroperoxide.

References: (1) Howard and Ingold (1967b). (2) Tsepalov and Shlyapintokh (1962). (3) Howard and Ingold (1968b). (4) Bamford and Dewar (1949). (5) Howard and Ingold (1968a). (6) Melville and Richards (1954).

TABLE 12

Absolute rate constants for autoxidation of aralkenes at 30°([a]) and 40°([b]) (Howard and Ingold, 1965c, 1967b)

Aralkene	$k_p/(2k_t)^{1/2} \times 10^3$ ($M^{-1/2}$ sec.$^{-1/2}$)	k_p (M^{-1} sec.$^{-1}$)	k_{abs} (M^{-1} sec.$^{-1}$)	k_{add} (M^{-1} sec.$^{-1}$)	$2k_t \times 10^{-6}$ (M^{-1} sec.$^{-1}$)
Allylbenzene[a]	0·49	10	10	—	440
Crotylbenzene[a]	4·2	8·2	8·2	—	3·8
Styrene[a]	6·3	41	—	41	42
Pentadeuterostyrene ($C_6D_5CH{=}CH_2$)[b]	8·9	69	—	69	60
β-Methylstyrene[a]	9·0	51	—	51	32
3-Chlorostyrene[b]	9·7	101	—	101	108
4-Cyanostyrene[b]	9·7	91	—	91	88
Styrene-α-d_1[b]	10·4	57	—	57	30
4-Methylstyrene[b]	11·1	83	—	83	56
Styrene-β-d_2[b]	12·0	78	—	78	42
Styrene-α-d_1-β-d_2[b]	12·3	55	—	55	20
4-Chlorostyrene[b]	12·4	123	—	123	98
α-Methylstyrene[a]	13·0	10	—	10	0·6
4-Methoxystyrene[b]	18·5	123	—	123	44
1,2-Dihydronaphthalene[a]	27·5	395	104	291	230
Indene[a]	28·4	142	14	128	50
1,4-Dihydronaphthalene[a]	35·0	900	900	—	700

TABLE 13

Absolute rate constants for autoxidation of alkenes (30°)

Alkene	$k_p/(2k_t)^{1/2} \times 10^3$ ($\text{M}^{-1/2}$ sec.$^{-1/2}$)	k_p (M^{-1} sec.$^{-1}$)	$2k_t \times 10^{-6}$ (M^{-1} sec.$^{-1}$)	Ref.
Oct-1-ene	0·062	1·0	260	1
Penta-1,4-diene	0·42	14	1080	1
Hept-3-ene	0·54	1·4	6·4	1
Dec-5-yne	0·74	2·8	14	1
Methyl oleate	0·89	0·92	1·06	1
Cyclohexene	2·3	5·4	5·6	1
Cyclopentene	2·8	7·0	6·2	1
2,3-Dimethylbut-2-ene	3·2	2·6	0·64	1
2,5-Dimethylhex-3-ene	5·4	2·3	0·18	1
Methyl linoleate	21	62	8·8	1
Methyl linolenate	39	234	36	1
Cyclohexa-1,4-diene	39	1480	1260	1
Cyclohexa-1,3-diene	100	810	66	1
Cyclohexene[a]	0·67	0·65	0·95	2
1-Methylcyclohexene[a]	1·55	1·1	0·5	2
Dihydromyrcene[a]	0·5	0·4	0·65	2
Ethyl linoleate[b]	8·1	5·7	0·5	2

[a] At 15°. [b] At 11°.
References: (1) Howard and Ingold (1967b). (2) Bateman and Gee (1948).

partly due to steric effects associated with the alkyl moiety of the peroxyl radical since termination rate constants appear to increase as the size of the olefin is decreased. Thus $2k_t$ increases from 10^6 M^{-1} sec.$^{-1}$ for methyl oleate to 6×10^6 M^{-1} sec.$^{-1}$ for hept-3-ene. Cyclic hydrocarbons have termination rate constants in the range $2\text{–}8 \times 10^6$ M^{-1} sec.$^{-1}$. Hydrocarbons giving tertiary peroxyl radicals have overall termination rate constants in the range $2\text{–}44 \times 10^4$ M^{-1} sec.$^{-1}$. Part of this difference is due to variations in $2k_{12}$ and it has been assumed that part is due to differences in the contribution from the 'crossed' termination step (reaction (16)).

Polar effects transmitted through a benzene ring appear to have little or no effect on the self reaction of secondary peroxyl radicals since peroxyl radicals from 4-methoxystyrene, styrene and 4-cyanostyrene have virtually the same termination rate constant.

Similarly, Howard, Ingold, and Symonds (1968) found little variation in termination rate constants for 4-methoxycumene

($2k_t = 4{\cdot}0 \times 10^4$ M^{-1} sec.$^{-1}$), cumene ($2k_t = 1{\cdot}5 \times 10^4$ M^{-1} sec.$^{-1}$), 3-methoxycumene ($2k_t = 6{\cdot}0 \times 10^4$ M^{-1} sec.$^{-1}$), and 4-carbomethoxycumene ($2k_t = 3{\cdot}0 \times 10^4$ M^{-1} sec.$^{-1}$).

Termination rate constants for α-deuterostyrenes substantiate the occurrence of a deuterium isotope effect for the bimolecular chain termination reaction of secondary peroxyl radicals. The value of the ratio $(k_t)_H/(k_t)_D$ is between 1·8 and 3·0 and provides support for the theory that the α-hydrogen of a secondary peroxyl radical is abstracted in the rate determining chain termination step (Russell, 1957).

Propagation rate constants for autoxidation of hydrocarbons have been found to depend to some extent on the nature of the attacking peroxyl radical as well as the nature of the substrate. Thus cumene is less reactive towards the cumylperoxyl radical than ethylbenzene is towards the 1-phenylethylperoxyl radical. This is because the cumylperoxyl radical is less reactive than the 1-phenylethylperoxyl radical and when cumene and ethyl benzene undergo reaction with the same peroxyl radical (see Tables 25 and 26) cumene is only slightly less reactive than ethylbenzene. Since cumene is generally more reactive towards free radicals than ethylbenzene, part of the low reactivity of cumene towards peroxyl radicals must be due to steric effects.

It has also been concluded from propagation rate constants that a phenyl group and a double bond have about the same activating effect on hydrogen atoms attached to the adjacent carbon, and thus ethylbenzene and oct-1-ene have similar propagation rate constants. Cyclic compounds such as cyclohexene and tetralin were found to be 3 to 4 times as reactive as acyclic compounds such as ethylbenzene. It has been suggested that the secondary hydrogens of ethylbenzene have a low reactivity because of steric inhibition to resonance in the incipient 1-phenylethyl radical arising from an interaction between the CH_3 group and the *ortho*-hydrogens on the ring. It is, however, quite likely that the difference in reactivity is due to the loss of fewer degrees of freedom by a cyclic than by an acyclic compound on entering the rigid transition state for hydrogen atom transfer.

When a hydrogen atom is activated by two groups it has a very high reactivity towards peroxyl radicals. For example, 9,10-dihydroanthracene and cyclohexa-1,4-diene are considerably more reactive than tetralin or cyclohexene.

Propagation rate constants for oxidation of *meta*- and *para*-substituted styrenes did not give a linear correlation with the Hammett equation using either σ or σ^+ substituent constants (Howard and Ingold, 1965c). This contrasts with the results obtained for oxidation of ring substituted toluenes and cumenes which give relatively good correlations with σ^+ substituent constants (Russell, 1956; Russell and Williamson, 1964a).

Alkanes. Alkanes (with the exception of those with labile tertiary hydrogens, such as 2,4,6-trimethylheptane) are too resistant to autoxidation at moderate temperatures for absolute rate constants to be measured by the rotating sector technique. Many alkane hydroperoxides are, however, readily available and oxidation of a reactive substrate such as cumene or tetralin in the presence of an alkane hydroperoxide enables termination rate constants for alkylperoxyl radicals to be measured (Howard and Ingold, 1968a, d; Howard, Adamic and Ingold, 1969). These rate constants are presented in Table 14 together with rate constants for autoxidation of 2,4,6-trimethylheptane.

Addition of sec-butyl-2-d_1 hydroperoxide and diphenylmethyl-α-d_1 hydroperoxide to hydrocarbon autoxidations has been used to give termination rate constants for deuterated peroxyl radicals and an average isotope effect of $1{\cdot}37 \pm 0{\cdot}14$ for the mutual termination reaction of secondary peroxyl radicals (Howard and Ingold, 1968d).

Ethers. Absolute rate constants for reactions of some α-alkoxyalkylperoxyl radicals determined from autoxidation of neat ethers or autoxidation of a reactive substrate in the presence of an ether hydroperoxide (Howard and Ingold, 1969b, 1970) are given in Table 15.

It would appear from the rate constants in Table 15 that α-alkoxyalkylperoxyl radicals have termination rate constants similar to structurally analogous alkylperoxyl radicals (see Table 11).

Rate constants for reaction of cyclic ethers towards their own peroxyl radical (and towards the t-butylperoxyl radical, see Table 27) increase in the order 1,4-dioxan < tetrahydropyran < tetrahydrofuran < 2,5-dimethyltetrahydrofuran < phthalan. The low reactivity of 1,4-dioxan compared with tetrahydropyran has been assumed to be due to deactivation of the hydrogen atoms by β-oxygens. Tetrahydrofuran is more reactive than tetrahydropyran towards peroxyl radicals. This is different from the situation found in the oxidation of hydrocarbons where 5- and 6-membered cyclic compounds, e.g. indan and tetralin, have about the same reactivity towards peroxyl radicals. The high reactivity of phthalan is expected since the labile

TABLE 14

Absolute rate constants for alkylperoxyl radicals (30°)

Alkane	$k_p/(2k_t)^{1/2} \times 10^3$ ($M^{-1/2}$ sec.$^{-1/2}$)	k_p (M^{-1} sec.$^{-1}$)	$2k_t \times 10^{-6}$ (M^{-1} sec.$^{-1}$)	$2k_{12} \times 10^{-6}$ (M^{-1} sec.$^{-1}$)	Ref.
Isobutane	—	—	—	0·0013	1, 2
1,1-Dimethylpropane	—	—	—	0·00026	1
2,3-Dimethylbutane	—	—	—	0·0015	1
Methylcyclohexane	—	—	—	0·001	1
1,1,3,3-Tetramethylbutane	—	—	—	0·003	1, 2
Methylcyclopentane	—	—	—	0·0057	1
2,4,6-Trimethylheptane	0·087	0·052	0·35	—	3
n-Butane (primary peroxyl)	—	—	40	—	4
n-Butane (secondary peroxyl)	—	—	1·5	—	2, 4
Cyclohexane	—	—	2·0	—	4

References: (1) Howard, Adamic and Ingold (1969). (2) Howard and Ingold (1968a). (3) Buchachenko, Kaganskaya, Neiman and Petrov (1961). (4) Howard and Ingold (1968d).

TABLE 15

Absolute rate constants[a] *for autoxidation of cyclic and acylic ethers* (30°) (Howard and Ingold, 1969b, 1970)

Ether	$k_p/(2k_t)^{1/2} \times 10^3$ ($M^{-1/2}$ sec.$^{-1/2}$)	k_p (M^{-1} sec.$^{-1}$)	$2k_t \times 10^{-7}$ (M^{-1} sec.$^{-1}$)	k_{10} (M^{-1} sec.$^{-1}$)	k_{17} (sec.$^{-1}$)
1,4-Dioxan	0·07	0·48	5	—	—
Diethyl ether	—	—	14	—	—
Di-n-butyl ether	0·1	(1·2)	(14)	—	—
Isopropyl t-butyl ether	0·1	(0·02)	(0·0043)	—	—
Tetrahydropyran	0·12	0·56	2·1	—	—
Benzyl phenyl ether	0·3	1·5	2·4	—	—
Tetrahydrofuran	0·78	4·4	3·1	—	—
Benzyl t-butyl ether	2·75	11·5	2·1	—	—
Di-isopropyl ether	3·7	0·4	0·1	0·11–0·33	0·23–0·68
Dibenzyl ether	7·1	84	16	30–61·5	86–176
2,5-Dimethyltetrahydrofuran	22·5	48	0·0046	—	—
Phthalan	42·5	432	10·5	—	—

[a] Figures in parentheses are estimated rate constants.

hydrogens are activated both by an α-phenyl group and by an α-alkoxyl group. The propagation rate constant for oxidation of 2,5-dimethyltetrahydrofuran is larger than the value for tetrahydrofuran, in contrast to the situation found in hydrocarbon oxidation, where k_p for ethylbenzene is larger than k_p for cumene.

Some acyclic ethers, e.g., di-isopropyl ether and dibenzyl ether undergo both intramolecular and intermolecular propagation reactions since they react with molecular oxygen to give mixtures of mono- and di-hydroperoxides. The following propagation and termination reactions have, therefore, to be used to describe the oxidation of these ethers (Howard and Ingold, 1970).

$$\text{Propagation: } ROO\cdot + RH \xrightarrow{k_{10}} ROOH + R\cdot \quad (10)$$

$$R\cdot + O_2 \longrightarrow ROO\cdot \quad (9)$$

$$\underset{(ROO\cdot)}{-\overset{|}{\underset{\underset{O\cdot}{\underset{|}{O}}}{\underset{|}{C}}}-O-\overset{|}{\underset{H}{\underset{|}{C}}}-} \xrightarrow{k_{17}} \underset{(D\cdot)}{-\overset{|}{\underset{\underset{OH}{\underset{|}{O}}}{\underset{|}{C}}}-O-\overset{|}{\underset{\cdot}{C}}-} \quad (17)$$

$$D\cdot + O_2 \longrightarrow DOO\cdot$$

$$DOO\cdot + RH \xrightarrow{k_{18}} DOOH + R\cdot \quad (18)$$

$$\left.\begin{array}{ll} \text{Termination: } ROO\cdot + ROO\cdot \xrightarrow{2k_{12}} & (12) \\ ROO\cdot + DOO\cdot \xrightarrow{4k_{19}} & (19) \\ DOO\cdot + DOO\cdot \xrightarrow{2k_{20}} & (20) \end{array}\right\} \text{Non-radical products}$$

R$\cdot$ represents an α-alkoxyalkyl radical; ROO$\cdot$ an α-alkoxylalkylperoxyl radical; ROOH a monohydroperoxide; D$\cdot$ an α'-hydroperoxy-α-alkoxyalkylperoxyl radical; and DOOH a dihydroperoxide. The overall rate of oxidation can be represented in the usual way:

$$R_s = \frac{k_p[RH]\, R_i^{1/2}}{(2k_t)^{1/2}},$$

where k_p is an overall propagation rate constant made up of contributions from k_{10}, k_{17} and k_{18}, and $2k_t$ is an overall termination rate constant made up of contributions from $2k_{12}$, $4k_{19}$ and $2k_{20}$.

From a kinetic analysis and a study of the products of autoxidation

TABLE 16

Rate constants for autoxidations of some aldehydes

Aldehyde	Temp. (°)	$k_p/(2k_t)^{1/2}$ ($M^{-1/2}$ sec.$^{-1/2}$)	$k_p \times 10^{-3}$ (M^{-1} sec.$^{-1}$)	$2k_t \times 10^{-6}$ (M^{-1} sec.$^{-1}$)	Ref.
Acetaldehyde	0	0·265	2·7	104	1
Heptanal	0	0·39	3·1	54	1
Octanal	0	0·47	3·9	69	1
Decanal	5	0·45	2·7	34	2
Cyclohexanecarbaldehyde	0	0·44	1·1	6·8	1
Pivalaldehyde	0	0·96	2·5	6·6	1
Benzaldehyde	0	0·29	12	1760	1
Benzaldehyde	5	0·13	1·9	210	3

References: (1) Zaikov, Howard and Ingold (1969). (2) Cooper and Melville (1951). (3) Ingles and Melville (1953).

of di-isopropyl ether and dibenzyl ether, absolute values of k_3 and k_{10} have been estimated, and are given in Table 15.

Aldehydes. Absolute rate constants for autoxidation of some aldehydes at either 0° (Zaikov, Howard and Ingold, 1969) or at 5° (Cooper and Melville, 1951; Ingles and Melville, 1953) are given in Table 16.

Values of the ratio $k_p/(2k_t)^{1/2}$ are quite similar for the different aldehydes ($0{\cdot}3-1{\cdot}0$ M$^{-1/2}$ sec.$^{-1/2}$) even though there is a wide variation in the absolute values of k_p ($1-12\times 10^3$ M^{-1} sec.$^{-1}$) and $2k_t$ ($7-1800\times 10^6$ M^{-1} sec.$^{-1}$). Termination rate constants for autoxidation of aldehydes are quite high and the value for benzaldehyde is close to the diffusion controlled limit for a bimolecular reaction. Acylperoxyl radicals appear to have termination rate constants about 10^2 times lower than aroylperoxyl radicals which may be because acyltetroxides decompose irreversibly more slowly than aroyltetroxides.

Variations in propagation rate constants for aldehyde autoxidation are apparently due to large differences in peroxyl radical reactivity as well as substrate reactivity (see Table 28).

α-Monosubstituted toluenes. The absolute rate constants for autoxidation of some α-substituted toluenes at 30° (Howard and Korcek, 1970) are summarised in Table 17. Termination rate constants for benzyl alcohol, benzyl chloride, and t-butyl phenylacetate are in the

TABLE 17

Absolute rate constants[a] *for autoxidation of some α-substituted toluenes* (30°) (Howard and Korcek, 1970)

α-Substituted toluene	$k_p/(2k_t)^{1/2}\times 10^3$ (M$^{-1/2}$ sec.$^{-1/2}$)	k_p (M^{-1} sec.$^{-1}$)	$2k_t\times 10^{-6}$ (M^{-1} sec.$^{-1}$)
Benzyl alcohol	0·85	4·8	32
Dibenzyl ketone	0·55	3·28	(36)
Benzyl cyanide	0·52	3·1	(36)
Benzyl chloride	0·42	3·0	50
Benzyl benzoate	0·36	5·2	220
Benzyl acetate	0·31	4·6	220
t-Butyl phenylacetate	0·28	1·6	33
Benzyl bromide	0·2	(1·2)	(36)

[a] Figures in parentheses are estimated constants.

TABLE 18

Arrhenius parameters for propagation and termination reactions of peroxyl radicals determined by the rotating sector technique

Substrate	log (A_p/M^{-1} sec.$^{-1}$)	E_p (kcal. mole^{-1})	log (A_t/M^{-1} sec.$^{-1}$)	E_t (kcal. mole^{-1})	Ref.
Ethylbenzene	6·0	8·5	7·3	~0	1
Tetralin	4·4	4·5	7·6	0·4	2
Tetralin	6·75	8·3	9·6	4·3	3
Cumene	—	6·7	—	<6	4
Cumene	8·2	12·4	10·7[a]	9·5[a]	5
Cumene	—	—	10·1[b]	9·2[b]	5
Styrene	7·7	8·4	8·7	1·8	6
Styrene-α-d_1	8·4	9·5	9·7	3·7	7
4-Methoxystyrene	7·1	7·2	7·3	~0	7
α-Methylstyrene	6·8	8·1	8·2	3·7	3
2,4,6-Trimethylheptane	5·3	9·1	5·5	~0	8

[a] Arrhenius parameters for $2k_5$. [b] Arrhenius parameters for $2k_t$.

References: (1) Tsepalov and Shylapintokh (1962). (2) Bamford and Dewar (1949). (3) Howard and Ingold (1966a). (4) Melville and Richards (1954). (5) Howard, Schwalm and Ingold (1968). (6) Howard and Ingold (1965b). (7) Howard and Ingold (1965c). (8) Buchachenko, Kaganskaya, Neiman and Petrov (1961).

range normally found for α-alkyltoluenes, such as ethylbenzene and bibenzyl, while benzyl acetate and benzyl benzoate give termination rate constants similar to those for toluene and benzyl ether. Propagation rate constants for the α-monosubstituted toluenes are similar and lie in the range 2–5 M^{-1} sec.^{-1}. These values are also complicated by differences in peroxyl radical reactivity as well as substrate reactivity (see Table 28).

Arrhenius parameters from rotating sector studies. Arrhenius parameters for the propagation and termination reactions for autoxidation of some hydrocarbons have been determined by the rotating sector technique and are given in Table 18. Most of the Arrhenius parameters are quite acceptable except for the termination activation energy for 2,4,6-trimethylheptane, which must be in error since it is inconsistent with data for the mutual reaction of other tertiary peroxyl radicals.

(iv) *Pre- and after-effect methods*

Several methods have been developed to determine termination rate constants for peroxyl radicals directly from studies of the non-stationary state period between one steady state autoxidation and another steady state autoxidation at a lower (or higher) rate of chain initiation.

The first pre- and after-effect method was developed by Bateman and Gee (1948, 1951) to separate propagation and termination rate constants for autoxidation of some olefins. This method involves following a photochemically initiated autoxidation, switching off the light and following the decay in the rate until a residual dark reaction is obtained. Extrapolation of the thermal rate back to the time the light was switched off gives a decay intercept, I_d, which is a function of the amount of oxygen absorbed during the non-stationary reaction. The decay intercept is given by the expression:

$$I_d = \frac{k_p[\text{RH}]}{2k_t} \ln\left(\frac{R_L + R_D}{2R_D}\right),$$

where R_L and R_D are the photochemical and thermal rates of oxidation respectively. The ratio of the rate constants $k_p/2k_t$ in conjunction with the ratio $k_p/(2k_t)^{1/2}$ enables the individual rate constants to be separated.

Alternatively, a growth intercept, I_g, can be obtained by following a thermal autoxidation, illuminating the reaction and following the growth in rate until a steady photochemical rate of oxidation is reached. Extrapolation of the photo-rate back to the time the light was switched on gives a growth intercept I_g which is given by:

$$I_g = \frac{k_p[\mathrm{RH}]}{2k_t} \ln\left(\frac{2R_L}{R_L + R_D}\right).$$

Although this method was originally used to determine termination rate constants for secondary peroxyl radicals derived from olefins it has recently been used with more accuracy for tertiary alkylperoxyl radicals (Hendry, 1967; Howard, Adamic and Ingold, 1969). This is because tertiary peroxyl radicals have relatively long lifetimes and therefore give easily measured growth and decay intercepts.

Robb and Shahin (1958) have developed a thermometric method of following non-stationary state autoxidations. Reaction between an organic compound and oxygen is exothermic and an adiabatic temperature rise can be detected by a thermocouple situated at the centre of the reaction vessel. This technique is sensitive enough to detect the induction period for a photochemically initiated autoxidation. Extrapolation of the steady state rate back to the time axis gives an intercept which is equal to $\tau \ln 2 + C$, where C is the lag in response of the recording system. This method seems best suited for rate constants in the region 10^5–10^6 M^{-1} sec.$^{-1}$ since rate constants higher than this give intercepts too small to be measured accurately and rate constants lower give intercepts which are too long with respect to the length of the adiabatic reaction.

Berger, Blaauw, Al and Smael (1968) have recently developed a method of disturbing a steady state thermally initiated oxidation by adding a small quantity of initiator or inhibitor to the reaction. Mathematical analysis of the induction period between the steady states gives the termination rate constant for the peroxyl radicals involved in the chain oxidation. A disadvantage of this method is that low rates of initiation and oxidation have to be used and the difficulties usually encountered in attempts to obtain reproducible results from radical reactions are magnified. Like other induction period methods this method is limited to substrates with low termination rate constants such as cumene although rate constants as high as 10^6 M^{-1} sec.$^{-1}$ have been determined with a fair degree of accuracy.

The weak chemiluminescence associated with autoxidation of organic compounds has been used to follow non-stationary state oxidations and to evaluate termination rate constants (Vichutinskii, 1964; Shlyapintokh, Karphukhin, Postnikov, Tsepalov, Vichutinskii and Zakharov, 1968). The occurrence of chemiluminescence during autoxidation has been attributed to the self reaction of peroxyl radicals, this reaction being exothermic by about 115–150 kcal. mole^{-1}. This phenomenon can be illustrated by considering the autoxidation of methyl ethyl ketone. This oxidation is believed to be terminated by reaction of two methyl ethyl ketone peroxyl radicals to give singlet 1-hydroxyethyl methyl ketone, triplet oxygen and triplet biacetyl.

Light is emitted by the triplet ketone as it decays to its singlet state, and the intensity, I, is proportional to the rate of termination.

$$I \propto 2k_t[\mathrm{ROO}\cdot]^2$$

Kellogg (1969) has recently suggested that the luminescence produced by autoxidations is weak because triplet ketone is rapidly quenched by triplet oxygen in the solvent cage and only one excited ketone molecule in 10^8 emits luminescence. Singlet oxygen is produced in this reaction and it has been identified as a reaction product formed from the self reaction of sec-butylperoxyl radicals (Howard and Ingold, 1968c).

During a non-stationary state oxidation the concentration of peroxyl radicals, and therefore the intensity of chemiluminescence, is changing. Thus when photochemical initiation is stopped the chemiluminescence intensity falls and:

$$(I_0/I)^{1/2} - 1 = t(2k_t R_i)^{1/2},$$

where I_0 is the intensity of the steady state chemiluminescence and I is the intensity after time t. Absolute values of termination rate

TABLE 19

Absolute rate constants and Arrhenius parameters for hydrocarbon autoxidation measured by pre- and after-effect methods

Hydrocarbon	Temp. (°)	k_p (M^{-1} sec.$^{-1}$)	log (A_p/M^{-1} sec.$^{-1}$)	E_p (kcal. mole^{-1})	$2k_t \times 10^{-6}$ (M^{-1} sec.$^{-1}$)	log (A_t/M^{-1} sec.$^{-1}$)	E_t (kcal. mole^{-1})	Ref.
Oct-1-ene	25	0·03	—	—	0·3	—	—	1
Ethylbenzene	36–60	—	—	—	40–200	—	—	2
Dihydromyrcene	25	0·26	—	—	0·52	—	—	1
Digeranyl	25	0·55	—	—	0·28	—	—	1
Digeranyl	25	3·0	—	—	9	—	—	3
Rubber	40	1·8	—	—	0·16	—	—	1
Gutta-percha	40	1·9	—	—	0·18	—	—	1
Ethyl linoleate	25	6·6	—	—	0·3	—	—	1
Ethyl linoleate	25	60	—	—	30	—	—	3
Cyclohexene	30	6·7	—	—	2·65	—	—	4
Cyclohexene	40	3·7	5·46	7·0	1·6	6·2	0	5
Cyclohexene	56	7·0	5·41	7·5	2·0	6·3	0	6
1-Methylcyclohexene	40	5·5	4·93	6·0	0·86	5·93	0	7
4-Methylcyclohexene	40	7·0	7·48	9·5	3·3	6·52	0	7
4,5-Dimethylcyclohexene	40	18·2	7·06	8·3	2·7	6·43	0	7
Tetralin	40	9·2	—	—	9·5	—	—	4
Cumene	35	0·2	6·5	10	0·01	8·3	6·0	8
Cumene	40	0·41	—	—	0·024	—	—	4
Cumene	56	1·1	4·0	6·5	0·18	5·25	0	6
Cumene	20	—	—	—	0·033	—	—	2
Cumene	60	—	—	—	0·048	6·2	2·3	9
p-Cymene	56	0·5	3·35	6·0	0·2	5·3	0	6
Decalin	60	—	—	—	0·15	—	—	9
Cyclohexanol	60	0·17	7·05	11·9	0·18	6·7	2·2	10

References: (1) Bateman and Gee (1951). (2) Vichutinskii (1964). (3) Bateman, Gee, Morris and Watson (1951). (4) Berger, Blaauw, Al and Smael (1968). (5) Robb and Shahin (1958). (6) Howard and Robb (1963). (7) Robb and Shahin (1959). (8) Hendry (1967). (9) Shlyapintokh, Karpukhin, Postnikov, Tsepalov, Vichutinskii and Zakharov (1968). (10) Aleksandrov and Denisov (1966).

constants have been obtained from the slope of a plot of $(I_0/I)^{1/2}-1$ against t.

As well as switching on and off the initiating light, non-stationary conditions for chemiluminescence studies have been achieved by the rapid addition of oxygen or an initiator to the reaction system.

Absolute rate constants and Arrhenius parameters for peroxyl radicals determined by pre- and after-effect methods are listed in Table 19. Agreement is quite good for data obtained by these methods and by the rotating sector technique. Exceptions are oct-1-ene, which gave a termination rate constant 10^3 larger by the rotating sector method, and esters of linoleic acid for which $2k_t$ has been reported to be in the range $3\times10^5 - 3\times10^7$ M^{-1} sec.$^{-1}$. Except for Hendry's data for cumene, activation energies for the termination reaction of tertiary peroxyl radicals must be too low in view of our present knowledge of this reaction.

The chemiluminescence method has been used extensively to study solvent effects on propagation and termination rate constants for autoxidation of ketones and hydrocarbons. Rate constants and Arrhenius parameters for autoxidation of methyl ethyl ketone, methyl propyl ketone, methyl isopropyl ketone and cyclohexanone, both as the pure materials and as very dilute solutions in the various solvents, are shown in Table 20.

Termination rate constants for the pure cyclic and acylic ketones are about 2×10^6 M^{-1} sec.$^{-1}$ at 30°, in good agreement with values for other secondary peroxyl radicals. Propagation rate constants for the ketones are in the range 0·07–0·75 M^{-1} sec.$^{-1}$, with the value of k_p increasing in the order methyl n-propyl < methyl ethyl < methyl isopropyl < cyclohexanone. This order of reactivities compares favourably with the reactivities of structurally analogous hydrocarbons.

As methyl ethyl ketone is diluted with solvents of lower dielectric constant, ϵ, such as benzene, n-decane, p-dichlorobenzene, carbon tetrachloride, or acetic acid, both the propagation and termination rate constants decrease according to Kirkwood's equation:

$$\log k_p \text{ or } \log(2k_t) = \log k_0 - \frac{1}{kT}\cdot\frac{\epsilon-1}{2\epsilon+1}\left(\frac{\mu_1^2}{r_1^3}+\frac{\mu_2^2}{r_2^3}-\frac{\mu_3^2}{r_3^3}\right)$$

where k_0 is the rate constant at $\epsilon = 1$, k is the Boltzmann constant, μ_1, μ_2, r_1 and r_2 are the dipole moments and effective radii of the reactants, and μ_3 and r_3 refer to the activated complex.

TABLE 20

Absolute rate constants and Arrhenius parameters for autoxidations of ketones in various solvents

Ketone	Solvent	log (A_p/M^{-1} sec.$^{-1}$)	E_p (kcal. mole^{-1})	log (A_t/M^{-1} sec.$^{-1}$)	E_t (kcal. mole^{-1})	k_p^a (M^{-1} sec.$^{-1}$)	$2k_t^a \times 10^{-6}$ (M^{-1} sec.$^{-1}$)	Ref.
Methyl ethyl	None	5·10	8·4	7·30	1.6	0·11	1·42	1
Methyl ethyl	n-Decane	—	—	—	—	0·05	0·04	2
Methyl ethyl	Carbon tetra-chloride	—	—	—	—	0·24[b]	0·13[b]	2
Methyl ethyl	*p*-Dichloro-benzene	—	—	—	—	0·052	0·05	2
Methyl ethyl	Benzene	8·4	13·7	6·78	2·7	0·031	0·068	3
Methyl ethyl	Chlorobenzene	6·25	10·3	7·14	3·0	0·067	0·095	2
Methyl ethyl	Nitrobenzene	5·08	8·1	8·04	1·2	0·17	15	4
Methyl ethyl	Acetic acid	5·80	9·6	7·85	2·9	0·075	0·575	5
Methyl ethyl	Water	9·11	16·5	7·86	4·6	0·0016	0·035	3, 6
Methyl n-propyl	None	5·92	9·8	7·48	1·7	0·071	1·8	7
Methyl isopropyl	None	4·75	7·2	7·48	1·6	0·365	2·1	7
Cyclohexanone	None	7·38	10·4	7·43	1·6	0·75	2·0	8

[a] Propagation and termination rate constants have been calculated at 30° for comparison with data for the oxidation of hydrocarbons, ethers, aldehydes, and α-monosubstituted toluenes. [b] at 60°.

References: (1) Zaikov, Vichutinskii and Maizus (1967). (2) Zaikov and Maizus (1969c). (3) Zaikov and Maizus (1968a). (4) Zaikov and Maizus (1968b). (5) Zaikov (1968a). (6) Zaikov, Andronov, Maizus and Émanuél (1967). (7) Zaikov (1968b). (8) Aleksandrov and Denisov (1969).

It has therefore been concluded that decreases in propagation and termination rate constants for oxidation of methyl ethyl ketone in these solvents depend solely on the dielectric constant of the medium, and that specific solvation, e.g., the formation of a π-complex between a peroxyl radical and an aromatic solvent, is unimportant.

Dilution of methyl ethyl ketone with water causes both the propagation and termination rate constants to decrease even though the polarity of the medium has increased. The rate constants decrease because the peroxyl radicals form hydrogen bonds with water. This lowers their reactivity by increasing the entropy and enthalpy of the reactions. The increase in entropy is due to a change in the degree of order on passing from initial solvated states to the activated complex, while the enthalpy increase is due to steric hindrance, associated with solvation of the peroxyl radicals by water.

Gradual increases in the dielectric constant of the medium for oxidation of cyclohexane by addition of either acetic acid or ethyl acetate have no effect on the propagation rate constant. However the termination rate constant is decreased because of a substantial increase in the activation energy, which overshadows an increase in the pre-exponential factor (Table 21).

t-Butyl alcohol influences the oxidation of cyclohexane both by altering the polarity of the medium and by changing the reactivity of peroxyl radicals by hydrogen bonding.

Acetic acid has the same influence on the oxidation of cumene as it has on the oxidation of cyclohexane. Thus acetic acid (78 %) increases the termination rate constant from $2{\cdot}4 \times 10^4$ to $6{\cdot}7 \times 10^4$ M^{-1} sec.$^{-1}$ at 40°, while the propagation rate constant remains unchanged (Vichutinskii, Guk, Tsepalov and Shlyapintokh, 1966).

Propagation and termination rate constants for autoxidation of 2-methylpent-2-ene increase on dilution with benzene and decrease on dilution with nitrobenzene (Table 22). In both cases changes in the rate constants with dilution obey the Kirkwood equation (Zaikov, Vichutinskii, Maizus and Émanuél, 1968). These results appear to conflict with the results obtained for oxidation of methyl ethyl ketone in non-polar solvents. It has, however, been suggested that the transition states for the propagation and termination steps for 2-methylpent-2-ene oxidation have dipole moments which are less than the sums of the dipole moments of the reactants. Dilution of 2-methylpent-2-ene with a less polar substance should, therefore, cause an increase in both k_p and $2k_t$.

TABLE 21

Absolute rate constants for autoxidations of cyclohexane (60°)

Solvent	k_p (M^{-1} sec.$^{-1}$)	$2k_t \times 10^{-6}$ (M^{-1} sec.$^{-1}$)	log (A_t/M^{-1} sec.$^{-1}$)	E_t (kcal. mole^{-1})	Ref.
Cyclohexane	0·53	5·25	7·77	1·6	1
Ethyl acetate	0·58	0·58	9·64	5·9	1
Acetic acid	0·54	0·52	9·72	6·1	1
t-Butyl alcohol	0·22	1·6	9·74	5·4	2
t-Butyl alcohol[a]	—	3·5	8·52	3·0	2

[a] Termination rate constant for reaction of a hydrogen bonded peroxyl radical with a non-hydrogen bonded peroxyl radical.
References: (1) Zaikov and Maizus (1969a). (2) Zaikov and Maizus (1969b).

TABLE 22

Rate constants and Arrhenius parameters for autoxidation of 2-methylpent-2-ene

Solvent	k_p at 60° (M^{-1} sec.$^{-1}$)	log (A_p/M^{-1} sec.$^{-1}$)	E_p (kcal. mole)	$2k_t \times 10^{-7}$ at 60° (M^{-1} sec.$^{-1}$)	log (A_t/M^{-1} sec.$^{-1}$)	E_t (kcal. mole)
2-Methylpent-2-ene	4·3	6·4	8·8	0·42	7·41	1·2
Benzene	11·5	5·08	6·1	0·76	—	—
Nitrobenzene	0·9	6·2	9·5	0·125	—	—

3. *Rate Constants for Reactions with Carbon Radicals*

At low oxygen pressures (usually < 10 torr) reaction between carbon radicals and peroxyl radicals (13) becomes an important

$$R\cdot + ROO\cdot \xrightarrow{4k_{13}} ROOR \tag{13}$$

termination process in autoxidations at ambient temperatures. The importance of this process increases with an increase in the stability of the carbon radical (R·) since the greater the stability the more slowly it reacts with oxygen. Absolute values of $4k_{13}$ have been estimated by Bateman, Gee, Morris and Watson (1951) and Bateman, Bolland and Gee (1951) using a photochemical pre- and after-effect method at 'low' oxygen pressures. These workers found that for oxidation of ethyl linoleate, $4k_{13} = 5{\cdot}0 \times 10^7$ M^{-1} sec.$^{-1}$ and for digeranyl $4k_{13} = 9 \times 10^6$ M^{-1} sec.$^{-1}$ at 25°. Autoxidation of triphenylmethane depends on the oxygen concentration at atmospheric pressure, and absolute values of $4k_{13}$ for reaction of triphenylmethyl radicals with triphenylmethylperoxyl radicals have been estimated to be $1{\cdot}5 \times 10^8$ M^{-1} sec.$^{-1}$ (Howard and Ingold, 1968a).

The equilibrium constant k_9/k_{-9} has been estimated from the measured rates of oxidation of tetralin and 9,10-dihydroanthracene in the presence of triphenylmethyl hydroperoxide (Howard and Ingold, 1968a), and was found to be 60 atm.$^{-1}$ at 30° for reaction of triphenylmethyl radicals with oxygen.

$$2(C_6H_5)_3C\cdot + O_2 \underset{k_{-9}}{\overset{k_9}{\rightleftharpoons}} (C_6H_5)_3COO\cdot$$

Janzen, Johnston and Ayers (1967) have also estimated the equilibrium constant for this reaction to be 25 atm.$^{-1}$ at 27° in a solid crystal lattice permeable to oxygen.

4. *'Crossed' Propagation and Termination Rate Constants*

Many 'crossed' propagation rate constants k_p^{ab}, for reaction of a peroxyl radical AOO·, with a substrate BH, and 'crossed' termination rate constants, $4k_t^{ab}$ for mutual reaction of unlike peroxyl radicals have been determined either by a hydroperoxide method or by co-oxidations of binary mixtures of organic compounds and use of 'uncrossed' rate constants.

(i) *Hydroperoxide method*

Autoxidation of a substrate BH in the presence of moderate concentrations of the hydroperoxide of another substrate AOOH usually means that all the peroxyl radicals derived from BH undergo chain transfer with the added hydroperoxide,

$$\mathrm{BOO\cdot + AOOH} \xrightarrow{k_{trans.}} \mathrm{BOOH + AOO\cdot}$$

and the rate controlling propagation and termination steps for the oxidation are

$$\mathrm{AOO\cdot + BH} \xrightarrow{k_p^a} \mathrm{AOOH + B\cdot}$$

$$\mathrm{AOO\cdot + AOO\cdot} \xrightarrow{2k_t^{aa}} \text{Non-radical products.}$$

The overall rate of oxidation is given by,

$$\frac{-\mathrm{d}[\mathrm{O_2}]}{\mathrm{d}t} = \frac{k_p^{ab}[\mathrm{BH}]\,R_i^{1/2}}{(2k_t^{aa})^{1/2}},$$

and determination of the absolute rate constants gives the 'crossed' propagation constant k_p^{ab}.

In Table 23 rate constants for reaction of some hydrocarbons with cumylperoxyl radicals, k_p^{cr}, are listed and compared with the reactivities of the hydrocarbons towards their own peroxyl radicals,

TABLE 23

Rate constants for reactions of cumylperoxyl radicals with hydrocarbons (30°) (Howard, Ingold and Symonds, 1968)

Hydrocarbon	k_p^{cr} ($\mathrm{M^{-1}\ sec.^{-1}}$)	k_p^{rr}/k_p^{cr}
Toluene	0·034	7·0
Oct-1-ene	0·27	3·7
Bibenzyl	0·20	2·8
Ethylbenzene	0·21	6·2
Diphenylmethane	0·50	4·0
Dec-5-yne	0·60	4·7
Tetralin	1·65	4·0
Indan	1·7	2·8
9,10-Dihydroanthracene	70	2·5
sec-Butylbenzene	0·07	1·14
sec-Amylbenzene	0·07	1·0
Cumene	0·18	1·0
1,1-Diphenylethane	0·46	0·74
2,3-Dimethylbut-2-ene	1·7	1·0
2,5-Dimethylhex-3-ene	2·1	1·1
Styrene	2·1	19
β-Methylstyrene	4·2	12
α-Methylstyrene	3·9	2·6

k_p^{rr}. Hydrocarbons with labile primary and secondary hydrocarbons are less reactive towards cumylperoxyl radicals than towards their own peroxyl radicals, while reactivities of hydrocarbons with labile tertiary hydrogens towards their own peroxyl radical and towards the cumylperoxyl radical are similar.

Rate constants for reaction of deuterocumenes and ring-substituted cumenes with cumylperoxyl radicals are shown in Table 24. The relative reactivities of the deuterocumenes and unsubstituted cumenes towards this radical indicate that the primary isotope effect for abstraction of a tertiary hydrogen is about 9 and the secondary isotope effect is about $1{\cdot}06 \pm 0{\cdot}03$ per deuterium.

Cumene is also more reactive towards t-butylperoxyl radicals ($k_p = 0{\cdot}22$ M^{-1} sec.$^{-1}$) than cumene-α-d_1 ($k_p = 0{\cdot}011$ M^{-1} sec.$^{-1}$) or cumene-β-d_6 ($k_p = 0{\cdot}19$ M^{-1} sec.$^{-1}$). Similarly, tetralin is more reactive towards t-butylperoxyl radicals ($k_p = 1{\cdot}55$ M^{-1} sec.$^{-1}$) than perdeutero-tetralin ($k_p = 0{\cdot}1$ M^{-1} sec.$^{-1}$).

Addition of cumene hydroperoxide to oxidising cumene-β-d_6 and cumene-α-d_1-β-d_6 was found to have no effect on the rate of oxidation,

TABLE 24

Rate constants ($k_p^{cc'}$) for reactions of cumyl peroxyl radicals with substituted cumenes (30°) (Howard, Ingold and Symonds, 1968)

Substituted cumene	$k_p^{cc'}$ (M^{-1} sec.$^{-1}$)
Cumene-α-d_1	0·02
Cumene-α-d_1-β-d_6	0·013
Cumene-β-d_6	0·11
p-Methoxycumene	0·33
p-Methylcumene	0·22
p-Isopropylcumene	0·24[a]
Cumene	0·18
m-Methoxycumene	0·18
p-Chlorocumene	0·15
m-Bromocumene	0·12
p-Carbomethoxycumene	0·13
p-Cyanocumene	0·13
m-Nitrocumene	0·12
p-Nitrocumene	0·12

[a] Statistically corrected.

implying that $2k_t$ for cumene-β-d_6-peroxyl radicals is about 6×10^3 M^{-1} sec.$^{-1}$.

Rate constants for reaction of ring-substituted cumenes with cumylperoxyl radicals can be correlated by the Hammett equation using σ^+ substitutent constants and $\rho^+ = -0{\cdot}29$. This correlation with σ^+ substituent constants supports Russell's suggestion (1957) that a structure with some charge separation ($RO\overset{\delta-}{O}{:}\overset{\delta+}{H{\cdot}}R$) plays an important role in the transition state for the rate controlling propagation reaction of autoxidations.

Rate constants for reaction of hydrocarbons with some tertiary peroxyl radicals are given in Table 25 and with some primary and

TABLE 25

Rate constants (M^{-1} *sec.*$^{-1}$) *for reactions of tertiary peroxyl radicals with hydrocarbons* (30°) (Howard and Ingold, 1968a)

	Peroxyl radical			
Hydrocarbon	t-Butyl	1,1,3,3-Tetra-methylbutyl	2-Phenyl-2-butyl	1,1-Diphenylethyl
Toluene	0·05	0·04	0·05	—
Ethylbenzene	0·2	0·32	0·22	—
Cumene	0·22[a]	0·14	0·15	0·18
Tetralin	2·0	2·0	1·3	—
Styrene	1·05	—	—	1·7

[a] Hendry (1967) has reported a value of 0·1 M^{-1} sec.$^{-1}$ for this rate constant at 35°.

secondary peroxyl radicals in Table 26. It is apparent that increasing the size of a tertiary peroxyl radical from t-butyl to 1,1-diphenylethyl has little effect on the reactivity of the radical towards hydrogen atom abstraction. Similarly, primary peroxyl and cyclic and acyclic secondary peroxyl radicals have about the same reactivity and they are all somewhat more reactive than tertiary peroxyl radicals.

Rate constants for reactions of cyclic and acylic ethers with t-butylperoxyl radicals and with some secondary ether and hydrocarbon peroxyl radicals are given in Table 27. Ethers show the expected reactivity towards t-butylperoxyl radicals. Thus the aliphatic ethers are less reactive than the benzylic ethers. However, α-methylbenzyl

TABLE 26

Rate constants (M^{-1} *sec.*$^{-1}$) *for reactions of primary and secondary peroxyl radicals with hydrocarbons* (30°) (Howard and Ingold, 1968b)

Hydrocarbon	Peroxyl Radical					
	n-Butyl	sec-Butyl	Cyclohexyl	α-Tetralyl	Diphenyl-methyl	9,10-Dihydro-9-anthracyl
Toluene	0·1	0·1	—	0·1	—	—
Ethylbenzene	0·55	0·5	0·5	0·5	0·8	0·6
Cumene	0·45	0·4	—	0·5	0·5	—
Tetralin	6·4	4·2	4·5	6·3	6·6	6·5
Styrene	14	6·3	6·8	8·8	16	21
α-Methylstyrene	25	12	14	16	25	33
9,10-Dihydroanthracene	310	140	~160	240	250	330

TABLE 27

Rate constants for the reactions of ethers with some peroxyl radicals (30°) (Howard and Ingold, 1970)

	k_p (M^{-1} sec.$^{-1}$)				
Ether	t-Butyl-peroxyl	α-Ethoxyethyl-peroxyl	α-Tetrahydro-furanyl-peroxyl	α-Tetrahydro-pyranyl-peroxyl	α-Tetralyl-peroxyl
Di-n-butyl ether	0·064	—	—	—	—
Isopropyl t-butyl ether	0·02	—	—	—	—
Tetrahydropyran	0·024	—	—	—	—
Di-α-methylbenzyl ether	0·084	—	—	—	—
Benzyl phenyl ether	0·20	—	—	—	—
Tetrahydrofuran	0·34	4·5	4·3	4·4	3·0
Benzyl t-butyl ether	1·10	—	—	—	—
Di-isopropyl ether	0·11	—	—	—	—
Dibenzyl ether	1·3	36	—	25·7	19
2,5-Dimethyltetrahydrofuran	0·8	—	—	—	—

ether has a low reactivity towards t-butylperoxyl radicals, when compared with the reactivity of dibenzyl ether, and this has been attributed to steric inhibition to hydrogen atom transfer. Steric hindrance by the α-t-butoxyl group has been invoked to explain the lower reactivity of isopropyl t-butyl ether as compared with di-isopropyl ether. 2,5-Dimethyltetrahydrofuran is considerably less reactive towards t-butylperoxyl radicals than it is towards its own peroxyl radical and it has been suggested that the propagation rate constant for the pure ether is an overall value containing a contribution from an intramolecular reaction.

Howard and Korcek (1970) recently measured rate constants for reaction of some α-monosubstituted toluenes with t-butylperoxyl radicals (k_p^{br}) and their results are compared with propagation rate constants for oxidation of the pure compounds (k_p^{rr}) in Table 28. It was concluded from the data in this Table that the reactivity of a peroxyl radical derived from an α-monosubstituted toluene depends to some extent on the nature of the group attached to the α-carbon.

TABLE 28

Rate constants for reactions of α-monosubstituted toluenes with t-butylperoxyl radicals (30°) (Howard and Korcek, 1970)

α-Monosubstituted toluene	k_p^{br} (per α-H) (M^{-1} sec.$^{-1}$)	k_p^{rr}/k_p^{br}
Benzaldehyde	0·85	~40000
Dibenzyl ether	0·3	25
Benzyl t-butyl ether	0·55	10·5
Benzyl alcohol	0·065	37
Benzyl ketone	0·045	18
Benzyl cyanide	0·01	156
Benzyl chloride	0·008	190
Benzyl benzoate	0·0085	306
Benzyl acetate	0·0075	307
Benzyl phenyl ether	0·1	7·5
t-Butyl phenylacetate	0·03	27
Ethylbenzene	0·1	6·5
Benzyl bromide	0·006	100
Toluene	0·009	9

A reasonable correlation was found between the logarithm of the ratio k_p^{rr}/k_p^{br} and σ_m substituent constants. This indicates that the reactivity of a peroxyl radical increases as the electron-withdrawing capacity of the α-substituent increases. Ratios of k_p^{rr}/k_p^{br} for oxidation of cumene, 1,1-diphenylethane, and α-methylbenzyl alcohol were found to correlate with $\Sigma\sigma_m$ substituent constants and Howard and Korcek (1970) suggested that the difference in reactivity between tertiary and secondary peroxyl radicals is due mainly to polar effects. The logarithms of the rate constants for reaction of α-monosubstituted toluenes with t-butylperoxyl radicals were found to correlate with σ_p^+ substituent constants with a value of ρ^+ equal to $-1{\cdot}2$. It was therefore suggested that both inductive and resonance effects influence the reactivity of the labile hydrogens of α-monosubstituted toluenes.

(ii) *Co-oxidation*

Co-oxidation of binary mixtures of organic compounds has been used to determine ratios of 'uncrossed' and 'crossed' propagation rate constants. Oxidation of two substrates AH and BH can be represented by the following rate controlling propagation and termination reactions.

$$\text{Propagation: } AOO\cdot + AH \xrightarrow{k_p^{aa}} AOOH + A\cdot$$

$$AOO\cdot + BH \xrightarrow{k_p^{ab}} AOOH + B\cdot$$

$$BOO\cdot + BH \xrightarrow{k_p^{bb}} BOOH + B\cdot$$

$$BOO\cdot + AH \xrightarrow{k_p^{ba}} BOOH + A\cdot$$

$$\left.\begin{array}{l} \text{Termination: } AOO\cdot + AOO\cdot \xrightarrow{2k_t^{aa}} \\ AOO\cdot + BOO\cdot \xrightarrow{4k_t^{ab}} \\ BOO\cdot + BOO\cdot \xrightarrow{2k_t^{bb}} \end{array}\right\} \text{Non-radical products}$$

The steady state treatment of these equations, assuming a long chain, yields the following rate expression.

$$-\frac{d[O_2]}{dt} = \frac{(r_a[AH]^2 + 2[AH][BH] + r_b[BH]^2)\, R_i^{1/2}}{(r_a^2\,\delta_a^2[AH]^2 + \phi r_a r_b \delta_a \delta_b[AH][BH] + r_b^2\,\delta_b^2[BH]^2)^{1/2}}, \quad (21)$$

where

$$r_a = k_p^{aa}/k_p^{ab},$$

$$\delta_a = (2k_t^{aa})^{1/2}/k_p^{aa},$$

$$r_b = k_p^{bb}/k_p^{ba},$$

$$\delta_b = (2k_t^{bb})^{1/2}/k_p^{bb},$$

$$\phi = k_t^{ba}/(k_t^{bb}\,k_t^{aa})^{1/2}.$$

The ratios δ_a and δ_b are obtained from oxidation of pure AH and BH. Rates of oxidation of mixtures of AH and BH of different compositions enable the ratios r_a, r_b, and ϕ to be calculated from equation (21) by using a computer to match predicted and experimental curves of oxidation rate against composition.

Alternatively, the concentration of AH and BH consumed during oxidation, or the concentrations of reaction products formed from AH and BH at low conversions can be used to calculate r_a and r_b directly from the differential form of the co-oxidation composition equation (22), where Δ[AH] and Δ[BH] are the concentrations of

$$\frac{\Delta[\mathrm{AH}]}{\Delta[\mathrm{BH}]} = \frac{r_a[\mathrm{AH}]/[\mathrm{BH}]+1}{r_b[\mathrm{BH}]/[\mathrm{AH}]+1} \tag{22}$$

AH and BH consumed and [AH]/[BH] is the average ratio of the concentrations of AH and BH during the reaction.

Equation (22) was modified by Fineman and Ross (1950) to give:

$$\frac{\rho-1}{R} = r_a - \frac{r_b\,\rho}{R^2},$$

where $\rho = \Delta[\mathrm{AH}]/\Delta[\mathrm{BH}]$ and $\mathrm{R} = [\mathrm{AH}]/[\mathrm{BH}]$. Graphical solution of this equation gives r_a and r_b from the intercept and slope of the plot of $\rho - 1/R$ against ρ/R^2.

Mayo, Syz, Mill and Castleman (1968) have discussed experimental data obtained from co-oxidation studies in a recent article and have emphasised that such data should be used only with extreme caution. Though there are small differences in the reactivity of peroxyl radicals they appear to have the same selectivity towards hydrogen atom abstraction and the product $r_a r_b$ should be about 1. (Although recent work on the oxidation of α-monosubstituted toluenes (Howard and Korcek, 1970) has indicated quite large differences in the reactivity of peroxyl radicals towards hydrogen atom abstraction, there

TABLE 29

'Crossed' propagation rate constants for hydrocarbon oxidations from co-oxidation data (30°)[a]

Substrate A	Substrate B	r_a	r_b	$r_a r_b$	k_p^{ab} (M^{-1} sec.$^{-1}$)	k_p^{ba} (M^{-1} sec.$^{-1}$)	Ref.
Cumene	Toluene	4·0	0·25	1·0	0·045	1·0	1
Cumene	Butadiene	0·03	23	0·7	6·0	—	2
Cumene	Tetralin	0·044	16	0·7	4·2	0·4	3
Cumene	Tetralin	0·13	4·6	0·6	1·4	1·4	4
Cumene	Tetralin	0·21	5·0	1·05	0·86	1·3	5
Cumene	*cis*-Decalin	0·58	1·56	0·90	0·31	—	4
Cumene	α-Methylstyrene	0·1	9·3	0·93	1·8	1·07	6
Cumene	Styrene	0·1	13	1·3	1·75	3·1	7
Cumene	Styrene	0·05	20	1·0	3·5	2·05	2
α-Methylstyrene	Tetralin	1·3	0·85	1·1	7·7	7·5	4
α-Methylstyrene	Styrene	1·2	0·9	1·08	8·3	45·5	4
Styrene	Butadiene	0·45	2·2	1·0	91	—	2
Styrene	Tetralin	2·3	0·43	1·0	17·8	15	2
Styrene	Tetralin	2·7	0·49	1·3	15·2	13·1	5
Tetralin	Butadiene	0·22	3·5	0·77	26·7	—	2

[a] It has been assumed that r_a and r_b are virtually independent of temperature.

References: (1) Alagy, Clement and Balaceanu (1961). (2) Mayo, Syz, Mill and Castleman (1968). (3) Russell (1955). (4) Russell and Williamson (1964b). (5) Niki, Kamiya and Ohta (1969). (6) Niki and Kamiya (1967). (7) Chevriau, Naffa and Balaceanu (1964).

are only small differences in reactivity between hydrocarbon peroxyl radicals for which most of the co-oxidation data are available.) Some reactivity ratios obtained from co-oxidation studies and calculated absolute 'crossed' propagation rate constants are given in Table 29. Values of $r_a r_b$ have been used only if the product $r_a r_b$ is in the range 0·5–1·5.

'Crossed' propagation rate constants obtained from co-oxidation studies and the hydroperoxide method agree quite well, as can be seen from the comparison given in Table 30.

TABLE 30

A comparison of 'crossed' propagation rate constants as measured by the hydroperoxide and the co-oxidation methods

Peroxyl radical from	Substrate	'Crossed' propagation rate constants (M^{-1} sec.$^{-1}$)	
		Hydroperoxide method	Co-oxidation method
Cumene	Styrene	2·1	1·75–3·5
Cumene	Tetralin	1·65	0·86–4·2
Cumene	α-Methylstyrene	3·9	1·8
Tetralin	Styrene	8·8	13·1–15
Tetralin	Cumene	0·5	0·4– 1·4
Tetralin	α-Methylstyrene	16	7·7

(iii) *The estimation of termination rate constants from co-oxidation studies*

French workers at the Institut Francais du Petrole have used co-oxidation studies extensively to estimate rate constants for autoxidation, and this work has been reviewed recently by Sajus (1968). This method is based on observations by Russell (1955) who found that small quantities of hydrocarbons which produce secondary peroxyl radicals, e.g., tetralin, retard the oxidation of cumene. Russell suggested, and it has since been confirmed (Howard, Schwalm and Ingold, 1968), that this retardation occurs because secondary peroxyl radicals are introduced into the reaction chain, and they have termination rate constants about three orders of magnitude larger than the termination rate constant for cumylperoxyl radicals. This means that the overall termination rate constant for oxidation of cumene in the presence of a small quantity of tetralin is much higher than for pure cumene, and the concentration of peroxyl radicals, and hence the rate of oxidation, must be lower. Thus if the rate–composition curve for co-oxidation of two substrates AH and BH shows a minimum (i.e., AH retards the oxidation of BH) then $2k_t^{aa} > 2k_t^{bb}$, but if it is a straight line, $2k_t^{aa} = 2k_t^{bb}$. By this method a wide variety of organic compounds have been classified according to their termination rate constants, and propagation rate constants have been estimated from rates of oxidation. The data has recently been summarised by Sajus (1968) and will not be repeated here. This co-oxidation method in many cases gives a good estimation of termination rate constants. It does, however, miss small and interesting differences between rate constants. For example, tetralin and ethylbenzene have been assumed to have the same termination rate constant, whereas rotating sector measurements indicate that tetralylperoxyl radicals are somewhat less reactive than 1-phenylethylperoxyl radicals towards termination. A much greater discrepancy occurs in the case of sec-butylbenzene, for which co-oxidation studies indicate a termination rate constant equivalent to that of ethylbenzene, whereas the actual measured value is about a hundred times lower.

(iv) *'Crossed' termination rate constants from co-oxidation studies*

Measurement of the rate of oxidation of a mixture of hydrocarbons AH and BH of known concentrations, at a known rate of initiation, and substitution of r_a, r_b, δ_a and δ_b into equation (21) enables a ϕ-factor, which is a measure of the preference for the mutual reaction of unlike peroxyl radicals, to be calculated. Some examples of ϕ-factors and 'crossed' termination rate constants, $4k_t^{ab}$, for substrates with known 'uncrossed' termination rate constants are given in Table 31. Again data have been considered only if the product $r_a r_b$ is in the range 0·5–1·5.

Values of ϕ for reaction of tertiary peroxyl radicals with primary and secondary peroxyl radicals are quite high, and 'crossed' termination rate constants are only slightly lower than termination rate constants for 'uncrossed' primary and secondary peroxyl radicals.

TABLE 31

'Crossed' termination rate constants for hydrocarbon oxidations

Substrate A	Substrate B	ϕ	$2k_t^{aa} \times 10^{-6}$ (M^{-1} sec.$^{-1}$)	$2k_t^{bb} \times 10^{-6}$ (M^{-1} sec.$^{-1}$)	$4k_t^{ab} \times 10^{-6}$ (M^{-1} sec.$^{-1}$)	Ref.
Cumene	Tetralin	11·8	0·015	7·6	4·0	1
Cumene	Tetralin	5·2	0·015	7·6	1·8	2
Cumene	Tetralin	12	0·015	7·6	4·0	3
Cumene	Ethylbenzene	13	0·015	40	10	2
Cumene	Styrene	26	0·015	42	20	2
Cumene	Styrene	21	0·015	42	16	4
Cumene	α-Methylstyrene	2	0·015	0·6	0·2	2
Cumene	Diphenylmethane	12·8	0·015	28	8·2	2

References: (1) Howard, Schwalm and Ingold (1968). (2) Niki, Kamiya and Ohta (1968). (3) Russell (1955). (4) Chevriau, Naffa and Balaceanu (1964).

5. *Reactions with Hydroperoxides*

'Crossed' and 'uncrossed' propagation and termination rate constants can be used to yield absolute rate constants for reaction of peroxyl radicals with hydroperoxides.

$$\text{AOO}\cdot + \text{BOOH} \xrightarrow{k_{trans.}} \text{AOOH} + \text{BOO}\cdot$$

This has been accomplished by kinetic analysis of the oxidation of a substrate AH in the presence of low concentrations of a hydroperoxide BOOH. The kinetic expression for the rate of oxidation is given by:

$$-\frac{\mathrm{d}[\mathrm{O}_2]}{\mathrm{d}t} = (k_p^{aa}[\mathrm{AH}] + k_{trans.}[\mathrm{BOOH}]) \times \left[\frac{R_i}{2k_t^{aa} + 4k_t^{ab}\dfrac{k_{trans.}[\mathrm{BOOH}]}{k_p^{ba}[\mathrm{AH}]} + 2k_t^{bb}\left(\dfrac{k_{trans.}[\mathrm{BOOH}]}{k_p^{ba}[\mathrm{AH}]}\right)^2}\right]^{1/2}.$$

TABLE 32

Rate constants for reactions of peroxyl radicals with hydroperoxides ($k_{trans.}$)

Peroxyl radical	Hydroperoxide	Temp. (°)	$k_{trans.}$ (M^{-1} sec.$^{-1}$)	Ref.
Cumyl	Tetralin	30	600	1
Cumyl	Tetralin	56	1100	1
Cumyl	Tetralin (OOD)	30	12	1
Cumyl	Tetralin[a]	30	10	1
Cumyl	Tetralin	56	300[b]	2
Cumyl	Tetrahydropyran	30	47	3
Cumyl	Tetrahydrofuran	30	38	3
Cumyl	α-Ethoxyethyl	30	66	3
α-Tetralyl	Cumene	30	2500	1
α-Tetralyl	Cumene	56	2800	1
α-Tetralyl	Cumene (OOD)	30	140	1
Poly(peroxystyryl)	t-Butyl	60	645[c]	4

[a] In the presence of 7·3 M t-butanol. [b] Calculated from Thomas and Tolman's (1962) results using more reliable rate constants. [c] From a value of 4·5 for the ratio $k_{trans.}/k_p$ and an absolute value for k_p for styrene of 143 M^{-1} sec.$^{-1}$ (from Table 18).

References: (1) Howard, Schwalm and Ingold (1968). (2) Thomas and Tolman (1962). (3) Howard and Ingold (1969b). (4) Hiatt, Gould and Mayo (1964).

Measurement of the rate, and knowledge of the rate constants k_p^{aa}, k_p^{ba}, $2k_t^{aa}$, $4k_t^{ab}$ and $2k_t^{bb}$ enables the rate constant, $k_{trans.}$, to be calculated from predicted and experimental rates of oxidation. Transfer rate constants for a number of peroxyl radicals and hydroperoxides are listed in Table 32. Absolute values of $k_{trans.}$ are relatively high and reflect the ease with which this completely symmetric hydrogen atom transfer reaction occurs. This reaction is also characterised by a very high deuterium isotope effect (17–30) though this isotope effect may be somewhat enhanced by stronger hydrogen bonding for the deuterated hydroperoxides. The importance of hydrogen bonding in this reaction is reflected by the low value of $k_{trans.}$ in t-butyl alcohol, a hydrogen bonding solvent. Increased hydrogen bonding has also been invoked to explain the lower values of $k_{trans.}$ for ether hydroperoxides than for alkyl hydroperoxides.

6. *Calculation of Absolute Rate Constants from Relative Rate Constants*

Propagation and termination rate constants determined by the methods described above have been used to calculate many other rate constants for reactions involving peroxyl radicals. For example, propagation rate constants for oxidation of boron compounds can be calculated from rates of oxidation once the chain-carrying peroxyl radical has been identified. Similarly, rate constants for reaction of peroxyl radicals with various antioxidants can be calculated once the kinetics and mechanism of inhibition have been established.

(i) *Boron compounds*

Alkyl derivatives of boron readily react with molecular oxygen to give peroxyl compounds by a free radical chain process which can be described by the following propagation and termination reactions (Davies and Roberts, 1967, 1969; Allies and Brindley, 1967, 1968).

$$\text{Propagation: } R\cdot + O_2 \longrightarrow ROO\cdot$$

$$ROO\cdot + RB \xrightarrow{k_p} ROOB + R\cdot$$

$$\text{Termination: } ROO\cdot + ROO\cdot \xrightarrow{2k_t} \text{Non-radical products}$$

where RB is a boron compound, R· is an alkyl radical and ROO· an alkylperoxyl radical. It has recently been shown that the rates of oxidation of many boron compounds obey the general kinetic expression for autoxidation (Ingold, 1969b; Davies, Ingold, Roberts and Tudor, 1970).

$$R_s = \frac{-\mathrm{d}[O_2]}{\mathrm{d}t} = \frac{k_p[\mathrm{RB}]\,R_i^{1/2}}{(2k_t)^{1/2}}$$

and measurement of the rate of oxidation, R_s, gives the rate constant ratio $k_p/(2k_t)^{1/2}$, where k_p is the rate constant for displacement of an alkyl radical on boron by a peroxyl radical and $2k_t$ is the termination rate constant for the alkylperoxyl radicals. Since termination rate constants for many alkylperoxyl radicals are known, absolute propagation rate constants for oxidation of a number of boron compounds have been calculated (Table 33).

TABLE 33

Propagation rate constants for autoxidation of some organo-boron compounds (30°)

	k_p (M^{-1} sec.$^{-1}$)[a]		
Compound	R = Bu^n	R = Bu^s	R = Bu^t
$RB(OR')_2$	Small	4	0·3
$(RBO)_3$	1×10^3	5×10^4	3×10^4
R_2BOR'	5×10^3	2×10^3	—
R_2BOOR	3×10^4	1×10^4	—
R_2BOBR_2	3×10^5	7×10^4	—
R_3B	2×10^6	—	—
$2k_t$ (ROO•)	4×10^7	$1{\cdot}5 \times 10^6$	1×10^3

[a] These rate constants have not been statistically corrected for the number of equivalent R groups available for displacement.

Propagation rate constants appear to cover a very wide range, reaching almost to the limits of a diffusion controlled process for trialkylboranes. The factors which control the magnitude of k_p include:

(*a*) The electron density at the boron atom, the reaction being retarded by a high density produced by complex formation with Lewis bases such as pyridine or by $p\pi$—$p\pi$ bond formation between oxygen and boron as in oxy-boron compounds.

(*b*) The stability of the displaced alkyl radical, the rate increasing as the radical stability increases.
(*c*) Steric protection of the boron compound and of the attacking peroxyl radical, which appears to retard reaction.

(ii) *Phenols*

The kinetics and mechanism of the inhibition of liquid phase autoxidation by phenols will not be dealt with in detail since this has been the subject of two recent reviews (Ingold, 1968; Mahoney, 1969).

It has been demonstrated fairly conclusively from deuterium isotope studies that the rate-controlling termination step for many inhibited oxidations is abstraction of the labile hydrogen from a phenol, AH, by a peroxyl radical (Howard and Ingold, 1962),

$$ROO\cdot + AH \xrightarrow{k_{inh}} ROOH + A\cdot$$

Rate constants for this reaction can readily be obtained for systems which obey the kinetic expression (23), where R_{inh} is the inhibited rate of oxidation.

$$R_{inh} = \frac{k_p[RH]\,R_i}{2k_{inh}} \qquad (23)$$

Systems which obey this equation include inhibition of the oxidation of polymerisable olefins, e.g., styrene, by hindered and non-hindered phenols, and inhibition of the oxidation of substrates with a labile hydrogen atom, e.g., tetralin and 9,10-dihydroanthracene, by hindered phenols and hydroquinones.

Howard and Ingold (1963a, b) have measured absolute values of k_{inh} for reaction of many ring-substituted phenols with poly(peroxystyryl)peroxyl radicals. They found that rate constants for *meta*- and *para*-substituted hindered and non-hindered phenols could be correlated by the Hammett equation using σ^+ substituent constants. The relatively large negative ρ values found for these phenols were interpreted by Howard and Ingold as being due to a strong polar character in the activated complex for the hydrogen transfer reaction. Rate constants for reaction of these phenols with poly(peroxystyryl)peroxyl radicals are summarised in Table 34. Instead of listing a large number of rate constants for *meta*- and *para*-substituted phenols, rate constants are given for the phenols unsubstituted in these positions and the appropriate ρ^+ values are indicated.

Rate constants for substituted phenols which cannot be characterised in this way are collected in Table 35.

TABLE 34

Rate constants and ρ^+ values for reactions of phenols with poly(peroxystyryl)peroxyl radicals (65°)

Phenol	$k_{inh} \times 10^{-4}$ (M^{-1} sec.$^{-1}$)	ρ^+	Ref.
Phenol	0·5	−1·49 or −1·58	1
2,6-Dimethylphenol	2·0	−1·36	2
2-t-Butylphenol	1·7	−1·46	2
2,6-Di-t-butylphenol	0·5	−1·11	2

References: (1) Howard and Ingold (1963a). (2) Howard and Ingold (1963b).

TABLE 35

Rate constants for reactions of some phenols with poly(peroxystyryl)peroxyl radicals (65°) (Howard and Ingold, 1963a)

Phenol	$k_{inh} \times 10^{-4}$ (M^{-1} sec.$^{-1}$)
2,4-Dimethyl	4·1
2,4,5-Trimethyl	4·8
2,3-Dimethyl	2·1
2,3,5,6-Tetramethyl	3·0
2,3,4,5,6-Pentamethyl	8·6
2,6-Diethyl	2·2
2-Isopropyl-4-methyl	3·9
2,6-Di-isopropyl	2·1
2,6-Di-isopropyl-4-methyl	5·9
2-t-Amyl-4-methyl	4·2
2,6-Di-t-amyl-4-methyl	1·6
2-Oct^{tt}-4-methyl[a]	3·7
2,6-Di-oct^{tt}-4-methyl[a]	1·14
2-Et_3C-4-methyl	3·3
2,6-Et_3C-4-methyl	1·6
2-t-Butyl-5,6-dimethyl	3·6
2-t-Butyl-4-methyl-6-oct^{tt} [a]	1·5
4,4′-Bis-2,6-di-t-butyl	4·2
4,4′-methylene-bis-2,6-di-t-butyl	3·0
α,α′-Bis-4-methylene-2,6-di-t-butyl	3·4

[a] Oct^{tt} ≡ $(CH_3)_3C.CH_2(CH_3)_2C$

Reactions of tetralylperoxyl radicals with 2,6-di-t-butyl-4-substituted phenols have also been shown to correlate with σ^+substituent constants and $\rho^+ = -1{\cdot}36$ (Howard and Ingold, 1965a). The value of k_{inh} for reaction of tetralylperoxyl radicals with 2,6-di-t-butylphenol at 65° is $1{\cdot}15 \times 10^4$ M^{-1} sec.$^{-1}$ and replacement of the phenolic hydrogen with deuterium yields an isotope effect of 10, i.e., $(k_{inh})_H/(k_{inh})_D = 10$.

Bickel and Kooyman (1956) and Davies, Goldsmith, Gupta and Lester (1956) have studied the inhibited oxidation of 9,10-dihydroanthracene and tetralin, respectively, in the presence of several hindered and non-hindered phenols. Absolute values of the rate constant, k_{inh} lie in the range $1\text{–}10 \times 10^4$ M^{-1} sec.$^{-1}$, when the ratios of rate constants obtained by these workers are put on an absolute basis.

Berger, Blaauw, Al and Smael (1968) were able to determine rate constants for reaction of 2,6-di-t-butyl-4-methylphenol with cumylperoxyl radicals at 40° ($k_{inh} = 1{\cdot}2 \times 10^4$ M^{-1} sec.$^{-1}$), cyclohexenylperoxyl radicals at 30° ($k_{inh} = 3{\cdot}0 \times 10^4$ M^{-1} sec.$^{-1}$) and tetralylperoxyl radicals at 40° ($k_{inh} = 2{\cdot}75 \times 10^4$ M^{-1} sec.$^{-1}$) from their studies of non-stationary state autoxidations.

Mahoney and DaRooge (1970b) have reported rate constants at 30° for reaction of tetralylperoxyl radicals with 2,4,6-tri-t-butylphenol ($k_{inh} = 2{\cdot}86 \pm 0{\cdot}07 \times 10^4$ M^{-1} sec.$^{-1}$) and for reaction of 9,10-dihydroanthracyl-9-peroxyl radicals with 2,4,6-tri-t-butylphenol ($k_{inh} = 1{\cdot}28 \pm 0{\cdot}07 \times 10^4$ M^{-1} sec.$^{-1}$) and with 2,6-di-t-butyl-4-methylphenol ($k_{inh} = 1{\cdot}1 \pm 0{\cdot}06 \times 10^4$ M^{-1} sec.$^{-1}$). These workers concluded that the difference in k_{inh} for reaction of 2,4,6-tri-t-butylphenol with tetralylperoxyl and 9,10-dihydroanthracyl-9-peroxyl radicals is significant, and that the value of k_{inh} depends on the nature of the peroxyl radical, decreasing in a regular manner with increasing size of the hydrocarbon moiety of the peroxyl radical.

Other workers have also reported absolute values of rate constants for reaction of peroxyl radicals with phenols and these are presented in Table 36.

The chemiluminescence associated with autoxidations has been used to estimate the activity of antioxidants and to determine rate constants for reaction of peroxyl radicals with inhibitors (Karpukhin, Shlyapintokh, Rusina and Zolotova, 1963; Karpukhin, Shlyapintokh and Zolotova, 1963).

This technique has been used to estimate rate constants for reactions of 1-phenylethylperoxyl radicals with some substituted

TABLE 36

Rate constants for reactions of peroxyl radicals with phenols

Peroxyl radical from	Phenol	Temp. (°)	$k_{inh} \times 10^{-4}$ (M^{-1} sec.$^{-1}$)	Ref.
Ethyl linoleate	Hydroquinone	45	110	1
Decanal	Hydroquinone	5	160	2
Benzaldehyde	Hydroquinone	5	39	3
Tetralin	β-Naphthol	25	2·1	4
Cyclohexane	β-Naphthol	75	0·96	5
Ethylbenzene	α-Naphthol	60	44	6
Ethylbenzene	α-Naphthol	70	0·5	7
Ethylbenzene	β-Naphthol	60	9·4	6
Ethylbenzene	2,6-Di-t-butyl-4-methyl	60	2·5	6
Ethylbenzene	2,4,6-Tri-t-butyl	60	1·3	6
Ethylbenzene	2,6-Di-t-butyl	60	0·95	6
Cumene	α-Naphthol	60	14	6
p-Methoxystyrene	2,6-Di-t-butyl-4-methyl	40	0·85	8
p-Chlorostyrene	2,6-Di-t-butyl-4-methyl	40	1·48	8
α-Methylstyrene	2,6-Di-t-butyl-4-methyl	65	1·5	9
Allylbenzene	2,6-Di-t-butyl-4-methyl	65	2·2	10
Diphenylmethane	2,6-Di-t-butyl-4-methyl	65	0·52[a]	10

[a] Using $2k_t = 2{\cdot}8 \times 10^7$ M^{-1} sec.$^{-1}$ for diphenylmethane.

References: (1) Bolland and ten Have (1947a, b). (2) Cooper and Melville (1948). (3) Ingles and Melville (1953). (4) Bamford and Dewar (1949). (5) Denisov and Aleksandrov (1964). (6) Karpukhin, Shlyapintokh, Rusina and Zolotova (1963). (7) Tsepalov and Shlyapintokh (1959). (8) Howard and Ingold (1965b). (9) Howard and Ingold (1966a). (10) Howard and Ingold (1966b).

phenols and terpenophenols. Values in the range $1\text{–}10 \times 10^4$ M^{-1} sec.$^{-1}$ were found using a value of $4{\cdot}0 \times 10^7$ M^{-1} sec.$^{-1}$ for $2k_t$, (Zakharova, Kruglyakova, Bogdanov, Kheifits and Émanuél, 1967).

Similarly, rate constants for reaction of the peroxyl radical from methyl oleate with substituted phenols were found to be in the range $0{\cdot}8\text{–}14 \times 10^4$ M^{-1} sec.$^{-1}$ (Burlakova, Khrapova, Shtol'ko and Émanuél, 1966).

Oxidation of methyl ethyl ketone in water inhibited by hydroquinone, α-naphthol, and some 4-substituted 2,6-di-t-butylphenols gave values of k_{inh} in the range $10^3\text{–}10^4$ M^{-1} sec.$^{-1}$ (Andronov, Zaikov and Maizus, 1967).

The kinetic behaviour of non-hindered phenols in hydrogen donor substrates is complex because of chain transfer of the phenoxyl radical with the substrate and with the hydroperoxide formed during oxidation. Mahoney (1967) has recently analysed the kinetics for the inhibited oxidation of 9,10-dihydroanthracene by some non-hindered phenols, and was able to estimate ratios of k_{inh}/k_p by comparing predicted and experimental plots of inhibited rate against phenol concentration. Absolute values of k_{inh} for some non-hindered phenols determined in this manner are presented in Table 37.

TABLE 37

Rate constants for the inhibited oxidation of 9,10-*dihydroanthracene by non-hindered phenols* (60°) (Mahoney, 1967)

Phenol	k_{inh}/k_p	$k_{inh} \times 10^{-4}$ (M^{-1} sec.$^{-1}$)
Phenol	9·7	0·82
2-Hydroxynaphthalene	39·0	3·3
m-Cresol	1·4	0·12
4-Hydroxybiphenyl	78·0	6·6
4,4′-Dihydroxybiphenyl	97[a]	8·2
4-Methoxyphenol	284	24
Pyren-3-ol	4400	374
Hydroquinone	147[b]	12·5
1-Hydroxynaphthalene	163	13·8
1,5-Dihydroxynaphthalene	650[b]	55
3,10-Dihydroxypyrene	8000[b]	680
3,8-Dihydroxypyrene	10000[b]	850
3-Hydroxyfluoranthene	1090	93

[a] $k_p = 850$ M^{-1} sec.$^{-1}$. [b] Statistically corrected by a factor of 2.

Arrhenius parameters for the reaction of phenols with peroxyl radicals are summarised in Table 38. Most of the pre-exponential factors and activation energies are in the range 10^7–10^9 M^{-1} sec.$^{-1}$ and 5–7 kcal. mole^{-1}. Mahoney and DaRooge (1970b), however, found significantly lower values, which they suggested are due to stabilisation of the activated complex by hydrogen bonding between the peroxyl radical and the O—H bond of the phenol.

(iii) *Amines*

The kinetics and mechanism of inhibition of autoxidation by aromatic amines have been studied much less thoroughly than inhibition

TABLE 38

Arrhenius parameters for reactions of peroxyl radicals with phenols

Peroxyl radical from	Inhibitor	log (A/M^{-1} sec.$^{-1}$)	E (kcal. mole^{-1})	Ref.
Ethylbenzene	β-Naphthol	8·94	7·3	1
Ethylbenzene	α-Naphthol	9·0	6·8	1
Tetralin	2,4,6-Tri-t-butylphenol	4·46	0	2
Cyclohexene	2,6-Di-t-butyl-4-methylphenol	7·02	4·5	3
1-Methylcyclohexene	2,6-Di-t-butyl-4-methylphenol	7·59	5·0	4
4-Methylcyclohexene	2,6-Di-t-butyl-4-methylphenol	7·56	5·0	4
4,5-Dimethylcyclohexene	2,6-Di-t-butyl-4-methylphenol	8·30	5·5	4
Styrene	2,6-Di-t-butyl-4-methylphenol	7·9	5·6	5
Cumene	2,6-Di-t-butyl-4-methylphenol	6·8	4·0	6
p-Cymene	2,6-Di-t-butyl-4-methylphenol	5·6	3·0	6
Methyl ethyl ketone	α-Naphthol	6·0	5·5	7

References: (1) Tsepalov and Shlyapintokh (1959). (2) Mahoney and DaRooge (1970b). (3) Robb and Shahin (1958). (4) Robb and Shahin (1959). (5) Howard and Ingold (1965a). (6) Howard and Robb (1963). (7) Andronov, Zaikov and Maizus (1967).

by phenols. Deuterium isotope effects are, however, consistent with abstraction of the amino-hydrogen atom by peroxyl radicals (Brownlie and Ingold, 1966).

$$\mathrm{ROO\cdot + R_2NH \xrightarrow{k_{inh}} ROOH + R_2N\cdot}$$

Although the kinetics of inhibition by aromatic amines tend to be complex, the inhibited oxidation of styrene by some amines appears to obey equation (23). Certain amines, e.g., diphenylamines and naphthylamines do not react with two peroxyl radicals since reaction of a peroxyl radical with an amino-radical produces a nitroxide radical and an alkoxyl radical.

$$\mathrm{ROO\cdot + NR_2 \longrightarrow RO\cdot + \cdot ONR_2}$$

These amines have been assumed to react with one peroxyl radical and equation (23) takes the form:

$$R_{inh} = \frac{k_p[\mathrm{RH}]\,R_i}{k_{inh}[\mathrm{AH}]}.$$

TABLE 39

Rate constants for reactions of aromatic amines with poly(peroxystyryl)peroxyl radicals (65°) (Brownlie and Ingold, 1967a)

Amine	$k_{inh} \times 10^{-4}$ ($\mathrm{M^{-1}\ sec.^{-1}}$)
Diphenylamine	4·0
N-Phenyl-β-naphthylamine	10·0
N-Phenyl-α-naphthylamine	14·0
4,4′-Dimethoxydiphenylamine	33
4-Methoxydiphenylamine	20
4,4′-Dimethyldiphenylamine	10
3-Chlorodiphenylamine	1·8
4-Nitrodiphenylamine	0·6
4,4′-Dinitrodiphenylamine	0·16
Di-β-naphthylamine	18
N,*N*′-Diphenyl-*p*-phenylenediamine	~190
N,*N*′-Dimethyl-*p*-phenylenediamine	~250
N,*N*,*N*′-Trimethyl-*p*-phenylenediamine	~330
p-Phenylenediamine	75
4-Methyl-*N*-methylaniline	1·2
3-Methyl-*N*-methylaniline	0·5
N-Methylaniline	0·4
4-Carbomethoxy-*N*-methylaniline	0·09
4-Methoxyaniline	1·9

Rate constants, k_{inh}, for these systems are listed in Table 39. The rate constants for reaction of nuclear substituted diphenylamines and *N*-methylanilines with poly(peroxystyryl)peroxyl radicals can be correlated with σ^+ substituent constants to yield ρ^+ values equal to $-0{\cdot}89$ and $-1{\cdot}6$ respectively. Arrhenius parameters for reaction of aromatic amines with peroxyl radicals have been measured for the diphenylamine inhibited oxidation of 2,4,6-trimethylheptane. The values are: $\log(A_{inh}/\text{M}^{-1}\ \text{sec.}^{-1}) = 4{\cdot}78\text{–}5{\cdot}3$ and $E_{inh} = 3{\cdot}5$ kcal. mole^{-1} (Buchachenko, Kaganskaya and Neiman, 1961; Buchachenko, Neiman and Kaganskaya, 1961).

(iv) *Hydroxylamines.*

2,2,6,6-Tetramethyl-4-piperidonehydroxylamine and 4,4′-dimethoxydiphenylhydroxylamine are efficient inhibitors of autoxidation, and rate constants for their reaction with poly(peroxystyryl)peroxyl radicals have been estimated to be about 5×10^5 M^{-1} sec.$^{-1}$ at 65° (Brownlie and Ingold, 1967b).

(v) *Polynuclear aromatic compounds*

The relative reactivity of a polynuclear aromatic compound and a hydrocarbon towards the peroxyl radical derived from the hydrocarbon enable absolute values of rate constants for addition of a peroxyl radical to the *meso*-positions of polynuclear aromatic compounds to be calculated (Mahoney, 1964). Rate constants per equivalent position for reaction of cumylperoxyl radicals with anthracene and α-benzoanthracene are 31 and $4{\cdot}3$ M^{-1} sec.$^{-1}$ respectively. Tetralylperoxyl radicals were found to add to tetracene with a rate constant of 5×10^3 M^{-1} sec.$^{-1}$.

(vi) *Metal ions*

The salts of metal ions of variable valency are believed to inhibit the initial stages of autoxidations by reacting with peroxyl radicals by a one electron transfer process to give molecular products.

$$\text{ROO}\cdot + \text{M}^{n+} \xrightarrow{k} \text{ROO}^- + \text{M}^{(n+1)+}$$

Shcheredin and Denisov (1967) obtained values of k by following the disappearance of Co^{2+} during the reaction of 2-cyano-2-propylperoxyl radicals with cobalt acetylacetonate at 74°. Steady-state rates of reaction enabled the ratio $k/(2k_t)^{1/2}$ to be determined, and since

$2k_t = 1{\cdot}6 \times 10^7$ M^{-1} sec.$^{-1}$ for 2-cyano-2-propylperoxyl radicals at 74° (Gol'dberg and Obukhova, 1965), absolute values of k could be calculated. The ratio $k/(2k_t)^{1/2}$ was also obtained from the dependence of the rate of consumption of Co^{2+} on the rate of initiation, and by the chemiluminescence method. The value of k obtained from chemiluminescence studies was about ten times larger than values obtained from the consumption of Co^{2+}, and it has been suggested that it may be in error because of quenching of the chemiluminescence by the cobalt salts.

The chemiluminescence method has also been used by Tochina, Postnikov and Shlyapintokh (1968) to investigate the inhibiting action of manganese and cobalt ethylendiaminetetra-acetates on the oxidation of ethylbenzene.

The Co^{2+} inhibited oxidation of styrene also gave values of k from the ratio $k/(2k_t)^{1/2}$ and the absolute value of $2k_t$ for styrene (Howard and Ingold, 1965b). This system was studied over a temperature range and gave the following Arrhenius parameters for k: $\log(A/$M^{-1} sec.$^{-1}) = 9{\cdot}46$, and $E = 7800 \pm 1500$ kcal. mole^{-1}.

MacLachlan (1967) has studied the reaction of α-hydroxyethylperoxyl radicals with Cu^+ by pulse radiolysis techniques and has estimated an absolute value for k.

Absolute rate constants for reactions of peroxyl radicals with transition metal ions are summarised in Table 40.

TABLE 40

Rate constants for reactions of peroxyl radicals with transition metal ions

Peroxyl radical	Transition metal ion	Temp. (°)	$k \times 10^{-5}$ (M^{-1} sec.$^{-1}$)	Ref.
Cumyl	Mn^{2+}	35	0·43	1
2-Cyano-2-propyl	Co^{2+}	74	0·20–0·23	2
2-Cyano-2-propyl	Mn^{2+}	74	2·6	3
Poly(peroxystyryl)	Co^{2+}	74	0·82	2
α-Hydroxyethyl	Cu^{+}	25	$\geqslant$10000	4
1-Phenylethyl	Mn^{2+}	60	1·4	5
1-Phenylethyl	Co^{2+}	60	3·8	5
1-Phenylethyl	Mn^{3+}	60	0·19	5

References: (1) Hendry and Schuetzle (1969). (2) Shcheredin and Denisov (1967). (3) Gol'dberg and Obukova (1965). (4) MacLachlan (1967). (5) Tochina, Postnikov and Shlyapintokh (1968).

Cumylperoxyl radicals have been reported to undergo an electron transfer reaction with the sulphur atom of zinc di-isopropyl dithiophosphate and the rate constant has been estimated to be 42 M^{-1} sec.$^{-1}$ (Burn, 1968). Burn also reports a rate constant of 404 M^{-1} sec.$^{-1}$ for abstraction of a hydrogen atom from zinc di-isopropyl dithiophosphate by cumylperoxyl radicals.

7. *The Hydroperoxyl Radical*

To conclude Section E it is worth considering separately the simplest peroxyl radical, HOO·.

The hydroperoxyl radical is believed to be the transient species involved in the rate determining steps for the autoxidation of cyclohexa-1,4-diene and 1,4-dihydronaphthalene at 30° (Howard and Ingold, 1967a). It must also play an increasingly important role in the oxidation of other organic compounds at higher temperatures. This is because the reaction of an alkyl radical with oxygen to give an unsaturated compound and the hydroperoxyl radical (24) becomes thermodynamically more favoured than reaction (9) (75).

$$R\dot{C}HCH_2R' + O_2 \longrightarrow RCH{=}CHR' + HO_2\cdot \tag{24}$$

The temperature at which this reaction becomes important depends on the stabilities of the alkyl radical and the unsaturated product.

Some propagation and termination rate constants for hydroperoxyl radicals are given in Table 41. The rate constants for reactions of aldehydes with hydroperoxyl radicals are significantly lower than the propagation rate constants for the autoxidation of the pure aldehydes (see Table 16). The termination rate constant for hydroperoxyl radicals appears to depend on the polarity of the medium, since in non-polar solvents it is close to the diffusion controlled limit, while in polar solvents it is three orders of magnitude lower.

Hydroperoxyl radicals have also been generated by pulse radiolysis of oxygen-saturated aqueous solutions and from the induced decomposition of hydrogen peroxide. Radical decay has been monitored either by e.s.r. or ultraviolet spectroscopy. The hydroperoxyl radical has been reported to have a pK-value of about 4·4 (Czapski and Bielski, 1963; Czapski and Dorfman, 1964; Rabani, Mulac and Matheson, 1965), and consequently at pH > 4·4 this radical will also exist in its basic form.

$$HO_2\cdot \longrightarrow H^+ + O_2^-$$

TABLE 41

Rate constants for reactions of hydrocarbons and aldehydes with hydroperoxyl radicals(k_p) *and termination rate constants for hydroperoxyl radicals* ($2k_t$)

Substrate	Solvent	$k_p/(2k_t)^{1/2} \times 10^2$ ($M^{-1/2}$ sec.$^{-1/2}$)	k_p (M^{-1} sec.$^{-1}$)	$2k_t \times 10^{-6}$ (M^{-1} sec.$^{-1}$)
Cyclohexa-1,4-diene[a]	n-Decane	0·9	340	1340
	Carbon tetrachloride	3·6	3800	11200
	Chlorobenzene	3·9	1400	1260
	Acetonitrile	12	350	8·6
1,4-Dihydro-naphthalene[a]	Chlorobenzene	3·5	900	700
Heptanal[b]	Chlorobenzene	—	50	—
Octanal[b]	Chlorobenzene	—	50	—
Cyclohexane-carbaldehyde[b]	Chlorobenzene	—	186	—
Pivalaldehyde[b]	Chlorobenzene	—	228	—
Benzaldehyde[b]	Chlorobenzene	—	17	—

[a] At 30°. [b] At 0°.

Czapski and Bielski (1963) have estimated a rate constant of $\sim 10^6$ sec.$^{-1}$ for this reaction from the pK value, assuming that the reverse reaction has a rate constant of 10^{10} M^{-1} sec.$^{-1}$.

Rate constants for the mutual reaction of the hydroperoxyl radical from pH 0·5 to 8 are given in Table 42. Bielski and Schwarz (1968) have reported Arrhenius parameters for this reaction as follows:

at pH 0·3 to 2, $\log(k/M^{-1}\ \text{sec.}^{-1}) = 9{\cdot}45 - 4900/RT$;

and at pH 5, $\log(k/M^{-1}\ \text{sec.}^{-1}) = 8{\cdot}6 - 2100/RT$.

Bielski and Schwarz have assumed that the first equation represents the rate constant for the reaction $HO_2\cdot + HO_2\cdot$ and that the second equation represents the rate constant for the reaction $O_2^- + O_2^-$ (see however Behor, Czapski, Rabani, Dorfman and Schwarz, 1970).

Recent work has, however, suggested that the pK value for the hydroperoxyl radical is larger than the previous values of about 4·4 and is either 4·8 (Rabani and Nielson, 1969) or $4{\cdot}88 \pm 0{\cdot}1$ (Behor, Czapski, Rabani, Dorfman and Schwarz, 1970). These workers have also reported rate constants for the mutual reaction of hydroperoxyl radicals from pH 0 to 8. Behor, Czapski, Rabani, Dorfman and

Schwarz (1970) report a rate constant of $0{\cdot}76 \times 10^6$ M^{-1} sec.$^{-1}$ for the reaction $HO_2\cdot + HO_2\cdot$, a rate constant of $8{\cdot}5 \times 10^7$ M^{-1} sec.$^{-1}$ for the reaction $HO_2\cdot + O_2^-$ and a rate constant less than 100 M^{-1} sec.$^{-1}$ for the reaction $O_2^- + O_2^-$ at about 27°. These values are in good agreement with those reported by Rabani and Nielson (1969).

TABLE 42

Bimolecular decay constants for the hydroperoxyl radical in aqueous solutions at ambient temperatures (10–25°)

pH	Rate constant $\times 10^{-6}$ (M^{-1} sec.$^{-1}$)	Ref.
0·5–1·5	4·7	1
2·0–3·0	4·4	2
1·7–3·0	5·4	3
2·7	2·5	4
Acid	2·0	5
Acid	5·0	6
5·0–7·0	34	3
5·0–8·0	30	2
7·0	29	7
Neutral	50	5
Neutral	106	6
Neutral	240	4

References: (1) Bielski and Saito (1962). (2) Czapski and Bielski (1963). (3) Czapski and Dorfman (1964). (4) Baxendale (1962). (5) Currie and Dainton (1965). (6) Adams, Boag and Michael (1966). (7) Schmidt (1961).

Several investigators have reported termination rate constants for hydroperoxyl radicals in the gas phase and the 'best' values are in the range 1–10^{10} M^{-1} sec.$^{-1}$ (see, for example, Burgess and Robb, 1957; Foner and Hudson, 1962).

Baker, Baldwin and Walker (1970) have recently estimated upper limits for the rate constants for the reaction of the hydroperoxyl radical with propane ($k < 1{\cdot}4 \times 10^5$ M^{-1} sec.$^{-1}$), n-butane ($k < 3{\cdot}7 \times 10^5$ M^{-1} sec.$^{-1}$) and isobutane ($k < 3{\cdot}7 \times 10^5$ M^{-1} sec.$^{-1}$) at 480°.

Tetranitromethane reacts rapidly with O_2^- ($k = 1{\cdot}9 \times 10^9$ M^{-1} sec.$^{-1}$) and relatively slowly with $HO_2\cdot$ ($k < 2 \times 10^4$ M^{-1} sec^{-1}.) (Rabani, Mulac and Matheson, 1965). [The dissociated form of the peroxyl radicals from methanol, ethanol and propan-2-ol have also been reported to react rapidly with tetranitromethane ($k = 1{\cdot}2 \times 10^9$ M^{-1} sec.$^{-1}$) while the undissociated forms of these peroxyl radicals have rate constants in the range $2\text{–}5 \times 10^8$ M^{-1} sec.$^{-1}$ (Stockhausen, Fojtik and Henglein, 1970)].

F. Aroxyl Radicals

The preparation and chemical and physical properties of aroxyl radicals have been reviewed several times during the past few years (Altwicker, 1967; Forrester, Hay and Thomson, 1968a; Strigun, Vartanyan and Émanuél, 1968; Pokhodenko, Khizhnyi and Bidzilya, 1968). Interest in the reactions and reactivities of these radicals has been enhanced by the technological importance of phenols and naphthols as inhibitors of autoxidation (Ingold, 1961; Ingold, 1968; Mahoney, 1969). Highly hindered aroxyl radicals such as tri-t-butylphenoxyl and galvinoxyl (I) are relatively unreactive, that is, they

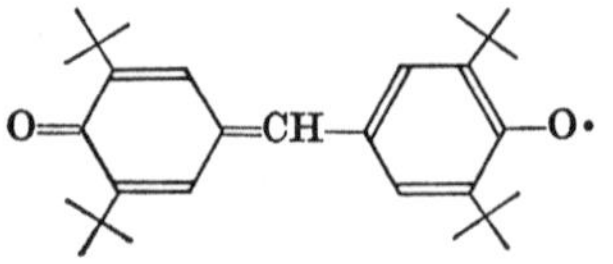

I

do not dimerise and they tend to react more slowly with hydrogen donors than do unhindered aroxyl radicals. However, these hindered aroxyl radicals are very reactive towards simple alkyl and alkoxyl radicals (Bartlett and Funahashi, 1962) and peroxyl radicals. This makes them very efficient inhibitors of a variety of free radical chain reactions (cf. for example, Davies and Roberts, 1967, 1968).

1. *Dimerisation and Disproportionation*

The earliest measurement of the rate of reaction between two phenoxyl radicals appears to have been made by Cook and Norcross (1956, 1959). These workers showed that 2,6-di-t-butyl-4-isopropyl-, -4-cyclohexyl- and -4-sec-butyl-phenoxyl radicals disproportionate to the corresponding quinone methides and the parent phenols by reactions second order in phenoxyl radical (see Table 43). At 25° the disproportionation rate constants decrease with increasing size

TABLE 43

Rate constants (25°) *and activation parameters for the disproportionation of* 2,6-*di-t-butyl*-4-*alkylphenoxyl radicals*

4-Substituent	Solvent	k (M^{-1} sec.$^{-1}$)	E (kcal. mole^{-1})	ΔH^* (kcal. mole^{-1})	ΔS^* (cal. deg.$^{-1}$ mole^{-1})
Isopropylphenoxyl	Benzene	2·2	6·2	5·7	−38
	Cyclohexane	3·8	5·5	4·9	−39
	Chlorobenzene	3·1	6·9	6·4	−35
	Anisole	3·2	7·2	6·6	−34
	Benzonitrile	5·6	8·2	7·6	−30
sec.-Butylphenoxyl	Benzene	0·62	7·0	6·5	—
Cyclohexylphenoxyl	Benzene	1·3	7·9	7·3	—
	Cyclohexane	—	7·5	6·8	—

of the 4-substituent along the series methyl > ethyl > isopropyl > cyclohexyl > sec.-butyl. Methyl and ethyl substituted radicals

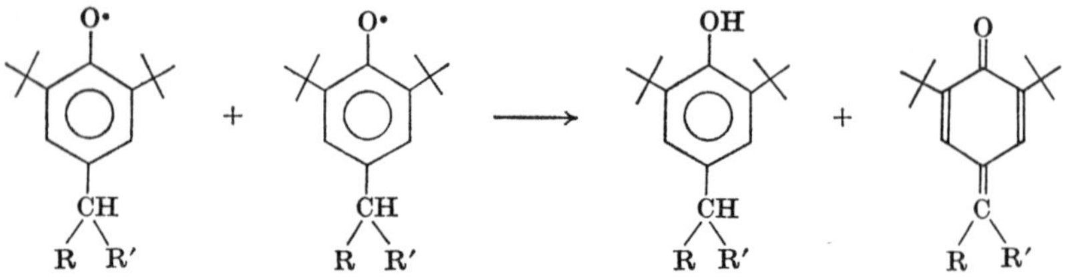

decayed too rapidly for the reaction to be followed and there has since been considerable disagreement as to the kinetics of their decay. However, a relatively slow bimolecular disproportionation similar to that for the other three phenoxyl radicals now seems to be firmly established (see below). The disproportionation may be preceded by a much faster, but reversible, association of these phenoxyl radicals.

In benzene the enthalpies of activation for the disproportionation of 2,6-di-t-butyl-4-alkylphenoxyl radicals increase in the order isopropyl < sec.-butyl < cyclohexyl (Cook and Norcross, 1959). The disproportionation rate constant for 2,6-di-t-butyl-4-isopropylphenoxyl varies little from one solvent to another. However, this is not due to the absence of interaction with the solvent but to the presence of a compensating effect with an isokinetic temperature of $-9°$. That is, the enthalpy of activation increases with increasing dielectric constant of the solvent, but there is a compensating change in the entropy of activation such that, at $-9°$, the rate constant would be approximately the same in all the solvents studied. The transition state is represented as follows.

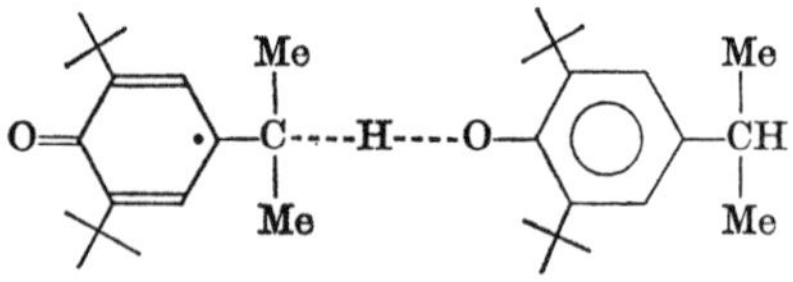

Desolvation occurs as the radicals enter the transition state. Thus, in a medium where the radicals are highly solvated, the enthalpy of activation would have to be relatively high to provide the necessary desolvation energy, and such cases would be accompanied by the largest entropy increases.

Brodskii, Pokhodenko, Khizhnyi and Kalibabchuk (1966) have reported that the decay of 2,6-di-t-butyl-4-ethylphenoxyl is second order at 47° ($k = 25{\cdot}4$ M^{-1} sec.$^{-1}$ in benzene) but first order at 26° ($k = 7{\cdot}8 \times 10^{-3}$ sec.$^{-1}$) (however, see below). Hubele, Suhr and Heilman (1962) obtained a rate constant of 4·2 M^{-1} sec.$^{-1}$ for the decay of 2,6-di-t-butyl-4-isopropylphenoxyl in cyclohexane at 25° (which is in excellent agreement with Cook and Norcross's (1959) value of 3·8 M^{-1} sec.$^{-1}$). They obtained a value of 2·2 M^{-1} sec.$^{-1}$ for the 2,6-di-t-butyl-4-diphenylmethylphenoxyl at 25° and 0·7 M^{-1} sec.$^{-1}$ for 2,6-di-t-butyl-4-(α-benzyloxy)benzylphenoxyl. However, 2,6-di-t-butyl-4-(α-methoxy)benzylphenoxyl apparently decayed with first order kinetics.

Flash photolysis has been used by Land and Porter (1963) and by Dobson and Grossweiner (1965) to study the decay of more reactive phenoxyl radicals (see also, Keene, Kemp and Salmon, 1965; Joschek and Grossweiner, 1966; Joschek and Miller, 1966). The initial decay was second order in all cases. However, Land and Porter found that decays of 2,6-di-t-butyl-4-methyl- and -4-ethyl-phenoxyl deviated significantly from second order kinetics as the reaction progressed, and appreciable quantities of radicals were observable at 'equilibrium'. It was suggested that two processes are superimposed, a rapid dimerisation and a slow isomerisation, as follows.

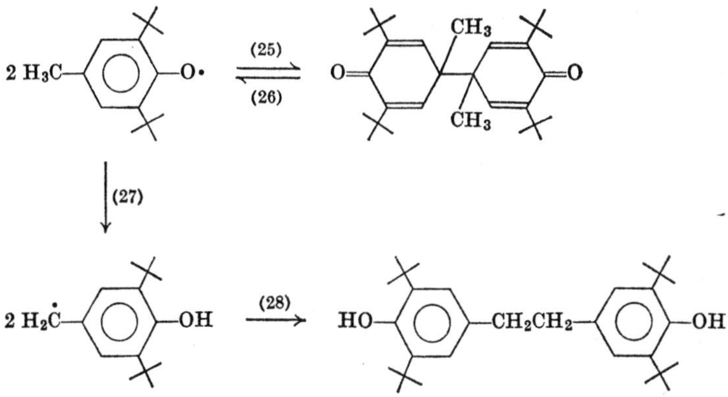

The same scheme applies to the 4-ethyl substituted radical at room temperature. Becker (1965) has shown that the rapidly formed dimer is actually a 'head to tail' dimer rather than a 'tail to tail' dimer as suggested by Land and Porter.

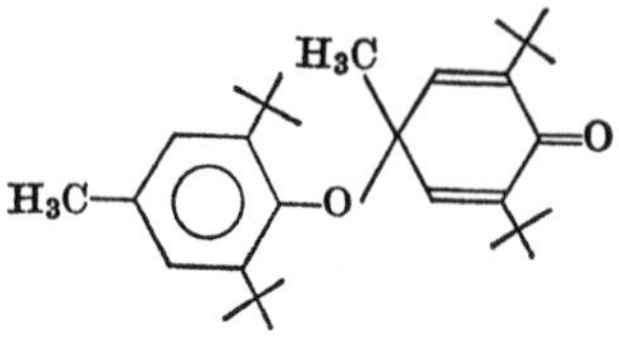

The rate constant, k_{25}, for the initial dimerisation of 2,6-di-t-butyl-4-methylphenoxyl in carbon tetrachloride at 25° was estimated by Land and Porter as $8{\cdot}2 \times 10^4/\epsilon$ cm. sec.$^{-1}$, where ϵ is the extinction coefficient of the absorption maximum at 4000 Å. This was taken as 1800, the same as for tri-t-butylphenoxyl in hexane, giving $k_{25} = 1{\cdot}5 \times 10^8$ M^{-1} sec.$^{-1}$. The equilibrium constant $K = k_{26}/k_{25} \sim 7 \times 10^{-7}$ M, and hence $k_{26} \sim 10^2$ sec.$^{-1}$. This estimation was only approximate since the 'equilibrium' concentration fell slowly, this latter reaction presumably being the process that has been followed by all the other workers who have studied the decay of 2,6-di-t-butyl-4-methylphenoxyl.

The results obtained by these other workers are conflicting. The isomerisation of 2,6-di-t-butyl-4-methylphenoxyl to the corresponding hydroxybenzyl radical (reaction 27) which was originally proposed by Cook, Nash and Flanagan (1955), implies that radical decay will be first order since benzyl radicals dimerise rapidly [Burkhart (1969) gives, for the unsubstituted benzyl radical, $k_{28} = 4 \times 10^9$ M^{-1} sec.$^{-1}$]. In 1960, Bennett reported first order kinetics with $k_{27} = 1{\cdot}3 \times 10^{-2}$ sec.$^{-1}$ in benzene at room temperature (for the 4-ethyl radical, $k_{27} = 7{\cdot}5 \times 10^{-3}$ sec.$^{-1}$). Land and Porter (1963) assumed rearrangement occurred by reaction 27. There was no spectroscopic evidence for the *p*-hydroxybenzyl radical, and this is consistent with a large value for k_{28}. In 1963, Bauer and Coppinger showed that the products formed by the decay of the phenoxyl radical were consistent with disproportionation to the corresponding phenol and quinone methide.

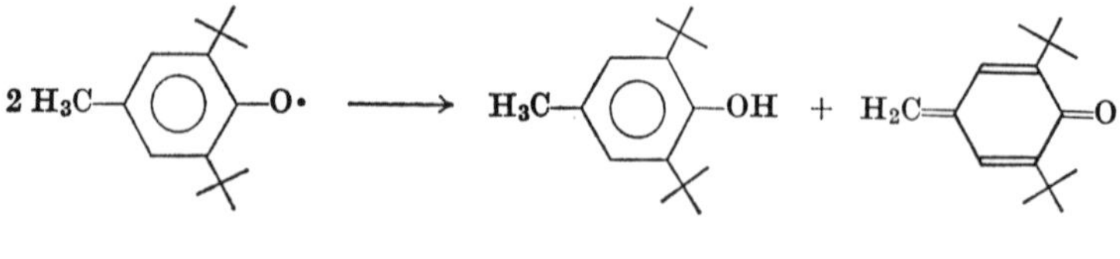

2 H₃C—Ar—O· ⟶ H₃C—Ar—OH + H₂C=quinone=O (29)

The dimeric products formed by further reactions of the quinone methide can be stoichiometrically represented by the following reaction.

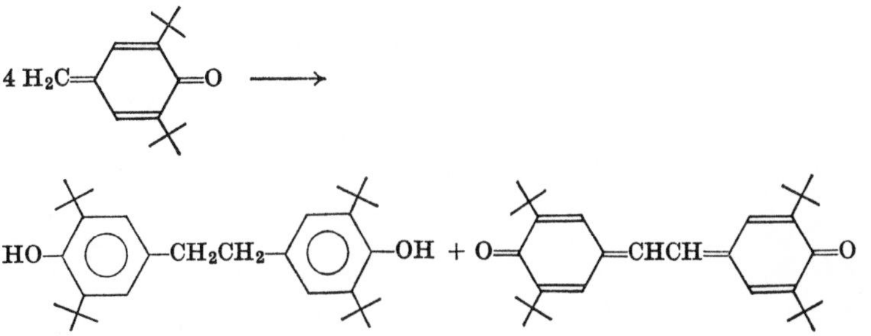

$$4\ H_2C{=}\langle\ \rangle{=}O \longrightarrow$$

$$HO{-}\langle\ \rangle{-}CH_2CH_2{-}\langle\ \rangle{-}OH + O{=}\langle\ \rangle{=}CHCH{=}\langle\ \rangle{=}O \qquad (30)$$

The postulated rearrangement of the phenoxyl radical to a *p*-hydroxybenzyl radical has been revived by Pokhodenko and coworkers on the basis of deuterium labelling experiments, e.s.r. studies and kinetics (Brodskii, Pokhodenko and Ganyuk, 1964; see also, Pokhodenko, Ganyuk and Brodskii, 1962; Pokhodenko, Ganyuk, Yakovleva, Shatenshtein and Brodskii, 1963; Pokhodenko and Bidzilya, 1965; Brodskii, Pokhodenko, Khizhnyi and Kalibabchuk, 1966; and, for related work on a similar system, Pokhodenko and Ganyuk, 1963). However, the migration of hydrogen from the 4-methyl group to oxygen, demonstrated by these workers, is readily accounted for by reaction (29) (see also, Filar and Winstein, 1960). Moreover, the e.s.r. spectrum assigned by the Russian workers to the *p*-hydroxybenzyl radical is probably due to some other radical (Starnes and Neureiter, 1967).

Stebbins and Sicilio (1970) have recently reinvestigated the kinetics of the decay of 2,6-di-t-butyl-4-methylphenoxyl. Their results show quite conclusively that in benzene and in tetrahydrofuran decay is second order, with rate constants of $4{\cdot}6 \times 10^2$ M^{-1} sec.^{-1} and $3{\cdot}3 \times 10^2$ M^{-1} sec.^{-1}, respectively, at 24°. The disproportionation mechanism (29) was confirmed by observation in both solvents of the intermediate quinone methide and the two dimeric products of reaction (30). The quinone methide build-up was followed by means of its strong absorption at 285 mμ. Stebbins and Sicilio concluded that the first order kinetics reported by Brodskii, Pokhodenko, Khizhnyi and Kalibabchuk (1966) arose from their failure to remove from their

TABLE 44

Rate constants for reactions of phenoxyl radicals at room temperature

Radical	Solvent	k_1/ϵ (cm. sec.$^{-1}$)	ϵ (M cm.$^{-1}$)	k_{25} (M^{-1} sec.$^{-1}$)	$k_{diff.}$ (M^{-1} sec.$^{-1}$)	K (M)	Ref.
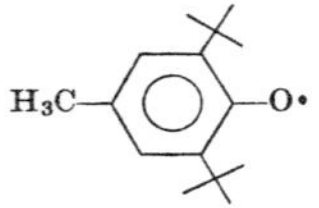 H_3C–(2,6-di-t-butyl ring)–O·	CCl_4	$8{\cdot}2\times10^4$	1800	$1{\cdot}5\times10^8$	$6{\cdot}7\times10^9$	7×10^{-7}	1
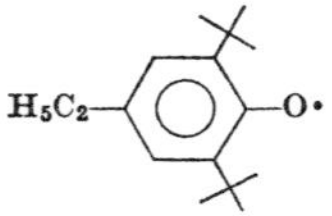 H_5C_2–(2,6-di-t-butyl ring)–O·	CCl_4	$7{\cdot}4\times10^3$	1800	$1{\cdot}3\times10^7$	$6{\cdot}7\times10^9$	1×10^{-6}	1
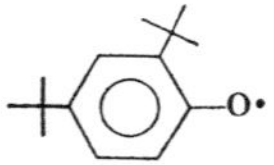 t-butyl–(2-t-butyl ring)–O·	Liquid paraffin	$1{\cdot}1\times10^4$	1800	$2{\cdot}0\times10^7$	$4{\cdot}6\times10^7$	—	1
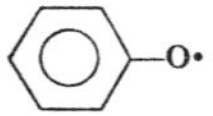 (ring)–O·	H_2O	$3{\cdot}0\times10^5$	1800	$5{\cdot}4\times10^8$	$6{\cdot}4\times10^9$	—	1
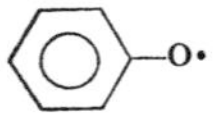 (ring)–O·	H_2O	$5{\cdot}1\times10^5$	11000	$5{\cdot}6\times10^9$	$6{\cdot}4\times10^9$	—	2
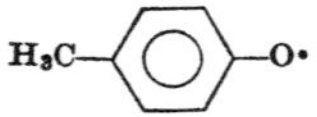 H_3C–(ring)–O·	H_2O	$4{\cdot}3\times10^5$	15000	$6{\cdot}5\times10^9$	$6{\cdot}4\times10^9$	—	2

References: (1) Land and Porter (1963). (2) Dobson and Grosswener (1965).

system all of the lead dioxide which was used to generate the phenoxyl radicals from the parent phenol. The rate constant for disproportionation of the hindered 4-methylphenoxyl radical is about one hundred

$$H_3C\text{–}C_6H_2(t\text{-}Bu)_2\text{–}OH \xrightarrow[2H^+]{PbO_2} H_3C\text{–}C_6H_2(t\text{-}Bu)_2\text{–}O\cdot + 2\,H_2O + Pb^{++}$$

times greater than that found by Cook and Norcross (1956, 1959) for the 4-isopropylphenoxyl radical.

Land and Porter's (1963) results for four phenoxyl radicals are given in Table 44. The rate constant, k_{25}, is calculated on their assumption that $\epsilon = 1800$ for all four radicals. This leads to the conclusion that the initial dimerisation rate is roughly 1–10% of the rate of a diffusion-controlled reaction. However, Dobson and Grossweiner (1965) have since shown that ϵ is actually 11000 for the unsubstituted phenoxyl radical. Both this radical and 4-methylphenoxyl couple at essentially the diffusion-controlled rate. The latter radical yields 2,2′-dihydroxy-4,4′-dimethylbiphenyl as the principal product.

$$O\cdot\text{–}C_6H_4\text{–}CH_3 + O\cdot\text{–}C_6H_4\text{–}CH_3 \longrightarrow HO(CH_3)C_6H_3\text{–}C_6H_3(CH_3)OH$$

Additional information on the products formed by the coupling of phenoxyl and substituted phenoxyl radicals has been obtained by Joschek and Miller (1966).

Williams and Kreilick (1967) have used nuclear magnetic resonance spectroscopy to measure the rate constant for the dimerisation of 2,6-di-t-butyl-4-acetylphenoxyl, k_{dim}, and for dissociation of the dimer, k_{diss}. The rate constant for a reaction which involves an interchange between the two halves of the dimer, k_{int}, was also measured. The complete process can be represented as follows.

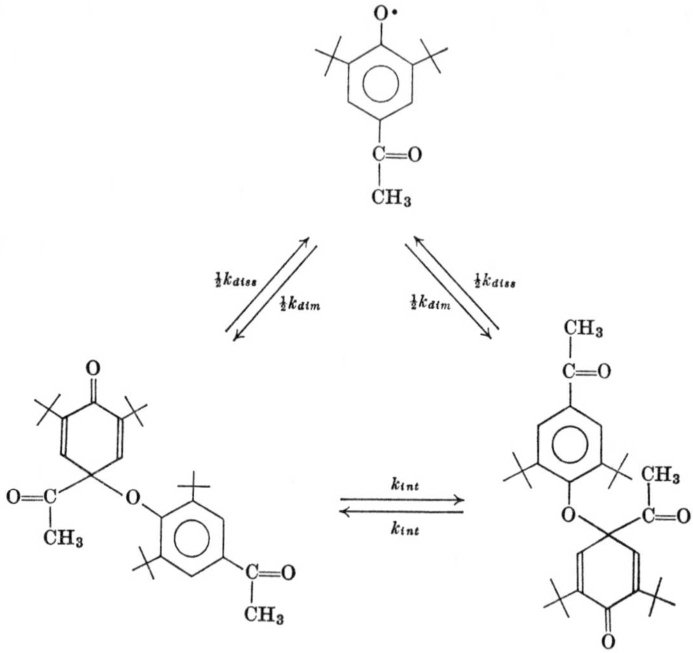

The rate constants and thermodynamic data at 243°K are summarised below.

$$k_{dim} = 2{\cdot}5 \times 10^5 \text{ M}^{-1} \text{ sec.}^{-1} \qquad \Delta H^{\ddagger}_{dim} < 1 \text{ kcal. mole}^{-1}$$

$$k_{diss} = 7{\cdot}6 \text{ sec.}^{-1} \qquad \Delta H_{diss} = 9{\cdot}1 \text{ kcal. mole}^{-1}$$

$$k_{int} = 2 \times 10^2 \text{ sec.}^{-1} \qquad \Delta H^{\ddagger}_{int} = 9{\cdot}1 \text{ kcal. mole}^{-1}$$

Bridge and Porter (1958) have used flash photolysis to study the disproportionation of the *p*-semiquinone radical.

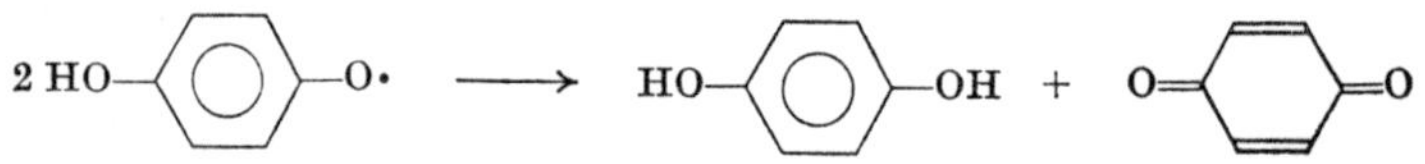

The rate constant for this reaction in ethanol–water at room temperature was found to be $\sim 8 \times 10^8$ M^{-1} sec.$^{-1}$.

Mahoney and DaRooge (1967) have estimated that the rate constant for the dimerisation of 4-methoxyphenoxyl radicals at 60° is $\leqslant 1{\cdot}7 \times 10^7$ M^{-1} sec.$^{-1}$.

2. *Hydrogen Atom Abstraction*

Kreilick and Weissman (1962, 1966) have used magnetic resonance techniques to measure the rate constants for symmetrical hydrogen atom exchange reactions between a number of hydroxy-compounds and their corresponding oxyl radicals. The compounds studied include 2,4,6-tri-t-butylphenol and 2,6,2′,6′-tetra-t-butylindophenol.

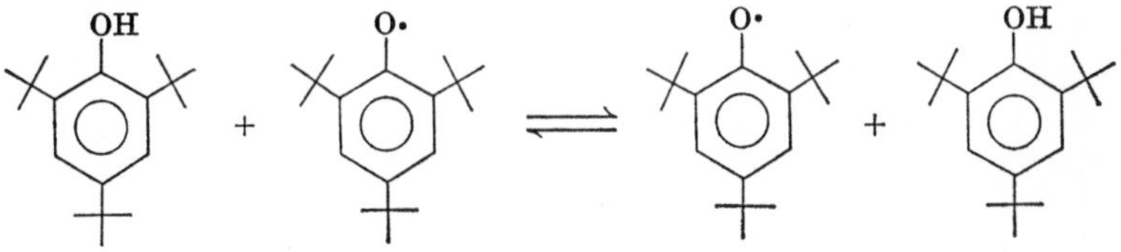

OH O· O· OH

At room temperature in carbon tetrachloride the rate constant for this reaction was found to be 330 M^{-1} sec.$^{-1}$ and the activation energy 1·2 kcal. mole^{-1}. For the indophenol, $k = 390$ M^{-1} sec.$^{-1}$ and $E = 2{\cdot}1$ kcal. mole^{-1}, with an isotope effect $k_{OH}/k_{OD} = 1{\cdot}77$. Subsequently, Arick and Weissman (1968) used a direct kinetic method to measure the rate constant for reaction of tri-t-butylphenoxyl with 3,5-di-deutero-2,4,6-tri-t-butylphenol and obtained $k = 220$ M^{-1} sec.$^{-1}$ and $k_{OH}/k_{OD} = 1{\cdot}24$.

Measurements on the reaction of 2,4,6-tri-t-butylphenoxyl with other phenols first appear to have been made by Cook, Depatie and English (1959). These workers measured only the equilibrium constant for the reaction of the phenoxyl radical with 2,6-di-t-butyl-4-t-butoxyphenol ($K_{25°} = 52$, $\Delta H = -3{\cdot}4$ kcal. mole^{-1}) and 2,6-di-t-butyl-4-methoxyphenol ($K_{25°} \sim 210$, $\Delta H \sim 5$ kcal. mole^{-1}).

Rate constants for the reactions of tri-t-butylphenoxyl with several nitrophenols have been measured by Buchachenko, Neiman, Sukhanova and Mamedova (1963). Sukhanova and Buchachenko (1965) have studied the effect of solvents on the reaction of this phenoxyl with *m*-nitrophenol and DaRooge and Mahoney (1967) Mahoney and DaRooge (1970a) have measured rate constants for its reaction with a variety of phenols in benzene and chlorobenzene (see Table 45). The total process can be represented by equations (31) and (32). The intramolecularly hydrogen bonded 2-nitrophenol and 2,4,6-trinitrophenol are very unreactive towards the hindered phenoxyl radical and the rate constants for these two phenols are

TABLE 45

Rate constants (24°, benzene solvent) and activation parameters for reactions of 2,4,6-tri-t-butylphenoxyl with unhindered phenols

Substituent in Phenol	k_{31} (M^{-1} sec.$^{-1}$)	log (A_{31}/ M^{-1} sec.$^{-1}$)	E_{31} (kcal. mole^{-1})	k_{32}/k_{-32} at 40°	log (A_{-32}/sec.$^{-1}$)	E_{-32} (kcal. mole^{-1})	$-\Delta H$ (kcal. mole^{-1})
None[a]	0·46[b]	5·3	7·6	—	—	—	—
3-Nitro-[a]	0·015[b]	5·7	10·2	—	—	—	—
4-Nitro-[a]	0·010[b]	5·7	10·5	—	—	—	—
2-Nitro-[a]	$7·4 \times 10^{-5}$	4·8	12·1	—	—	—	—
2,4,6-Trinitro-[a]	$4·6 \times 10^{-5}$	5·4	13·2	—	—	—	—
4-Methoxy-[c]	3030	—	—	—	12·2	21·0	20·2
4-Phenyl[c]	125	—	—	—	—	—	—
4-t-Butyl[c]	46	5·5	4·8	128	13·8	26·1	19·9
3,5-Dimethyl[c]	15	6·4	6·8	~63	14·5	28·0	20·1
3-t-Butyl[c]	10	—	—	—	—	—	—
4-Bromo-[c]	4·4	6·5	7·5	111	13·9	27·8	20·4
None[c]	3·1	—	—	—	—	—	—
3-Carboethoxy-[c]	0·60	5·9	8·2	—	13·9	29·5	20·5
4-Carbomethoxy-[c]	0·21	—	—	—	—	—	—
3-Cyano-[c]	0·081	—	—	—	—	—	—
4-Cyano-[c]	0·077	—	—	—	—	—	—
2,4,6-Trichloro-[d]	2·2	4·6	5·5	$> 10^3$	13·5	22·3	15·9
2,6-Dimethyl[d]	—	—	—	—	—	—	15·4
2,4,6-Trimethyl[d]	—	—	—	—	—	—	15·4

[a] Buchachenko, Neiman, Sukhanova and Mamedova (1963). [b] Obtained at concentrations where self-association is significant (see text). [c] Values of k_{31} from DaRooge and Mahoney (1967). Activation parameters and k_{32}/k_{-31} obtained in chlorobenzene (Mahoney and DaRooge, 1970a). [d] Results obtained in chlorobenzene (Mahoney and DaRooge, 1970a).

unaffected by a change in the solvent. In contrast, with 3- and 4-nitrophenols the rate constants decrease as the phenol becomes more

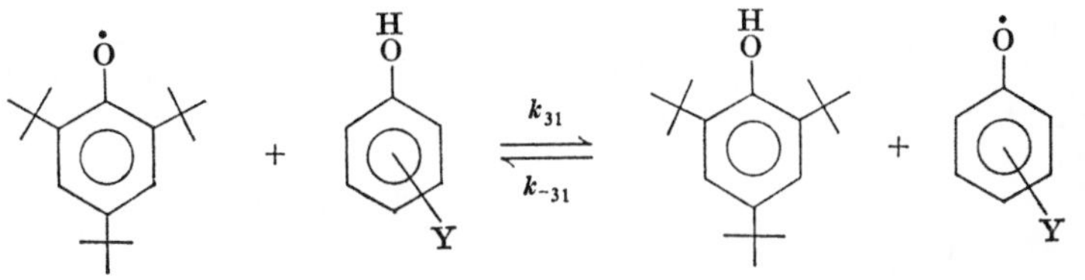

(31)

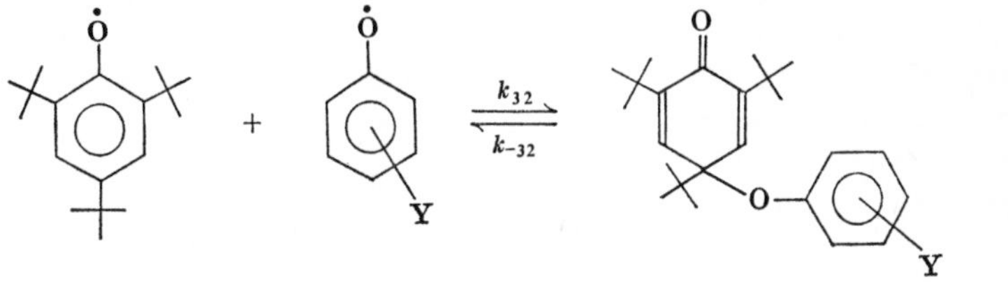

(32)

strongly associated with the solvent. Thus, with 3-nitrophenol $(k_{31})_{24°} = 4{\cdot}5 \times 10^{-4}$ M^{-1} sec.$^{-1}$ and $E = 14{\cdot}7$ kcal. mole^{-1} in acetone, and $(k_{31})_{24°} = 6{\cdot}6 \times 10^{-5}$ M^{-1} sec.$^{-1}$ and $E = 17{\cdot}3$ kcal. mole^{-1} in methanol (Buchachenko, Neiman, Sukhanova and Mamedova, 1963).

Unfortunately, the rate constants obtained by the Russian workers were measured at phenol concentrations of 0·1–0·2 M where self-association for phenol and for *m*- and *p*-nitrophenols is appreciable. In contrast, the data of DaRooge and Mahoney (1967) and Mahoney and DaRooge (1970a) were obtained at much lower phenol concentrations where self-association is unimportant. These workers showed that the deuterium isotope effect for 4-hydroxybiphenyl, $k_{OH}/k_{OD} \geqslant 7{\cdot}5$. The values of $\log A_{31}$ are rather small compared with those reported for most hydrogen atom transfer reactions in solution but are consistent with other reactions involving hydrogen transfer to this hindered phenoxyl radical (Kreilick and Weissman, 1962). The differences in the absolute values of k_{31} are due principally to differences in the values of the activation energies for this reaction. The effect of the substituents can be roughly correlated with the Hammett equation using σ^+ constants with $\rho = -2{\cdot}8$. The rate constants for the decomposition of the quinol ethers, k_{-32}, are affected by the substitution of the phenol in the same manner as are the values of k_{31}. A correlation with σ^+ gives $\rho = -3{\cdot}3$ at 60°. In contrast to the large effect of substituents on k_{31} and k_{-32}, the ratio k_{32}/k_{-31} appears to be essentially independent of the substituent. The values of the total equilibrium constants, $k_{31}k_{32}/k_{-31}k_{-32}$, at 40° ($2{\cdot}2 \pm 0{\cdot}4 \times 10^8$ M^{-1}) and the total heats of reaction, $-\Delta H$ at 25° ($-20{\cdot}3 \pm 0{\cdot}5$ kcal. mole^{-1}) are also independent of *meta-* and *para-*substituents. However, the observed heat of reaction for several phenols disubstituted at the *ortho*-positions were only $-15{\cdot}5 \pm 0{\cdot}5$ kcal. mole^{-1}. This 5 kcal. mole^{-1} difference is due to the presence of strain energy in the quinol ether compounds derived from the 2,6-substituted compounds. As a result of this strain energy the values of k_{-32} for such quinol ethers are extremely large.

Mahoney and DaRooge (1970a) were also able to estimate the differences in the heats of formation of some phenoxyl radicals and the parent phenols in chlorobenzene, i.e., $(\Delta H_f)^{\mathrm{PhCl}}_{\mathrm{ArO\cdot}} - (\Delta H_f)^{\mathrm{PhCl}}_{\mathrm{ArOH}}$. This quantity increased by a total of 8 kcal. mole^{-1} along the series 4-methoxyphenol < 4-t-butylphenol < 2,4,6-trichlorophenol < 4-bromophenol < 3,5-dimethylphenol < 3-carboethoxyphenol. It was concluded that hydrogen atom abstractions from *meta-* and *para-*substituted phenols can be correlated with σ^+ constants because these constants actually reflect the position of the hydrogen transfer *equilibrium* (e.g. k_{31}/k_{-31}) for reactions involving phenoxyl radicals rather than, as has usually been supposed, because they reflect the importance of polar contributions to the transition state of the following type.

$$\mathrm{Y{-}C_6H_4{-}\overset{\delta+}{O}\cdot\ H{:}\overset{\delta-}{R}}$$

Ayscough and Russell (1965, 1967) have studied the reaction of 2,2-diphenyl-1-picrylhydrazyl (DPPH) with 2,4,6-tri-t-butylphenol and 2,6-di-t-butyl-4-isopropylphenol, and the reverse reactions.

$$\mathrm{DPPH\cdot} + \text{2,6-di-t-Bu-4-R-C}_6\mathrm{H_2OH} \rightleftharpoons \mathrm{DPPH{-}H} + \text{2,6-di-t-Bu-4-R-C}_6\mathrm{H_2O\cdot} \quad (33)$$

The kinetic and thermodynamic data for R = t-butyl at 20° are,

$k_{33}^{\mathrm{OH}} = 0{\cdot}12\ \mathrm{M}^{-1}\ \mathrm{sec.}^{-1}$ $\quad E_{33}^{\mathrm{OH}} = 6{\cdot}5\ \mathrm{kcal.\ mole}^{-1}$

$k_{33}^{\mathrm{OD}} = 0{\cdot}010\ \mathrm{M}^{-1}\ \mathrm{sec.}^{-1}$ $\quad E_{33}^{\mathrm{OD}} = 7{\cdot}7\ \mathrm{kcal.\ mole}^{-1}$

$k_{-33} \sim 2\ \mathrm{M}^{-1}\ \mathrm{sec.}^{-1}$ $\quad E_{-33} = 4{\cdot}8\ \mathrm{kcal.\ mole}^{-1}$

$K_{33} = 0{\cdot}1$ $\quad \Delta H_{33}^{\circ} = 1{\cdot}6\ \mathrm{kcal.\ mole}^{-1}$

For R = isopropyl they are,

$k_{33}^{\mathrm{OH}} = 0{\cdot}18\ \mathrm{M}^{-1}\ \mathrm{sec.}^{-1}$ $\quad E_{33}^{\mathrm{OH}} = 6{\cdot}8\ \mathrm{kcal.\ mole}^{-1}$

$k_{-33} = 1{\cdot}3\ \mathrm{M}^{-1}\ \mathrm{sec.}^{-1}$ $\quad E_{-33} = 6{\cdot}0\ \mathrm{kcal.\ mole}^{-1}$

In addition to the disproportionation reaction between 2,6-di-t-butyl-4-isopropylphenoxyl radicals [$k = 1{\cdot}7\ \mathrm{M}^{-1}\ \mathrm{sec.}^{-1}$, $E = 6{\cdot}2$ kcal. mole^{-1} (Cook and Norcross, 1959)] there is a reaction of the phenoxyl with DPPH.

DPPH· + (2,6-di-t-butyl-4-isopropylphenoxyl, O·, HCMe₂) ⟶ DPPH—H + (2,6-di-t-butyl-4-isopropylidene-cyclohexadienone, O, CMe₂)

This reaction has a rate constant of $0{\cdot}68\ \mathrm{M}^{-1}\ \mathrm{sec.}^{-1}$ and an activation energy of 6·4 kcal. mole^{-1}.

Bidzilya, Pokhodenko and Brodskii (1966) have measured rate constants for reactions of tetra-t-butylindophenoxyl with phenols and anilines.

O=(quinone)=N—(aryl)—O· + {ArOH / ArNHR} ⟶ O=(quinone)=N—(aryl)—OH + {ArO· / ArṄR}

TABLE 46

Rate constants and activation energies for reactions of 2,2′,6,6′-tetra-t-butylindophenoxyl with phenols and amines in o-xylene

Substrate	Rate constant (M^{-1} sec.$^{-1}$)		E	log		
	30°	50°	(kcal. mole^{-1})	(A/M^{-1} sec.$^{-1}$)	k_H/k_D	Ref.
2,6-Bu_2^t-phenol	1·55					1
2,6-Bu_2^t-4-Me-phenol	0·49	0·7	5·5	3·54		1
2,6-Bu_2^t-4-Et-phenol	0·28					1
Phenol	0·04	0·12	10·5	6·20		1
p-Cresol	0·04					1
p-Nitrophenol	0·04					1
Diphenylamine	$2·5 \times 10^{-4}$	$0·47 \times 10^{-3}$	6·4	1·0		1
Aniline	2×10^{-2}	$43·5 \times 10^{-3}$	7·3	3·54		1
p-Nitroaniline		$12·0 \times 10^{-3}$				1
m-Nitroaniline		$10·5 \times 10^{-3}$				1
o-Nitroaniline		$7·5 \times 10^{-3}$				1
p-Phenylenediamine	1·41					2
p-Methoxyaniline	0·97					2
p-Toluidine	0·11					2
Aniline	0·019				1·6	2
p-Bromoaniline	0·012					2
Diphenylamine[a]	$3·7 \times 10^{-4}$				1·5	2

[a] At 40°.

References: (1) Bidzilya, Pokhodenko and Brodskii (1966). (2) Pokhodenko and Bidzilya (1966).

The original rate constants given for these reactions have been divided by two since the ArO• and ArṄR radicals are presumably able to react with a second indophenoxyl radical. These revised rate constants are listed in Table 46 together with the revised results obtained by Pokhodenko and Bidzilya (1966) for reaction with *para*-substituted anilines. The latter results were correlated with the Hammett equation using the σ constants of the substituents ($\rho = -2{\cdot}53$). A somewhat better correlation would have been obtained had σ^+ constants been used. The pre-exponential factors for the reactions with the hindered phenol, diphenylamine and aniline all seem far too small for simple bimolecular hydrogen atom transfers.

McGowan and Powell (1960) have reported a rate constant of 4 M^{-1} sec.$^{-1}$ for the reaction of 2,4,6-tri-t-butylphenoxyl with 3-chloroaniline in carbon tetrachloride at 23·5°. For reaction (34),

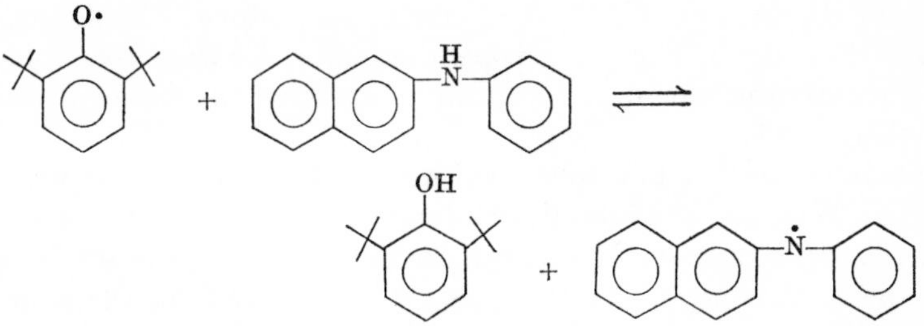

(34)

Karpukhina and Maizus (1969) have estimated that the rate constants for both the forward and reverse reaction are about 10^5 M^{-1} sec.$^{-1}$.

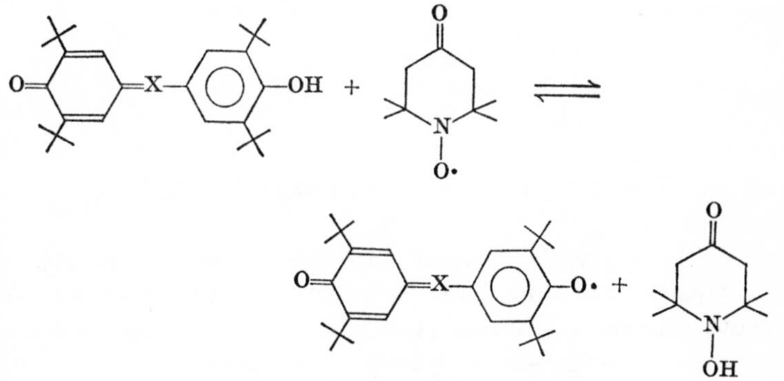

(35)

X = N, CH

The equilibrium constant for reaction of tetra-t-butylindophenol with 2,2,6,6-tetramethylpiperidone nitroxide (35) in heptane at 50–95° is given by $K_{35} = k_{35}/k_{-35} = 4{\cdot}5 \times 10^{-2}$ $\exp(-4100/RT)$ (Vasserman, Buchachenko, Nikiforov, Ershov and Neiman, 1967). For the analogous reaction with X = CH, $K = 5{\cdot}3 \times 10^{-2} \exp(-3900/T)$.

It has been realised for some time (Ingold, 1968; Mahoney, 1969) that chain transfer reactions between phenoxyl radicals and hydrocarbons and their hydroperoxides are very important in determining the kinetics of phenol-inhibited autoxidations and the overall efficiency of the inhibitor. However, there have been relatively few *absolute* (as opposed to relative) rate constant measurements on these reactions. Most of these relate to the highly hindered radical, 2,4,6-tri-t-butylphenoxyl, which is much less reactive in hydrogen atom abstractions than other less hindered phenoxyl radicals. The reaction of this radical with hydrocarbons is very slow, and has a high activation energy (see Table 47; Neiman, Mamedova, Blenke and Buchachenko, 1962). The pre-exponential factors for these reactions show a remarkable range and indicate how important steric effects are in this reaction.

The reaction of tri-t-butylphenoxyl with hydroperoxides has been studied by McGowan and Powell (1960) and by Mahoney and DaRooge (1967, 1969). The agreement between the different groups of workers is very poor. Mahoney and DaRooge's results appear to be the more accurate. The pre-exponential factors for these hydrogen atom transfers are remarkably low (cf. also Kreilick and Weissman, 1962, 1966). The equilibrium constant for the reaction:

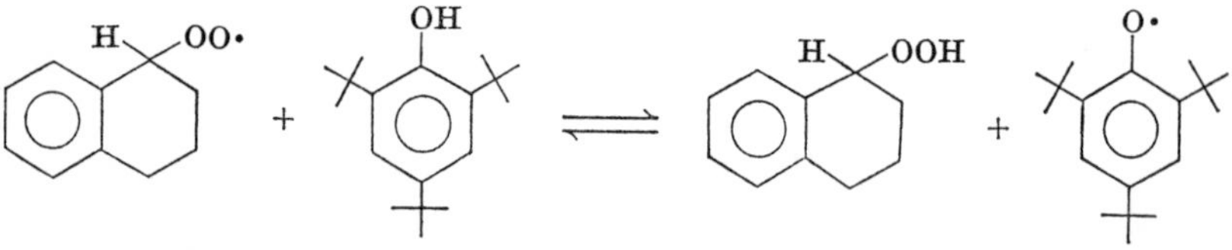

is given by $K = (1{\cdot}0 \pm 0{\cdot}2) \exp(7000 \pm 1700/RT)$ (Mahoney and DaRooge, 1969).

Although there have been many estimates of the rate constants for various reactions of unhindered phenoxyl radicals there are almost no absolute rate data. Thomas (1964) has estimated that for the abstraction of hydrogen from tetralin hydroperoxide by an unhindered phenoxyl radical, $k \geqslant 2 \times 10^{5}$ M^{-1} sec.$^{-1}$ (cf. however,

TABLE 47

Rate constants and activation parameters for reactions of 2,4,6-tri-t-butylphenoxyl with some hydrocarbons and hydroperoxides

Substrate	Solvent	Temp. (°)	Rate const. (M^{-1} sec.$^{-1}$)	E (kcal. mole^{-1})	log (A/M^{-1} sec.$^{-1}$)	Ref.
n-Decane	Substrate	60	$0{\cdot}22 \times 10^{-7}$	27·5	10·415	1
Iso-decane	Substrate	60	$0{\cdot}32 \times 10^{-7}$	25·5	9·25	1
Toluene	Substrate	60	$4{\cdot}5 \times 10^{-7}$	23	8·76	1
Ethylbenzene	Substrate	60	79×10^{-7}	18	6·72	1
Cumene	Substrate	60	$7{\cdot}1 \times 10^{-7}$	16·5	4·69	1
t-Butyl hydroperoxide	CCl_4	21	90	8·3	8·15	2
Tetralin hydroperoxide	C_6H_5Cl	30	0·34	5·5	3·61	3
Tetralin hydroperoxide	Tetralin, C_6H_5Cl	30	0·43	7·2	4·72	3
Cumyl hydroperoxide	Cumene	40	0·37	—	—	3
9,10-Dihydroanthracene-9-hydroperoxide	C_6H_5Cl	30	0·15	—	—	3
9,10-Dihydroanthracene-9-hydroperoxide	C_6H_5Cl	24	0·088	8·7	5·69	4

References: (1) Neiman, Mamedova, Blenke and Buchachenko (1962). (2) McGowan and Powell (1960). (3) Mahoney and DaRooge (1969). (4) Mahoney and DaRooge (1967).

Mahoney, 1967). Howard and Ingold (1964, 1965a) have listed rate constant *ratios* for various reactions involving a variety of phenoxyl radicals in the autoxidation of tetralin at 65°. The reactions for which rate constant ratios were measured include hydrogen abstraction from tetralin and its hydroperoxide, reaction with peroxyl radicals and reaction with a second phenoxyl radical.

Mahoney (1967) has listed rate constant ratios for similar reactions at 60° involving 9,10-dihydroanthracene as the oxidising substrate. Chain termination can involve both a peroxyl-phenoxyl radical reaction and a bimolecular phenoxyl radical reaction.

$$\mathrm{ROO\cdot + ArO\cdot} \longrightarrow \text{Non-radical products.} \quad (36)$$

$$\mathrm{ArO\cdot + ArO\cdot} \longrightarrow \text{Non-radical products.} \quad (37)$$

There is a very large variation in the rate constant ratio $(k_{36}/k_{37})^{1/2}$ for different phenols. This was attributed to a decrease in the bimolecular phenoxyl termination rate constant as the phenoxyl radical becomes more resonance stabilised. The relative rates of hydrogen abstraction from the hydroperoxide and from the hydrocarbon by the phenoxyl radical were found to be relatively insensitive to the phenolic structure, indicating that both reactions are affected to the same extent by the stability of the phenoxyl radical. Mahoney (1967) also re-examined the results obtained by Thomas (1964) and by Howard and Ingold (1964, 1965a) on tetralin. He concluded that the reactivity ratios for hydrogen abstraction from 9,10-dihydroanthracene and from tetralin at 60° are 24 for the peroxyl radical, 70 for the unsubstituted phenoxyl radical and > 200 for the *p*-methoxyphenoxyl radical. In a subsequent paper, Mahoney and DaRooge

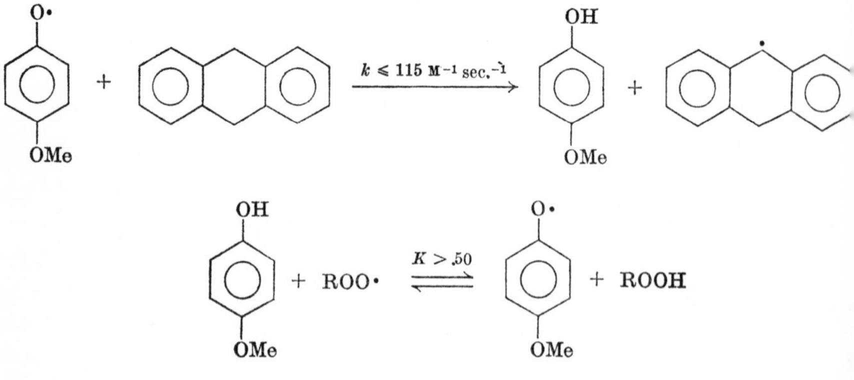

O·, OMe + [structure] $\xrightarrow{k \leqslant 115\ \mathrm{M}^{-1}\ \mathrm{sec.}^{-1}}$ OH, OMe + [structure]

OH, OMe + ROO· $\underset{}{\overset{K > 50}{\rightleftharpoons}}$ O·, OMe + ROOH

(1967) listed additional rate constant ratios for the oxidation of 9,10-dihydroanthracene at 60° in the presence of phenols. They also estimated the following rate and equilibrium constants for hydrogen abstraction reactions involving the *p*-methoxyphenoxyl radical.

3. *Reactions with Other Radicals*

The rate constant for the reaction between galvinoxyl and methyl radicals has been estimated to be 2×10^7 M^{-1} sec.$^{-1}$ (Schuler, 1964). This is probably a lower limit because galvinoxyl is a very efficient inhibitor of chain reactions (cf. Bartlett and Funahashi, 1962; Davies and Roberts, 1967, 1968).

Thomas (1964) has estimated that the rate constant for reaction of an unhindered phenoxyl radical with a tetralylperoxyl radical is $> 10^9$ M^{-1} sec.$^{-1}$. Mahoney and DaRooge (1967) concluded that $k_{38} > 8{\cdot}7 \times 10^7$ M^{-1} sec.$^{-1}$ for reaction (38). Chien and Boss (1967a)

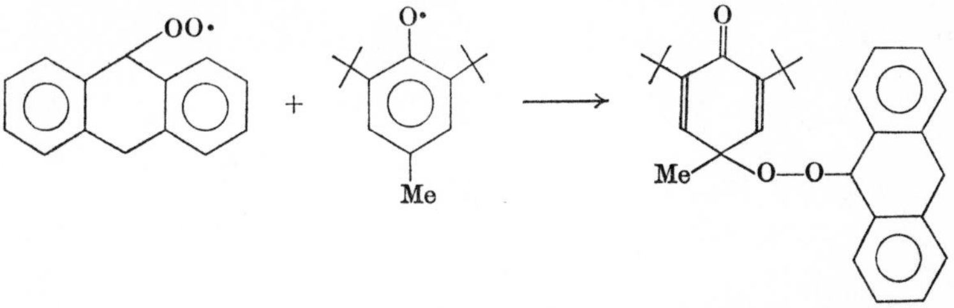

(38)

gave $k = 7 \times 10^7$ M^{-1} sec.$^{-1}$ at 130° for the reaction of the same hindered phenoxyl radical with the peroxyl radicals present during the autoxidation of polypropylene. However, at 130°, cyclohexadienone peroxides must be quite unstable and therefore the efficiency of the coupling process as a chain-terminating step could be quite low.

4. *Bimolecular Homolytic Substitution* (S_H2) *at Oxygen*

Prokof'ev, Solodovnikov, Bogdanov, Nikiforov and Ershov (1967) have shown that 2,6-di-t-butyl-4-substituted phenoxyl radicals induce the decomposition of t-butyl peroxide by the reaction:

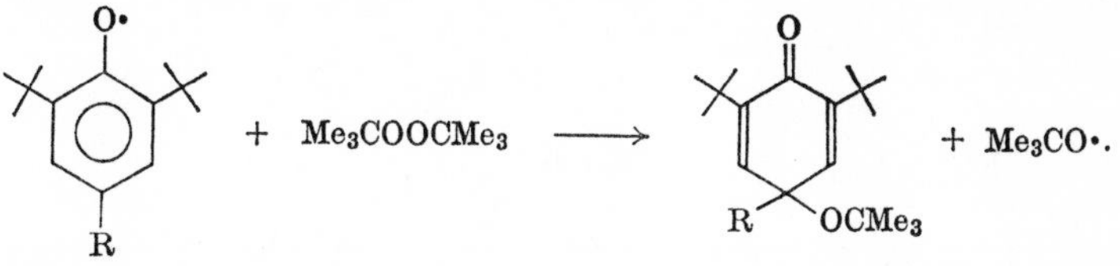

$$\text{ArO}\cdot + \text{Me}_3\text{COOCMe}_3 \longrightarrow \text{R(OCMe}_3\text{)-cyclohexadienone} + \text{Me}_3\text{CO}\cdot.$$

The reaction is first order in the phenoxyl radical and in the peroxide. Pseudo-first order rate constants were determined over a temperature range from 10 to 50° for a number of different substituents. The Arrhenius parameters for these pseudo-first order reactions are given in Table 48. The progressive increase in activation energy shown by the radicals in Table 48 is paralleled by a decrease in the unpaired

TABLE 48

Arrhenius parameters for reactions of 4-*substituted* 2,6-*di-t-butyl-phenoxyl radicals with t-butyl peroxide*

4-Substituent	*A* (sec.$^{-1}$)	*E* (kcal. mole^{-1})
Me_3C—	4×10^6	9·9
$MeC(=O)$—C_6H_4— (4-)	7×10^6	10·2
$C_{10}H_7$— (1-naphthyl)	12×10^6	10·4
Cl—C_6H_4— (4-)	9×10^6	10·5
C_6H_5—	13×10^6	11·0
Me—C_6H_4— (4-)	12×10^6	11·2
OMe—C_6H_4— (2-)	20×10^6	11·4
MeO—C_6H_4— (4-)	26×10^6	11·9

electron spin density at their *para*-position, i.e., by a decrease in spin density at the reaction centres of the radicals.

G. Nitroxides

Although the first organic nitroxide was isolated in 1901 (Piloty and Schwerin, 1901), there have been very few quantitative measurements of the reactivity of nitroxides towards one another and towards other compounds. While the electron spin resonance spectra of several hundred nitroxides have been recorded, the information on their reactivity is usually confined to qualitative statements that a specific radical is stable, moderately stable or unstable. Nitroxides have an inherently stable electronic arrangement around the oxygen and nitrogen atoms (Forrester, Hay and Thomson, 1968b). Steric and resonance effects which arise from the groups attached to the nitrogen play little or no part in preventing dimerisation at the O—N centre. However, these factors may be important in preventing the nitroxide from reacting with itself in some other way such as disproportionation, for example, as follows:

$$R(H)N{-}O\cdot + R(H)N{-}O\cdot \longrightarrow R(H)NOH + R{-}N{=}O$$

$$R(R'CH_2)N{-}O\cdot + R(R'CH_2)N{-}O\cdot \longrightarrow R(R'CH_2)NOH + R(R'CH){=}N{\rightarrow}O,$$

or dimerisation at some other centre in the molecule, e.g.,

$$2\; Bu^t{-}N(C_6H_5){-}O\cdot \longrightarrow [\text{dimer: } Bu^t{-}N(\rightarrow O){=}C_6H_5(H){-}O{-}N(Bu^t)C_6H_5] \longrightarrow Bu^t{-}N(\rightarrow O){=}C_6H_4{=}O + Bu^tN(H)C_6H_5.$$

Absolute rate constants for the bimolecular decay of diethyl nitroxide radicals in difluorodichloromethane have been measured recently (Adamic, Bowman and Ingold, 1970; Adamic, Bowman, Gillan and Ingold, 1971). At room temperature the measured rate constant was $1{\cdot}5 \pm 0{\cdot}5 \times 10^4$ M^{-1} sec.$^{-1}$. The rate constants over a temperature range from 27° to −137° can be represented by $k = 10^{(5\pm1)}\exp(-1000 \pm 500/RT)$ M^{-1} sec.$^{-1}$. Although dialkyl nitroxides do not form dimers at room temperature, diethyl nitroxide was found to form a dimer reversibly at temperatures from −100° to −145°. Because the irreversible decomposition of the nitroxide is quite rapid even at these low temperatures, it was not possible to form very much of the dimer/nitroxide equilibrium mixture. Under equilibrium conditions the variation in the nitroxide concentration with temperature gave $\Delta H = 7{\cdot}0 \pm 2{\cdot}0$ kcal. mole^{-1} and $\Delta S = 25 \pm 15$ cal. deg.$^{-1}$ mole^{-1}. The total process was represented by reactions (39) and (40).

$$2(CH_3CH_2)_2NO\cdot \rightleftharpoons [(CH_3CH_2)_2NO]_2 \quad (39)$$

$$[(CH_3CH_2)_2NO]_2 \longrightarrow \left[\begin{array}{c} CH_3CH_2N\overset{O}{\frown}O{-}N(CH_2CH_3)_2 \\ HC \cdots H \\ CH_3 \end{array}\right]^{\ddagger} \longrightarrow \begin{array}{c} CH_3CH_2N{\rightarrow}O \\ \| \\ HC \\ | \\ CH_3 \end{array} + HON(CH_2CH_3)_2 \quad (40)$$

The rate constant for irreversible decay of the dimer, k_{40}, was $10^{10{\cdot}5}$ $\exp(-8000/RT)$ M^{-1} sec.$^{-1}$. The pre-exponential factor for this reaction is within the range generally found for unimolecular decompositions proceeding by five- or six-centre cyclic transition states.

The bimolecular decay rate constant for diethyl nitroxide is very dependent on the solvent. Thus, the rate constant at room temperature decreases by a factor of 40 with increasing solvent polarity along the series: isopentane > benzene > dichlorodifluoromethane > methanol > water.

The reactions of nitroxides with alkyl radicals are very fast. Thus, Khloplyankina, Buchachenko, Neiman and Vasil'eva (1965), studying the inhibition of hydrocarbon autoxidation by 2,2,6,6-tetramethylpiperidine nitroxide at 60°, found that the rate constant ratio,

k_9/k_{41} for reactions (9) and (41) was 26 for RH = ethylbenzene and 1·4 for RH = diphenylmethane.

$$R\cdot + O_2 \longrightarrow ROO\cdot \qquad (9)$$

Me Me Me Me

$$R\cdot + \text{(2,2,6,6-tetramethylpiperidine)}\ N{-}O\cdot \longrightarrow \text{(2,2,6,6-tetramethylpiperidine)}\ NOR \qquad (41)$$

Me Me Me Me

For autoxidising styrene at 65°, the analogous ratio of rate constants is 10 for 2,2,6,6-tetramethyl-4-piperidone nitroxide (Brownlie and Ingold, 1967b). For styrene k_9 has been estimated to be 1×10^8 M^{-1} sec.$^{-1}$ (Mayo, 1958) and hence $k_{41} = 1 \times 10^7$ M^{-1} sec.$^{-1}$.

Aliphatic nitroxides, such as tetramethylpiperidone nitroxide, do not react with peroxyl radicals. Aromatic nitroxides such as diphenyl nitroxide react with both alkyl radicals (rapidly) and peroxyl radicals (more slowly). It is likely that the latter reaction involves an initial addition of the peroxyl radical to the aromatic ring to give eventually an *N*-phenyl-*p*- or -*o*-benzoquinoneimine *N*-oxide and an alcohol.

$$ROO\cdot + C_6H_5{-}N(O\cdot){-}C_6H_5 \longrightarrow$$

$$ROO(H)C_6H_4{=}N(\rightarrow O){-}C_6H_5 \longrightarrow ROH + O{=}C_6H_4{=}N(\rightarrow O){-}C_6H_5$$

The rate constant for the reaction of 4,4′-dimethoxydiphenylnitroxide with 1-phenylethylperoxyl radicals at 60° has been estimated as 6×10^5 M^{-1} sec.$^{-1}$ (Khloplyankina, Buchachenko, Neiman and Vasil'eva, 1965) and with poly(peroxystyryl)peroxyl radicals at 65° as 5×10^4 M^{-1} sec.$^{-1}$ (Brownlie and Ingold, 1967b). The lower value is to be preferred because the nitroxide is a less potent inhibitor at 1 atmosphere pressure of oxygen than the corresponding amine, which has a rate constant for its reaction with peroxyl radicals of $3{\cdot}3 \times 10^5$ M^{-1} sec.$^{-1}$ at 65°.

Very few absolute rate constants for hydrogen atom abstraction by stable nitroxides have been measured. Kreilick and Weissman (1966) have used magnetic resonance techniques to study the hydrogen atom exchange between some hydroxylamines and the corresponding nitroxides. The reaction is believed to proceed through a short-lived intermediate complex.

$$\begin{matrix}R'\\R\end{matrix}\!\!>\!NO\cdot + \begin{matrix}R'\\R\end{matrix}\!\!>\!NOH \rightleftharpoons \begin{matrix}R'\\R\end{matrix}\!\!>\!NOH\dot{O}N\!<\!\!\begin{matrix}R'\\R\end{matrix} \rightleftharpoons \begin{matrix}R'\\R\end{matrix}\!\!>\!NOH + \begin{matrix}R'\\R\end{matrix}\!\!>\!NO\cdot$$

In carbon tetrachloride at 27°, the rate constants for the hydrogen exchange process are: R = R′ = Ph, $k > 10^7$ M^{-1} sec.$^{-1}$; R = R′ = t-Bu, $k = 320$ M^{-1} sec.$^{-1}$; R = t-Bu, R′ = 2,6-dimethoxyphenyl, $k = 2000$ M^{-1} sec.$^{-1}$. The rate constants for the di-t-butyl and for the t-butyl-2,6-dimethoxyphenyl systems are decreased in more polar solvents. They also show an unusual dependence on temperature, having minimum values at about room temperature which increase when the temperature is raised or lowered.

Buchachenko, Sukhanova, Kalashnikova and Neiman (1965) studied the reactions of the four nitroxides listed in Table 49 with 2,6-di-t-butylphenol in a number of solvents over a range of temperatures.

$$>N\text{—}O\cdot + \text{2,6-di-t-butylphenol (OH)} \longrightarrow >NOH + \text{2,6-di-t-butylphenoxyl (O}\cdot) \xrightarrow{>N\text{—}O\cdot} \text{coupled products}$$

The solvents used were benzene, ethylbenzene and pyridine, which were supposed to be able to form π-complexes with the nitroxides, and n-heptane and dichloroethane, which cannot form π-complexes. The hydrogen donating phenol is sterically too hindered to form either intermolecular hydrogen bonds or π-complexes. Some of these workers' kinetic results are summarised in Table 49. The pre-exponential factors and activation energies show wide variations from one solvent to another. These values were interpreted (*sic*) in terms of π-complexes which could either decrease or increase the reactivity

Rate constants (30°), *pre-exponential factors and activation energies for reactions of nitroxides with* 2,6-*di-t-butylphenol in n-heptane, ethylbenzene and pyridine* (Buchachenko, Sukhanova and Neiman (1965))

Nitroxide	n-Heptane			Ethylbenzene			Pyridine		
	k (M^{-1} sec.$^{-1}$)	log (A/M^{-1} sec.$^{-1}$)	E (kcal. mole^{-1})	k (M^{-1} sec.$^{-1}$)	log (A/M^{-1} sec.$^{-1}$)	E (kcal. mole^{-1})	k (M^{-1} sec.$^{-1}$)	log (A/M^{-1} sec.$^{-1}$)	E (kcal. mole^{-1})
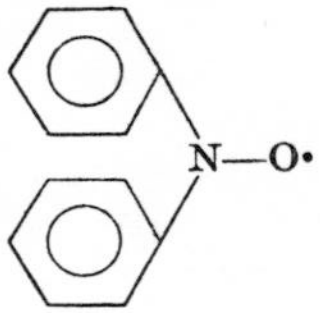 $(C_6H_5)_2N$—O•	$4{\cdot}2\times10^{-5}$	0·68	7·0	$2{\cdot}1\times10^{-6}$	3·78	12·6	—	—	—
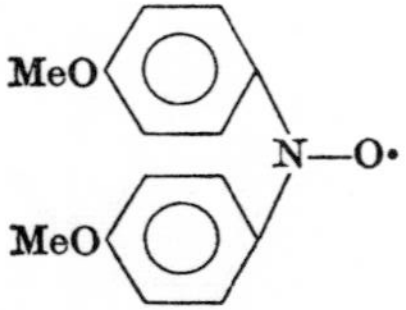 $(p\text{-}MeOC_6H_4)_2N$—O•	$5{\cdot}2\times10^{-5}$	4·38	12·0	$2{\cdot}2\times10^{-6}$	8·78	20·0	$5{\cdot}0\times10^{-3}$	4·78	9·8
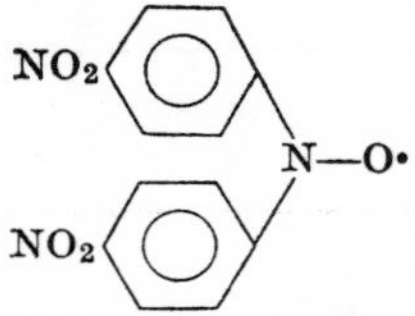 $(p\text{-}NO_2C_6H_4)_2N$—O•	$7{\cdot}5\times10^{-4}$	6·48	13·3	$1{\cdot}1\times10^{-4}$	8·68	17·5	—	—	—
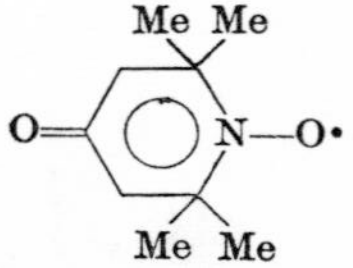 O=(2,2,6,6-tetramethyl ring, Me Me / Me Me)N—O•	$5{\cdot}2\times10^{-5}$	1·78	8·4	$5{\cdot}2\times10^{-5}$	1·78	8·4	$5{\cdot}5\times10^{-5}$	6·28	14·6

TABLE 50

Rate constants (100°), *pre-exponential factors and activation energies for reactions of nitroxides with ethylbenzene.*
(Mamedova, Buchachenko and Neiman, 1965)

Nitroxide	k (M^{-1} sec.$^{-1}$)	log (A/M^{-1} sec.$^{-1}$)	E (kcal. mole^{-1})	Nitroxide	k (M^{-1} sec.$^{-1}$)	log (A/M^{-1} sec.$^{-1}$)	E (kcal. mole^{-1})
2,2,6,6-tetramethyl-4-oxopiperidine-N–O• (Me Me, O=, N—O•, Me Me)	$6{\cdot}2 \times 10^{-7}$	2·89	15·5	C_6H_5–N(O•)–C_6H_5	$4{\cdot}8 \times 10^{-6}$	5·89	19·1
2,2,6,6-tetramethylpiperidine-N–O• (Me Me, N—O•, Me Me)	$1{\cdot}1 \times 10^{-6}$	3·15	15·5	MeO–C_6H_4–N(O•)–C_6H_4–OMe	$3{\cdot}4 \times 10^{-6}$	3·04	14·5
2,2,6,6-tetramethyl-4-oxopiperidine-N–O• (Me Me, O=, N—O•, Me Me)	$1{\cdot}1 \times 10^{-6}$	3·08	15·4	NO_2–C_6H_4–N(O•)–C_6H_4–NO_2	$2{\cdot}7 \times 10^{-6}$	8·98	24·8

of the nitroxide. The bonding energy of 4,4′-dimethoxydiphenyl nitroxide with benzene was estimated to be $6{\cdot}4 \pm 2{\cdot}0$ kcal. mole.$^{-1}$

Mamedova, Buchachenko and Neiman (1965) have measured absolute rate constants for the reactions of six stable nitroxides with ethylbenzene, their results are summarised in Table 50. The large

$$\gt N{-}O\cdot + C_6H_5CH_2CH_3 \longrightarrow \gt NOH + C_6H_5\dot{C}HCH_3 \xrightarrow{\gt N{-}O\cdot} \text{coupled products}$$

variation in the pre-exponential factors compared with the small change in rate constant at 100° raises serious doubts as to whether the reaction can be quite as simple as that shown above.

Acknowledgment

The author wishes to thank Dr. K. U. Ingold for the considerable help that he has given in the preparation of this article.

References

Adamic, K., Bowman, D. F. and Ingold, K. U. (1970) *J. Amer. Chem. Soc.*, **92**, 1093.
Adamic, K., Bowman, D. F., Gillan, T. and Ingold, K. U. (1971) *J. Amer. Chem. Soc.*, **93**, 902.
Adamic, K., Howard, J. A. and Ingold, K. U. (1969) *Canad. J. Chem.*, **47**, 3803.
Adams, G. E., Boag, J. W. and Michael, B. D. (1966) *Proc. Roy. Soc.*, **289A**, 321.
Adams, G. E. and Michael, B. D. (1967) *Trans. Faraday Soc.*, **63**, 1171.
Adams, G. E., Michael, B. D. and Land, E. J. (1966) *Nature*, **211**, 293.
Alagy, J., Clement, G. and Balaceanu, J. C. (1961) *Bull. Soc. chim. France*, 1303.
Aleksandrov, A. L. and Denisov, E. T. (1966) *Izvest. Akad. Nauk SSSR, Ser. Khim.*, 1737.
Aleksandrov, A. L. and Denisov, E. T. (1969) *Kinetika i Kataliz*, **10**, 904.
Allies, P. G. and Brindley, P. B. (1967) *Chem. and Ind.*, 319.
Allies, P. G. and Brindley, P. B. (1968) *Chem. and Ind.*, 1439.
Altwicker, E. R. (1967) *Chem. Rev.*, **67**, 475.
Anbar, M., Meyerstein, D. and Neta, P. (1966a) *J. Phys. Chem.*, **70**, 2660.
Anbar, M., Meyerstein, D. and Neta, P. (1966b) *J. Chem. Soc.* (*B*), 742.
Anbar, M. and Neta, P. (1967) *Internat. J. Appl. Radiation Isotopes*, **18**, 493.
Andronov, L. M., Zaikov, G. E. and Maizus, Z. K. (1967) *Teor. Eksp. Khim.* **3**, 620.

Arick, M. R. and Weissman, S. I. (1968) *J. Amer. Chem. Soc.*, **90**, 1654.
Ayscough, P. B. and Russell, K. E. (1965) *Canad. J. Chem.*, **43**, 3039.
Ayscough, P. B. and Russell, K. E. (1967) *Canad. J. Chem.*, **45**, 3019.
Bäckström, H. L. J. and Sandros, K. (1958) *Acta Chem. Scand.*, **12**, 823.
Bäckström, H. L. J. and Sandros, K. (1960) *Acta Chem. Scand.*, **14**, 48.
Baker, R. R., Baldwin, R. R. and Walker, R. W. (1970) *Trans. Faraday Soc.*, **66**, 2812.
Bamford, C. H. and Dewar, M. J. S. (1949) *Proc. Roy. Soc.* **198A**, 252.
Bartlett, P. D. and Funahashi, T. (1962) *J. Amer. Chem. Soc.*, **84**, 2596.
Bartlett, P. D. and Guaraldi, G. (1967) *J. Amer. Chem. Soc.*, **89**, 4799.
Bartlett, P. D. and Traylor, T. G. (1963) *J. Amer. Chem. Soc.*, **85**, 2407.
Bateman, L., Bolland, J. L. and Gee, G. (1951) *Trans. Faraday Soc.*, **47**, 274.
Bateman, L. and Gee, G. (1948) *Proc. Roy. Soc.*, **195A**, 391.
Bateman, L. and Gee, G. (1951) *Trans. Faraday Soc.*, **47**, 155.
Bateman, L., Gee, G., Morris, A. L. and Watson, W. F. (1951) *Discuss. Faraday Soc.*, 250.
Bauer, R. H. and Coppinger, G. M. (1963) *Tetrahedron*, **19**, 1201.
Baxendale, J. H. (1962) *Radiation Res.*, **17**, 312.
Becker, H.-D. (1965) *J. Org. Chem.*, **30**, 982.
Beckett, A. and Porter, G. (1963) *Trans. Faraday Soc.*, **59**, 2038.
Behar, D., Czapski, G., Rabani, J., Dorfman, L. M. and Schwarz, H. A. (1970) *J. Phys. Chem.*, **74**, 3209.
Bell, J. A. and Linschitz, H. (1963) *J. Amer. Chem. Soc.*, **85**, 528.
Bennett, J. E. (1960) *Nature*, **186**, 385.
Bennett, J. E., Brown, D. M. and Mile, B. (1970a) *Trans. Faraday Soc.* **66**, 386.
Bennett, J. E., Brown, D. M. and Mile, B. (1970b) *Trans. Faraday Soc.*, **66**, 397.
Benson, S. W. and Shaw, R. (1968) Adv. Chem. Series, No. 75 (Amer. Chem. Soc.), p. 288.
Berger, H., Blaauw, A. M. W., Al, M. M. and Smael, P. (1968) Adv. Chem. Series, No. 75 (Amer. Chem. Soc.), p. 346.
Bickel, A. F. and Kooyman, E. C. (1956) *J. Chem. Soc.*, 2215.
Bidzilya, V. A., Pokhodenko, V. D. and Brodskii, A. I. (1966) *Doklady Akad. Nauk SSSR*, **166**, 1099.
Bielski, B. H. J. and Saito, E. (1962) *J. Phys. Chem.*, **66**, 2266.
Bielski, B. H. J. and Schwarz, H. A. (1968) *J. Phys. Chem.*, **72**, 3836.
Black, G. and Porter, G. (1962a) *Proc. Roy. Soc.*, **266A**, 185.
Black, G. and Porter, G. (1962b) *Discuss. Faraday Soc.*, **33**, 284.
Blanchard, H. S. (1959) *J. Amer. Chem. Soc.*, **81**, 4548.
Bolland, J. L. and ten Have, P. (1947a) *Trans. Faraday Soc.*, **43**, 201.
Bolland, J. L. and ten Have, P. (1947b) *Trans. Faraday Soc.*, **43**, 252.
Boss, C. R. and Chien, J. C. W. (1969) *Amer. Chem. Soc.*, Div. of Petroleum Chem. Preprints, 14(2), A122.
Bridge, N. K. and Porter, G. (1958) *Proc. Roy. Soc.*, **244A**, 259, 276.
Briers, F., Chapman, D. L. and Walters, E. (1926) *J. Chem. Soc.*, 562.
Brodskii, A. I., Pokhodenko, V. D. and Ganyuk, L. N. (1964) *Roczniki Chem.*, **38**, 105.

Brodskii, A. I., Pokhodenko, V. D., Khizhnyi, V. A. and Kalibabchuk, N. N. (1966) *Doklady Akad, Nauk SSSR*, **169**, 339.
Brook, J. H. T. (1957) *Trans. Faraday Soc.*, **53**, 327.
Brook, J. H. T. and Glazebrook, R. W. (1960) *Trans. Faraday Soc.*, **56**, 1014.
Brownlie, I. T. and Ingold, K. U. (1966) *Canad. J. Chem.*, **44**, 861.
Brownlie, I. T. and Ingold, K. U. (1967a) *Canad. J. Chem.*, **45**, 2419.
Brownlie, I. T. and Ingold, K. U. (1967b) *Canad. J. Chem.*, **45**, 2427.
Buchachenko, A. L., Kaganskaya, K. Ya. and Neiman, M. B. (1961) *Kinetika i Kataliz*, **2**, 161.
Buchachenko, A. L., Kaganskaya, K. Ya., Neiman, M. B. and Petrov, A. A. (1961) *Kinetika i Kataliz*, **2**, 44.
Buchachenko, A. L., Neiman, M. B., Sukhanova, O. P. and Mamedova, Ya. G. (1963) *Zhur. fiz. Khim.*, **37**, 221.
Buchachenko, A. L., Sukhanova, O. P., Kalashnikova, L. A. and Neiman, M. B. (1965) *Kinetika i Kataliz*, **6**, 601.
Burgess, R. H. and Robb, J. C. (1957) *Chem. Soc. Special Publ. No. 9*, p. 167.
Burkhart, R. D. (1969) *J. Phys. Chem.*, **73**, 2703.
Burlakova, E. B., Khrapova, N. G., Shtol'ko, V. N. and Émanuél, N. M. (1966) *Doklady Akad. Nauk SSSR*, **169**, 688.
Burn, A. J. (1968) Adv. Chem. Series, No. 75 (Amer. Chem. Soc.), p. 323.
Burnett, G. M. and Melville, H. W. (1963) *Technique of Organic Chemistry*, Ed. Freiss S. L., Lewis, E. S. and Weissberger, A., Interscience, New York, **8**, Pt. II, 1107.
Caldwell, J. and Back, R. A. (1965) *Trans. Faraday Soc.*, **61**, 1939.
Carlsson, D. J., Howard, J. A. and Ingold, K. U. (1966) *J. Amer. Chem. Soc.*, **88**, 4725.
Carlsson, D. J. and Ingold, K. U. (1967a) *J. Amer. Chem. Soc.*, **89**, 4885.
Carlsson, D. J. and Ingold, K. U. (1967b) *J. Amer. Chem. Soc.*, **89**, 4891.
Carlsson, D. J. and Ingold, K. U. (1968) *J. Amer. Chem. Soc.*, **90**, 7047.
Carlsson, D. J., Ingold, K. U. and Bray, L. C. (1969) *Int. J. Chem. Kinetics*, **1**, 315.
Cercek, B. (1968) *J. Phys. Chem.*, **72**, 3832.
Chevriau, C., Naffa, P. and Balaceanu, J. C. (1964) *Bull Soc. chim. France*, 3002.
Chien, J. C. W. (1967) *J. Amer. Chem. Soc.*, **89**, 1273.
Chien, J. C. W. and Boss, C. R. (1967a) *J. Polymer Sci.* (*A*), **5**, 1683.
Chien, J. C. W. and Boss, C. R. (1967b) *J. Polymer Sci.* (*A*), **5**, 3091.
Clark, W. D. K., Litt, A. D. and Steel, C. (1969a) *J. Chem. Soc.* (*D*), 1087.
Clark, W. D. K., Litt, A. D. and Steel, C. (1969b) *J. Amer. Chem. Soc.*, **91**, 5413.
Cohen, S. G. and Baumgarten, R. J. (1965) *J. Amer. Chem. Soc.*, **87**, 2996.
Cohen, S. G. and Chao, H. M. (1968) *J. Amer. Chem. Soc.*, **90**, 165.
Cohen, S. G., Laufer, D. A. and Sherman, W. V. (1964) *J. Amer. Chem. Soc.*, **86**, 3060.
Cohen, S. G. and Litt, A. D. (1970) *Tetrahedron Letters*, 837.
Cohen, S. G. and Sherman, W. V. (1963) *J. Amer. Chem. Soc.*, **85**, 1642.
Cohen, S. G. and Stein, N. (1969) *J. Amer. Chem. Soc.*, **91**, 3690.
Cook, C. D., Depatie, C. B. and English, E. S. (1959) *J. Org. Chem.*, **24**, 1356.

Cook, C. D., Nash, N. G. and Flanagan, H. R. (1955) *J. Amer. Chem. Soc.*, **77**, 1783.
Cook, C. D. and Norcross, B. E. (1956) *J. Amer. Chem. Soc.*, **78**, 3797.
Cook, C. D. and Norcross, B. E. (1959) *J. Amer. Chem. Soc.*, **81**, 1176.
Cooper, H. R. and Melville, H. W. (1951) *J. Chem. Soc.*, 1994.
Cullis, C. F., Francis, J. M., Raef, Y. and Swallow, A. J. (1967) *Proc. Roy. Soc.*, **300A**, 443.
Currie, D. J. and Dainton, F. S. (1965) *Trans. Faraday Soc.*, **61**, 1156.
Czapski, G. and Bielski, B. H. J. (1963) *J. Phys. Chem.*, **67**, 2180.
Czapski, G. and Dorfman, L. M. (1964) *J. Phys. Chem.*, **68**, 1169.
DaRooge, M. A. and Mahoney, L. R. (1967) *J. Org. Chem.*, **32**, 1.
Davies, A. G., Griller, D., Roberts, B. P. and Tudor, R. (1970) *J. Chem. Soc. (D)*, 640.
Davies, A. G. and Roberts, B. P. (1967) *J. Chem. Soc. (B)*, 17.
Davies, A. G. and Roberts, B. P. (1968) *J. Chem. Soc. (B)*, 1074.
Davies, A. G. and Roberts, B. P. (1969) *J. Chem. Soc. (B)*, 311.
Davies, A. G., Ingold, K. U., Roberts, B. P. and Tudor, R. (1971) *J. Chem. Soc. (B)*, 699.
Davies, D. S., Goldsmith, H. L., Gupta, A. K. and Lester, G. R. (1956) *J. Chem. Soc.*, 4926.
Del Greco, F. P. and Kaufman, F. (1962) *Discuss. Faraday Soc.*, **33**, 128.
Denisov, E. T. and Aleksandrov, A. L. (1964) *Zhur. fiz. Khim.*, **38**, 491.
Dobson, G. and Grossweiner, L. I. (1965) *Trans. Faraday Soc.*, **61**, 708.
Fineman, M. and Ross, S. D. (1950) *J. Polymer Sci.*, **5**, 259.
Foner, S. N. and Hudson, R. L. (1962) Adv. Chem. Series, No. 36, (Amer. Chem. Soc.), p. 34.
Forrester, A. R., Hay, J. M. and Thompson, R. H. (1968a) *Organic Chemistry of Stable Free Radicals*, Academic Press, London and New York, Chapter 7.
Forrester, A. R., Hay, J. M. and Thompson, R. H. (1968b) *Organic Chemistry of Stable Free Radicals*, Academic Press, London and New York, Chapter 5.
Filar, L. and Winstein, S. (1960) *Tetrahedron Letters*, 9.
Franklin, J. L. (1967) *Ann. Rev. Phys. Chem.*, **18**, 261.
Fry, A. J., Liu, R. S. H. and Hammond, G. S. (1966) *J. Amer. Chem. Soc.*, **88**, 4781.
Gall, B. L. and Dorfman, L. M. (1969) *J. Amer. Chem. Soc.*, **91**, 2199.
Gol'dberg, V. H. and Obukhova, L. K. (1965) *Doklady Akad. Nauk SSSR*, **165**, 860.
Gray, P., Shaw, R. and Thynne, J. C. J. (1967) *Progress in Reaction Kinetics*, Ed. G. Porter, Pergamon, Oxford, **4**, 63.
Greiner, N. R. (1968a) *J. Phys. Chem.*, **72**, 406.
Greiner, N. R. (1968b) *J. Chem. Phys.*, **48**, 1413.
Guk, A. F., Tsepalov, V. F., Shuvalov, V. F. and Shlyapintokh, V. Ya. (1968) *Izvest. Akad. Nauk SSSR. Ser. Khim.*, 2250.
Guttenplan, J. and Cohen, S. G. (1969) *J. Chem. Soc. (D)*, 247.
Hammond, G. S., Baker, W. P. and Morre, W. M. (1961) *J. Amer. Chem. Soc.*, **83**, 2795.
Hammond, G. S. and Foss, R. P. (1964) *J. Phys. Chem.*, **68**, 3739.
Hammond, G. S. and Leermakers, P. A. (1962) *J. Phys. Chem.*, **66**, 1148.

Hendry, D. G. (1967) *J. Amer. Chem. Soc.*, **89**, 5433.
Hendry, D. G. and Schuetzle, D. (1969) *Amer. Chem. Soc., Div. of Petroleum Chem. Preprints*, 14(2), A31.
Herkstroeter, W. G. and Hammond, G. S. (1966) *J. Amer. Chem. Soc.*, **88**, 4769.
Hiatt, R., Gould, C. W. and Mayo, F. R. (1964) *J. Org. Chem.*, **29**, 3461.
Howard, J. A., Adamic, K. and Ingold, K. U. (1969) *Canad. J. Chem.*, **47**, 3793.
Howard, J. A. and Ingold, K. U. (1962) *Canad. J. Chem.*, **40**, 1851.
Howard, J. A. and Ingold, K. U. (1963a) *Canad. J. Chem.*, **41**, 1744.
Howard, J. A. and Ingold, K. U. (1963b) *Canad. J. Chem.*, **41**, 2800.
Howard, J. A. and Ingold, K. U. (1964) *Canad. J. Chem.*, **42**, 2324.
Howard, J. A. and Ingold, K. U. (1965a) *Canad. J. Chem.*, **43**, 2724.
Howard, J. A. and Ingold, K. U. (1965b) *Canad. J. Chem.*, **43**, 2729.
Howard, J. A. and Ingold, K. U. (1965c) *Canad. J. Chem.*, **43**, 2737.
Howard, J. A. and Ingold, K. U. (1966a) *Canad. J. Chem.*, **44**, 1113.
Howard, J. A. and Ingold, K. U. (1966b) *Canad. J. Chem.*, **44**, 1119.
Howard, J. A. and Ingold, K. U. (1967a) *Canad. J. Chem.*, **45**, 785.
Howard, J. A. and Ingold, K. U. (1967b) *Canad. J. Chem.*, **45**, 793.
Howard, J. A. and Ingold, K. U. (1968a) *Canad. J. Chem.*, **46**, 2655.
Howard, J. A. and Ingold, K. U. (1968b) *Canad. J. Chem.*, **46**, 2661.
Howard, J. A. and Ingold, K. U. (1968c) *J. Amer. Chem. Soc.*, **90**, 1056.
Howard, J. A. and Ingold, K. U. (1968d) *J. Amer. Chem. Soc.*, **90**, 1058.
Howard, J. A. and Ingold, K. U. (1969a) *Canad. J. Chem.*, **47**, 3797.
Howard, J. A. and Ingold, K. U. (1969b) *Canad. J. Chem.*, **47**, 3809.
Howard, J. A. and Ingold, K. U. (1970) *Canad. J. Chem.*, **48**, 873.
Howard, J. A., Ingold, K. U. and Symonds, M. (1968) *Canad. J. Chem.*, **46**, 1017.
Howard, J. A. and Korcek, S. (1970) *Canad. J. Chem.*, **48**, 2165.
Howard, J. A. and Robb, J. C. (1963) *Trans. Faraday Soc.*, **59**, 1590.
Howard, J. A., Schwalm, W. J. and Ingold, K. U. (1968) Adv. Chem. Series, No. 75 (Amer. Chem. Soc.), p. 6.
Hubele, A., Suhr, H. and Heilmann, U. (1962) *Chem. Ber.*, **95**, 639.
Ingles, T. A. and Melville, H. W. (1953) *Proc. Roy. Soc.*, **218A**, 175.
Ingold, K. U. (1961) *Chem. Rev.*, **61**, 563.
Ingold, K. U. (1963) *Canad. J. Chem.*, **41**, 2807, 2816.
Ingold, K. U. (1968) Adv. Chem. Series, No. 75 (Amer. Chem. Soc.), p. 296.
Ingold, K. U. (1969a) *Accounts of Chem. Res.*, **2**, 1.
Ingold, K. U. (1969b) *J. Chem. Soc.* (*D*), 911.
Janzen, E. G., Johnston, F. J. and Ayers, C. L. (1967) *J. Amer. Chem. Soc.*, **89**, 1176.
Jarvie, J. M. and Laufer, A. H. (1964) *J. Phys. Chem.*, **68**, 2557.
Jefcoate, C. R. E. and Norman, R. O. C. (1968) *J. Chem. Soc.* (*B*), 48.
Johnston, K. M. and Williams, G. H. (1958) *Chem. and Ind.* (*London*), 328.
Johnston, K. M. and Williams, G. H. (1960) *J. Chem. Soc.*, 1446.
Joschek, H.-I. and Grossweiner, L. I. (1966) *J. Amer. Chem. Soc.*, **88**, 3261.
Joschek, H.-I. and Miller, S. I. (1966) *J. Amer. Chem. Soc.*, **88**, 3269, 3273.
Karpukhin, O. N., Shlyapintokh, V. Ya., Rusina, I. F. and Zolotova, N. V. (1963) *Zhur. Analit. Khim.*, **18**, 1021.
Karpukhin, O. N., Shlyapintokh, V. Ya., and Zolotova, N. V. (1963) *Izvest. Akad. Nauk, SSSR*, 1718.

Karpukhina, G. V. and Maizus, Z. K. (1969) *Ivzest. Akad. Nauk SSSR, Ser. Khim.*,1253.
Keene, J. P., Kemp, T. J. and Salmon, G. A. (1965) *Proc. Roy. Soc.*, **287A**, 494.
Kellogg, R. E. (1969) *J. Amer. Chem. Soc.*, **91**, 5433.
Kerr, J. A. (1966) *Chem. Rev.*, **66**, 465.
Khloplyankina, M. A., Buchachenko, A. L., Neiman, M. B. and Vasil'eva, A. G. (1965) *Kinetika i Kataliz*, **6**, 394.
Kreilick, R. W. and Weissman, S. I. (1962) *J. Amer. Chem. Soc.*, **84**, 306.
Kreilick, R. W. and Weissman, S. I. (1966) *J. Amer. Chem. Soc.*, **88**, 2645.
Land, E. J. and Ebert, M. (1967) *Trans. Faraday Soc.*, **63**, 1181.
Land, E. J. and Porter, G. (1963) *Trans. Faraday Soc.*, **59**, 2016.
Lebedev, Ya. S., Tsepalov, V. F. and Shlyapintokh, V. Ya. (1961) *Doklady Akad. Nauk SSSR*, **139**, 1409.
MacCarthey, R. L. and MacLachlan, A. (1961a) *J. Chem. Phys.*, **35**, 1625.
MacCarthey, R. L. and MacLachlan, A. (1961b) *Trans. Faraday Soc.*, **57**, 1107.
MacLachlan, A. (1965) *J. Amer. Chem. Soc.*, **87**, 960.
MacLachlan, A. (1967) *J. Phys. Chem.*, **71**, 4132.
Maguire, W. J. and Pink, R. C. (1967) *Trans. Faraday Soc.*, **63**, 1097.
Mahoney, L. R. (1964) *J. Amer. Chem. Soc.*, **86**, 444.
Mahoney, L. R. (1967) *J. Amer. Chem. Soc.*, **89**, 1895.
Mahoney, L. R. (1969) *Angew. Chem. Internat. Ed.*, **8**, 547.
Mahoney, L. R. and DaRooge, M. A. (1967) *J. Amer. Chem. Soc.*, **89**, 5619.
Mahoney, L. R. and DaRooge, M. A. (1969) *Amer. Chem. Soc., Division of Petroleum Chem. Preprints*, **14(2)**, A7.
Mahoney, L. R. and DaRooge, M. A. (1970a) *J. Amer. Chem. Soc.*, **92**, 890.
Mahoney, L. R. and DaRooge, M. A. (1970b) *J. Amer. Chem. Soc.*, **92**, 4063.
Mamedova, Yu. G., Buchachenko, A. L. and Neiman, M. B. (1965) *Izvest. Akad. Nauk SSSR, Ser. Khim.*, 911.
Mayo, F. R. (1958) *J. Amer. Chem. Soc.*, **80**, 2465.
Mayo, F. R., Syz, M. G., Mill, T. and Castleman, J. K. (1968) Adv. Chem. Series, No. 75 (Amer. Chem. Soc.), p. 38.
McGowan, J. C. and Powell, T. (1960) *J. Chem. Soc.*, 238.
Melville, H. W. and Richards, S. (1954) *J. Chem. Soc.*, 944.
Neiman, M. B., Mamedova, Yu. G., Blenke, P. and Buchachenko, A. L. (1962) *Doklady Akad. Nauk SSSR*, **144**, 392.
Niki, E. and Kamiya, Y. (1967) *Kogyo Kaguku Zasshi*, **70**, 42.
Niki, E., Kamiya, Y. and Ohta, N. (1968) *Kogyo Kaguku Zasshi*, **71**, 1187.
Niki, E., Kamiya, Y. and Ohta, N. (1969) *Bull. Chem. Soc. Jap.*, **42**, 512.
Norman, R. O. C. and Pritchett, R. J. (1967) *J. Chem. Soc. (B)*, 926.
Padwa, A. (1964) *Tetrahedron Letters*, 3465.
Pagsberg, P., Christensen, H., Rabani, J., Nilsson, G., Fenger, J. and Nielsen, S. O. (1969) *J. Phys. Chem.*, **73**, 1029.
Patmore, E. L. and Gritter, R. J. (1962) *J. Org. Chem.*, **27**, 4196.
Pearson, D. E. and Moss, M. Y. (1967) *Tetrahedron Letters*, 3791.
Piette, L. H., Bulow, G. and Loeffler, K. (1964) *Amer. Chem. Soc. Div. of Petroleum Chem. Preprints*, **9**, C9.
Piloty, O. and Schwerin, B. G. (1901) *Ber.* **34**, 1870, 2354.

Pokhodenko, V. D. and Bidzilya, V. A. (1965) *Teor. i Eksperim. Khim., Akad. Nauk Ukr. SSSR*, **1**, 801 (Chem. Abs. (1966) **64**, 9557).
Pokhodenko, V. D. and Bidzilya, V. A. (1966) *Teor. Eksp. Khim.*, **2**, 691.
Pokhodenko, V. D. and Ganyuk, L. N. (1963) *Dopovidi Akad. Nauk Ukr. SSR*, **73**, (*Chem. Abs.* (1963) **59**, 6232).
Pokhodenko, V. D., Ganyuk, L. N. and Brodskii, A. I. (1962) *Doklady Akad. Nauk SSSR*, **145**, 815.
Pokhodenko, V. D., Ganyuk, L. N., Yakovleva, E. A., Shatenshtein, A. I. and Brodskii, A. I. (1963) *Doklady Akad. Nauk SSSR*, **148**, 1314.
Pokhodenko, V. D., Khizhnyi, V. A. and Bidzilya, V. A. (1968) *Uspekhi Khim.*, **37**, 998.
Prokof'ev, A. I., Solodovnikov, S. P., Bogdanov, G. N., Nikiforov, G. A. and Eroshov, V. V. (1967) *Teor. Eksp. Khim.*, **3**, 416.
Rabani, J., Mulac, W. A. and Matheson, M. S. (1965) *J. Phys. Chem.*, **69**, 53.
Rabani, J. and Matheson, M. S. (1966) *J. Phys. Chem.*, **70**, 761.
Rabani, J. and Nielsen, S. O. (1959) *J. Phys. Chem.*, **73**, 3736.
Robb, J. C. and Shahin, M. (1958) *J. Inst. Petrol.*, **44**, 283.
Robb, J. C. and Shahin, M. (1959) *Trans. Faraday Soc.*, **55**, 1753.
Rodgers, A. S., Golden, D. M. and Benson, S. W. (1967) *J. Amer. Chem. Soc.*, **89**, 4578.
Russell, G. A. (1955) *J. Amer. Chem. Soc.*, **77**, 4583.
Russell, G. A. (1956) *J. Amer. Chem. Soc.*, **78**, 1047.
Russell, G. A. (1957) *J. Amer. Chem. Soc.*, **79**, 3871.
Russell, G. A. and Williamson, Jr., R. C. (1964a) *J. Amer. Chem. Soc.*, **86**, 2357.
Russell, G. A. and Williamson, Jr., R. C. (1964b) *J. Amer. Chem. Soc.*, **86**, 2364.
Sajus, L. (1968) Adv. Chem. Series, No. 75 (Amer. Chem. Soc.), p. 59.
Sakurai, H. and Hosomi, A. (1967) *J. Amer. Chem. Soc.*, **89**, 458.
Sandros, K. (1964) *Acta Chem. Scand.*, **18**, 2355.
Sandros, K. and Bäckström, H. L. J. (1962) *Acta Chem. Scand.*, **16**, 956.
Schmidt, K. (1961) *Z. Naturforsch.*, **16B**, 206.
Schuler, R. H. (1964) *J. Phys. Chem.*, **68**, 3873.
Schwetlick, K., Karl, R. and Jentzsch, J. (1963) *J. Prakt. Chem.*, **22**, 113.
Sehested, K., Rasmussen, O. L. and Fricke, H. (1968) *J. Phys. Chem.*, **72**, 626.
Shcheredin, V. P. and Denisov, E. T. (1967) *Izvest. Akad. Nauk SSSR, Khim.*, 1428.
Shlyapintokh, V. Ya., Karpukhin, O. N., Postnikov, L. M., Tsepalov, V. F., Vichutinskii, A. A. and Zakharov, I. V. (1968) *Chemiluminescence Techniques in Chemical Reactions*, Consultants Bureau, New York, Chapter 8.
Simonaites, R., Cowell, C. W. and Pitts, J. N. (1967) *Tetrahedron Letters*, 3751.
Smaller, B., Remko, J. R. and Avery, E. C. (1968) *J. Chem. Phys.*, **48**, 5174.
Starnes, W. H. Jr. and Neureiter, N. P. (1967) *J. Org. Chem.*, **32**, 333.
Stebbins, R. and Sicilio, F. (1970) *Tetrahedron*, **26**, 291.
Stockhausen, K., Fojtik, A. and Henglein, A. (1970) *Ber. Bunsenges. Phys. Chem.*, **74**, 34.
Strigun, L. M., Vartanyan, L. S. and Émanuél, N. M. (1968) *Uspekhi Khim.* **37**, 969.
Sukhanova, O. P. and Buchachenko, A. L. (1965) *Zhur. fiz. Khim.*, **39**, 2413.

Symons, M. C. R. (1969) *J. Amer. Chem. Soc.*, **91**, 5924.
Thomas, J. R. (1963) *J. Amer. Chem. Soc.*, **85**, 591.
Thomas, J. R. (1964) *J. Amer. Chem. Soc.*, **86**, 4807.
Thomas, J. R. (1965) *J. Amer. Chem. Soc.*, **87**, 3935.
Thomas, J. R. (1967) *J. Amer. Chem. Soc.*, **89**, 4872.
Thomas, J. R. and Ingold, K. U. (1968) Adv. Chem. Series, No. 75 (Amer. Chem. Soc.), p. 258.
Thomas, J. R. and Tolman, C. A. (1962) *J. Amer. Chem. Soc.*, **84**, 2079.
Thomas, J. K. (1967) *J. Phys. Chem.*, **71**, 1925.
Tochina, E. M., Postnikov, L. M. and Shlyapintokh, V. Ya. (1968) *Izvest. Akad. Nauk SSSR, Ser. Khim.*, 71.
Traylor, T. G. and Russell, C. A. (1965) *J. Amer. Chem. Soc.*, **87**, 3698.
Tsepalov, V. F. and Shlyapintokh, V. Ya. (1959) *Doklady Akad. Nauk SSSR*, **124**, 883.
Tsepalov, V. F. and Shlyapintokh, V. Ya. (1962) *Kinetika i Kataliz*, **3**, 870.
Turro, N. J. and Engel, R. (1969) *J. Amer. Chem. Soc.*, **91**, 7113.
Vasserman, A. M., Buchachenko, A. L., Nikiforov, G. A., Ershov, V. V. and Neiman, M. B. (1967) *Zhur. fiz. Khim.*, **41**, 705.
Vichutinskii, A. A. (1964) *Doklady Akad. Nauk SSSR*, **157**, 150.
Vichutinskii, A. A., Guk, A. F., Tsepalov, V. F., and Shlyapintokh, V. Ya. (1966) *Izvest. Akad. Nauk SSSR, Ser. Khim.*, 1672.
Wagner, P. J. (1969) *Mol. Photochem.* **1**, 71.
Wagner, P. J. and Hammond, G. S. (1968) *Adv. in Photochem.*, Wiley, New York, **5**, 21.
Wagner, P. and Walling, C. (1965) *J. Amer. Chem. Soc.*, **87**, 5179.
Wallace, T. J. and Gritter, R. J. (1963) *Tetrahedron*, **19**, 657.
Walling, C. and Gibian, M. J. (1965) *J. Amer. Chem. Soc.*, **87**, 3361.
Walling, C. and Jacknow, B. B. (1960) *J. Amer. Chem. Soc.*, **82**, 6108.
Walling, C. and Kurkov, V. (1966) *J. Amer. Chem. Soc.*, **88**, 4727.
Walling, C. and Kurkov, V. (1967) *J. Amer. Chem. Soc.*, **89**, 4895.
Walling, C. and McGuinness, J. A. (1969) *J. Amer. Chem. Soc.*, **91**, 2053.
Walling, C. and Mintz, M. J. (1967) *J. Amer. Chem. Soc.*, **89**, 1515.
Walling, C. and Padwa, A. (1963) *J. Amer. Chem. Soc.*, **85**, 1597.
Walling, C. and Papaioannou, C. G. (1968) *J. Phys. Chem.*, **72**, 2260.
Walling, C. and Thaler, W. (1961) *J. Amer. Chem. Soc.*, **83**, 3877.
Walling, C. and Wagner, P. J. (1963) *J. Amer. Chem. Soc.*, **85**, 2333.
Walling, C. and Wagner, P. J. (1964) *J. Amer. Chem. Soc.*, **86**, 3368.
Wander, R. Gall, B. L. and Dorfman, L. M. (1970) *J. Phys. Chem.*, **74**, 1819.
Wander, R., Neta, P. and Dorfman, L. M. (1968) *J. Phys. Chem.*, **72**, 2946.
Weeks, J. L. and Rabani, J. (1966) *J. Phys. Chem.*, **70**, 2100.
Weiner, S. and Hammond, G. S. (1969) *J. Amer. Chem. Soc.*, **91**, 2182.
Williams, D. J. and Kreilick, R. (1967) *J. Amer. Chem. Soc.*, **89**, 3408.
Williams, A. L. Oberright, E. A. and Brooks, J. W. (1956) *J. Amer. Chem. Soc.*, **78**, **1190**.
Zaikov, G. E. (1968a) *Kinetika i Kataliz*, **9**, 511.
Zaikov, G. E. (1968b) *Kinetika i Kataliz*, **9**, 1166.
Zaikov, G. E., Andronov, L. M., Maizus, Z. K. and Émanuél, N. M. (1967) *Doklady Akad. Nauk SSSR*, **174**, 127.

Zaikov, G. E., Howard, J. A. and Ingold, K. U. (1969) *Canad. J. Chem.*, **47**, 3017.

Zaikov, G. E. and Maizus, Z. K. (1968a) Adv. Chem. Series, No. 75 (Amer. Chem. Soc.), p. 150.

Zaikov, G. E. and Maizus, Z. K. (1968b) *Izvest Akad. Nauk SSSR. Ser. Khim.*, 47.

Zaikov, G. E. and Maizus, Z. K. (1969a) *Izvest Akad. Nauk SSSR. Ser. Khim.*, 311.

Zaikov, G. E. and Maizus, Z. K. (1969b) *Izvest Akad. Nauk. SSSR. Ser. Khim.*, 598.

Zaikov, G. E. and Maizus, Z. K. (1969c) *Zhur. fiz. Khim.*, **43**, 57.

Zaikov, G. E., Vichutinskii, A. A. and Maizus, Z. K. (1967) *Kinetika i Kataliz*, **8**, 675.

Zaikov, G. E., Vichutinskii, A. A., Maizus, Z. K. and Émanuél, N. M. (1968) *Izvest Akad. Nauk SSSR.*, 1743.

Zakharova, N. A., Kruglyakova, K. E., Bogdanov, N. G., Kheifits, L. A and Émanuél, N. M. (1967) *Zhur. obschchei Khim.*, **37**, 801.

Zwolenik, J. T. (1967) *J. Phys. Chem.*, **71**, 2464.

3

ALLYLIC HALOGENATION

A. Nechvatal

Department of Chemistry, University of Dundee, Dundee

A. Introduction

The overall reaction of allylic halogenation involves the replacement by halogen of hydrogen adjacent to a carbon–carbon double bond.

$$>C{=}C(-)-CH< \longrightarrow >C{=}C(-)-CX<$$

The reaction has been extensively investigated for its synthetic usefulness (Sosnovsky, 1964), producing reactive allylic halides, and also because the characteristic feature of being able to substitute hydrogen by halogen without simultaneous addition to the carbon–carbon double bond has caused much attention to be focused on the mechanism of the reaction.

Wohl (1919) first demonstrated that allylic bromides could be prepared without simultaneous addition to the double bond from the corresponding alkene by the use of *N*-bromoacetamide.

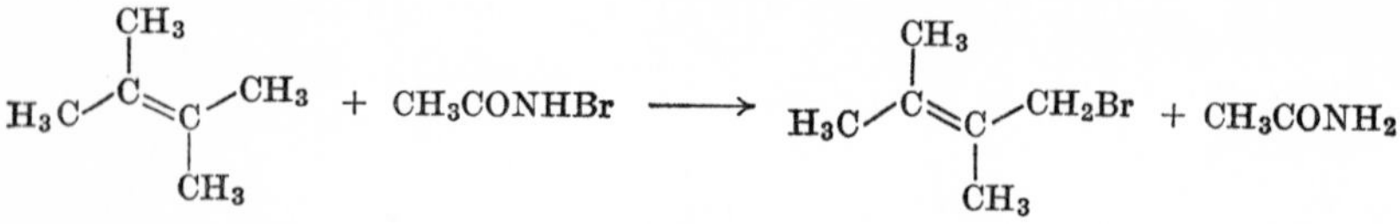

$$H_3C{-}C(CH_3){=}C(CH_3){-}CH_3 + CH_3CONHBr \longrightarrow H_3C{-}C(CH_3){=}C(CH_3){-}CH_2Br + CH_3CONH_2$$

Later it was discovered (Wendt, 1941) that even molecular bromine itself in the presence of strong light would brominate in the allylic position, provided the double bond was deactivated by, for example, a carboxyl group, as in the case of β-cyclogeranic acid (I).

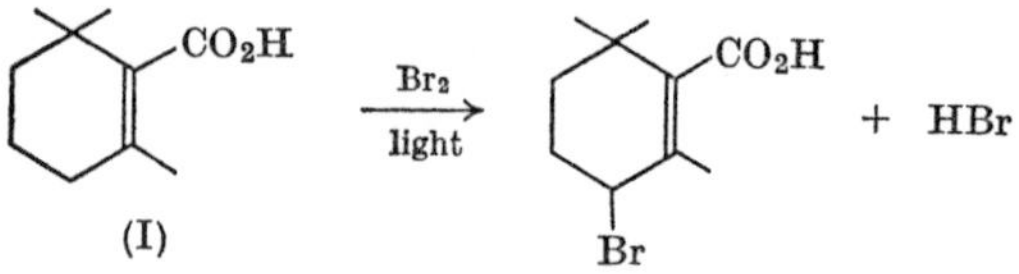

$$\underset{\text{(I)}}{C_9H_{15}{-}CO_2H} \xrightarrow[\text{light}]{Br_2} Br{-}C_9H_{14}{-}CO_2H + HBr$$

Ziegler *et al.* (1942) investigated a number of *N*-bromo-amides and found that *N*-bromosuccinimide (II) was a particularly effective allylic brominating reagent, and this bromo-compound has since been widely used in conjunction with light, or with radical initiators such as benzoyl peroxide (III) or azobisisobutyronitrile (IV) to form allylic bromides.

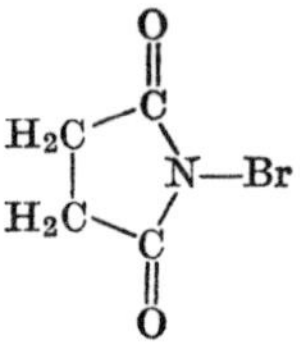

$$H_2C{-}C({=}O){-}N(Br){-}C({=}O){-}CH_2 \text{ (ring)}$$

(II)

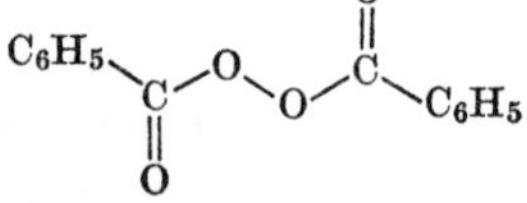

$$C_6H_5{-}C({=}O){-}O{-}O{-}C({=}O){-}C_6H_5$$

(III)

$$(CH_3)_2C(CN){-}N{=}N{-}C(CN)(CH_3)_2$$

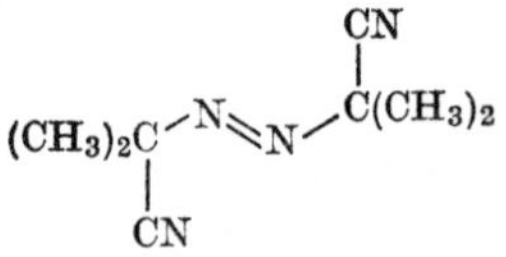

(IV)

The use of *N*-bromosuccinimide in synthesis has been reviewed (Djerassi, 1948; Horner and Winkelmann, 1959) and will not be further discussed here. Instead, attention will be concentrated on various methods of bringing about allylic halogenation and the underlying mechanistic features of these reactions.

Normally, the use of an equimolar ratio of halogenating agent to unsaturated compound results in the insertion of only one halogen atom. However, excess of *N*-bromosuccinimide with a compound such as methyl 3-methylbut-2-enoate (V) leads to the formation of methyl 4-bromo-3-bromomethylenebut-2-enoate (VI) rather than methyl 4,4-dibromo-3-methylbut-2-enoate (VII) (Gedye, unpublished observation).

$$\underset{\textbf{(V)}}{(CH_3)_2C{=}CHCO_2CH_3} \xrightarrow[\text{(excess)}]{N\text{-bromo-succinimide}} \underset{\textbf{(VI)}}{(BrCH_2)_2C{=}CHCO_2CH_3}$$

$$\underset{\text{(VII)}}{CH_3{-}\underset{\substack{|\\CHBr_2}}{C}{=}CHCO_2CH_3}$$

Allylic chlorination on a laboratory scale is less easy to effect with compounds such as *N*-chlorosuccinimide and can best be brought about with t-butyl hypochlorite in the presence of light or radical initiators (Walling and Thaler, 1961). On an industrial scale, however, the gas-phase preparation of allyl chloride from propene was first patented by Groll, Hearne, Burgin and LaFrance (1938) and is used in the first step in the manufacture of glycerol from propene (Charles, 1958).

$$CH_3CH{=}CH_2 \xrightarrow[500^\circ]{Cl_2} ClCH_2CH{=}CH_2 \xrightarrow{Cl_2/H_2O} ClCH_2\overset{\substack{OH\\|}}{C}HCH_2Cl + ClCH_2\overset{\substack{Cl\\|}}{C}HCH_2OH$$

$$\downarrow H_2O$$

$$HOCH_2\overset{\substack{OH\\|}}{C}HCH_2OH$$

B. General Features of the Reaction

From what has been stated in the Introduction about the use of light, high temperature, and peroxides, it will be appreciated that allylic halogenations occur by way of radical intermediates, and with this fact in mind, allylic halogenation can be regarded as a special case of hydrocarbon halogenation to which the usual chain reaction scheme applies (Poutsma, 1969).

Initiation $$X_2 \xrightarrow[\text{or heat}]{\text{light}} 2\,X\cdot \quad (1)$$

Propagation
$$RH + X\cdot \longrightarrow HX + R\cdot \quad (2)$$
$$R\cdot + X_2 \longrightarrow RX + X\cdot \quad (3)$$

Termination
$$2\,R\cdot \longrightarrow R_2 \quad (4)$$
$$R\cdot + X\cdot \longrightarrow RX \quad (5)$$
$$2\,X\cdot \xrightarrow{M} X_2 + M^* \quad (6)$$

The propagation steps in the reaction scheme for halogenation of an alkene such as propene are complicated by the fact that the halogen atom X· may either abstract hydrogen, as in reaction (2), or add to the double bond (reaction 7). The electron on the allyl radical is delocalised over three carbon atoms, i.e.

$$\dot{C}H_2{-}CH{=}CH_2 \leftrightarrow CH_2{=}CH{-}\dot{C}H_2$$

or

$$(CH_2{\cdots}CH{\cdots}CH_2)\cdot.$$

Chain propagation resulting in substitution
$$X\cdot + CH_3CH{=}CH_2 \xrightarrow{k_2} HX + \dot{C}H_2{-}CH{=}CH_2 \updownarrow CH_2{=}CH{-}\dot{C}H_2 \quad (2)$$
$$(CH_2{\cdots}CH{\cdots}CH_2)\cdot + X_2 \xrightarrow{k_3} XCH_2{-}CH{=}CH_2 + X\cdot \quad (3)$$

Chain propagation resulting in addition
$$X\cdot + CH_3CH{=}CH_2 \underset{k_{-7}}{\overset{k_7}{\rightleftharpoons}} CH_3\dot{C}HCH_2X \quad (7)$$
$$CH_3\dot{C}HCH_2X + X_2 \xrightarrow{k_8} CH_3CH(X)CH_2X \quad (8)$$

Rust and Vaughan (1940) showed that both addition and substitution occurred simultaneously during the chlorination of propene and that, as the temperature was raised, the proportion of substitution increased. Earlier, it had been found by Stewart, Dod and Stenmark (1937) that in the halogenation of a series of alkanes such as the pentenes, hex-1- and -2-ene, and hept-1- and -2-ene, a decrease in the concentration of chlorine resulted in an increase in the proportion of substitution as against addition product.

By applying a stationary state treatment to the steps involved in the chlorination of propene, Adam, Gosselain and Goldfinger (1953) derived the following expression for the ratio of the rate of substitution (r_S) to that of addition (r_A).

$$\frac{r_S}{r_A} = \frac{k_2}{k_{-7}}\left(1 + \frac{k_{-7}}{k_8[X_2]}\right)$$

The expression shows that as the concentration of halogen $[X_2]$ decreases, r_S/r_A increases, as observed by Stewart, Dod and Stenmark (1937). If it is assumed that the activation energies E_7, E_{-7} and E_2 are all small and in the order $E_{-7} > E_2 > E_7$, and that the pre-exponential terms are in the order $A_{-7} > A_7 > A_2$, it can be seen that, at low temperatures, k_7 is larger than k_2 or k_{-7}, but increasing temperature increases k_2 and k_{-7} more rapidly than k_7, thus accounting for the increasing amount of substitution product, as observed by Rust and Vaughan (1940). In qualitative terms, the effect of temperature can be related to the reversibility of the addition step (7) and the non-reversibility of the abstraction step (2). Reversal of the abstraction step is unfavourable due to the strength of the HX bond. Thus we have some theoretical justification for the experimental findings that allylic halogenation can be effected by the molecular halogens, provided that the reaction is carried out at a sufficiently high dilution and temperature.

In practice the method is not much used on a laboratory scale. In the case of fluorine, which reacts explosively with saturated hydrocarbons at ordinary temperatures, control of the reaction to produce allylic fluorides would be difficult, so that it is not surprising that there has been no report of a successful allylic fluorination with molecular fluorine. Allylic chlorination using molecular chlorine can be achieved with difficulty in the laboratory, but under ordinary conditions it is difficult to suppress the competing addition reaction. However, even under conditions not designed to favour allylic halogenation, cyclohexene reacts with chlorine at 25° to form a mixture of addition and substitution products containing 28% of the allylic 3-chlorocyclohexene (Poutsma, 1965a). It would be expected that compounds containing deactivated double bonds such as conjugated acids or esters, for example, methyl but-2-enoate, should be more amenable to direct allylic radical chlorination. Radical bromination by molecular bromine, a much less energetic and therefore more selective process than chlorination, is more frequently observed in the laboratory. The example of the bromination of β-cyclogeranic acid (Wendt, 1941) has already been mentioned (page 176) and other examples are shown below.

$$CH_3CH{=}CHCO_2CH_3 \xrightarrow[\text{u.v. light}]{Br_2} BrCH_2CH{=}CHCO_2CH_3 \quad \text{(Schaltegger, 1957)}$$

$$(CH_3)_2C{=}CHCO_2C_2H_5 \xrightarrow[CCl_4]{Br_2} BrCH_2\underset{\displaystyle CH_3}{\underset{|}{C}}{=}CHCO_2C_2H_5 \quad \text{(Sisido, } et\ al.\text{, 1960)}$$

It was shown by Sixma and Riem (1958) that the presence of a group such as carboxyl to deactivate the double bond was unnecessary when they prepared 3-bromocyclohexene in 80% yield by passing gaseous bromine diluted with nitrogen through cyclohexene in refluxing carbon tetrachloride.

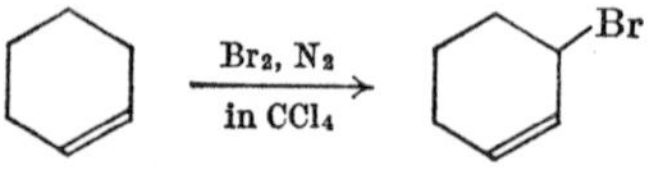

$$\text{cyclohexene} \xrightarrow[\text{in } CCl_4]{Br_2,\ N_2} \text{3-bromocyclohexene (Br)}$$

The energetics of allylic radical iodination of propene with molecular iodine are unfavourable owing to the low bond energy for hydrogen iodide [$D(\text{H—I}) = 297$ kJ. mole^{-1} (71 kcal.mole^{-1})]. Using the value of 335 kJ. mole^{-1} (85 kcal.mole^{-1}) for the bond dissociation energy of the allylic C—H bond (Benson, 1965; Kerr, 1966), ΔH for the overall chain process is approximately zero and the reaction is understandably not self-sustaining. Organic compounds such as *N*-iodosuccinimide do not effect allylic iodination, but will iodinate enol acetates (Djerassi and Lenk, 1953).

In practice, allylic halogenation is not usually effected directly, owing to the difficulty of suppressing the addition reaction. The majority of allylic halogenations are achieved using organic halogenating agents, and in particular the topic of allylic halogenation is closely connected with the development of compounds such as *N*-bromosuccinimide which are easy to handle and act in a selective manner on a variety of unsaturated compounds. The topic of allylic bromination has been discussed by Thaler (1969) in a review on radical bromination reactions.

C. Bromination

1. *Mechanism of Reaction with N-Bromosuccinimide*

It was at one time thought that the selectivity of *N*-bromosuccinimide was due to the abstraction step (2) (page 178) being brought about by the succinimidyl radical (Bloomfield, 1944).

However, Adam, Gosselain and Goldfinger (1953) argued that in the reaction of *N*-chlorosuccinimide with toluene, traces of impurities such as water led to the formation of hydrogen chloride, which reacts with *N*-chlorosuccinimide to form traces of molecular chlorine. The halogen was then thought to undergo thermal dissociation and thus effect the chlorination.

They suggested that impurities reacting with *N*-bromosuccinimide would release traces of bromine in a similar manner and that the specificity of the reagent towards halogenation at the allylic position was due merely to the provision of a sufficiently low concentration of bromine by *N*-bromosuccinimide. After traces of bromine have been formed by reaction of impurities with *N*-bromosuccinimide, the initiation step is the same as that of bromination by molecular bromine.

Initiation
$$Br_2 \xrightarrow[\text{heat or radical initiators}]{\text{light}} 2\,Br\cdot \qquad (1)$$

Propagation
$$Br\cdot + CH_3CH{=}CH_2 \longrightarrow HBr + (CH{\cdots}CH{\cdots}CH_2)\cdot \qquad (2)$$

$$HBr + \text{(N-bromosuccinimide) } N{-}Br \longrightarrow Br_2 + \text{(succinimide) } NH \qquad (9)$$

$$(CH_2{\cdots}CH{\cdots}CH_2)\cdot + Br_2 \longrightarrow BrCH_2{-}CH{=}CH_2 + Br\cdot \qquad (3)$$

The alternative fate of the bromine atom, that of addition to the double bond (7) does not lead to simultaneous addition since the

$$Br\cdot + CH_3CH{=}CH_2 \rightleftharpoons CH_3\dot{C}HCH_2Br \quad (7)$$

probability that the adduct radical, $CH_3\dot{C}HCH_2Br$, will react with molecular bromine is low, because of the low concentration of bromine; instead it will dissociate again to yield the unsaturated compound.

Experimental support for this mechanism came from the work already mentioned (page 180) of Wendt (1941) and Sixma and Riem (1958) who showed that low concentrations of molecular bromine could indeed bring about allylic bromination without much addition across the double bond. Evidence for the reversible nature of the addition step (7) was obtained by McGrath and Tedder (1961) who showed that during the bromination of *cis*-hex-3-ene by *N*-bromosuccinimide, a rapid isomerisation to the thermodynamically more stable *trans*-hex-3-ene occurred.

$$\underset{cis\text{-}}{(C_2H_5)(H)C{=}C(C_2H_5)(H)} \underset{-Br\cdot}{\overset{Br\cdot}{\rightleftharpoons}} C_2H_5\dot{C}H{-}CH(Br)C_2H_5 \underset{Br}{\overset{-Br\cdot}{\rightleftharpoons}} \underset{trans\text{-}}{(C_2H_5)(H)C{=}C(H)(C_2H_5)}$$

Investigation of the relative rates of isomerisation and bromination (Gedye and Nechvatal, 1964) of a pair of *cis-trans*-isomers, *trans*- and *cis*-methyl but-2-enoate [(VIII) and (IX) respectively],

$$\underset{trans\text{-}(VIII)}{(CH_3)(H)C{=}C(H)(CO_2CH_3)} \underset{-Br}{\overset{Br\cdot}{\rightleftharpoons}}{}_{fast}\ CH_3CH(Br){-}\dot{C}H{-}CO_2CH_3 \underset{Br\cdot}{\overset{-Br\cdot}{\rightleftharpoons}}{}_{fast}\ \underset{cis\text{-}(IX)}{(CH_3)(H)C{=}C(CO_2CH_3)(H)}$$

$$trans\text{-}(VIII) \xrightarrow{Br\cdot} \left((CH_2\cdots)(H)C{\cdots}C(H)(CO_2CH_3)\right)\cdot \xrightarrow{Br_2} \underset{\text{(Main product)}}{(BrCH_2)(H)C{=}C(H)(CO_2CH_3)}$$

$$cis\text{-}(IX) \xrightarrow{Br\cdot} \left((CH_2\cdots)(H)C{\cdots}C(CO_2CH_3)(H)\right)\cdot \xrightarrow{Br_2} \underset{\text{(Trace)}}{(BrCH_2)(H)C{=}C(CO_2CH_3)(H)}$$

showed that, while complete bromination required about 3 hours, the isomerisation of either of the two isomers to the equilibrium ratio of *cis*/*trans* = 1/10 was complete in about three minutes.

The rate of isomerisation and bromination were sensitive to the purity of the reagents. *N*-bromosuccinimide did not cause any isomerisation in the absence of an initiator such as azobisisobutyronitrile.

Dauben and McCoy (1959) showed that the rate of bromination of cyclohexene with *N*-bromosuccinimide could be increased by the presence of bromine, or of compounds such as hydrogen bromide, water, ethanol, and thiophenol, which react with *N*-bromosuccinimide to form bromine.

Incremona and Martin (1970) found that there was no difference in the relative reactivities of *N*-bromosuccinimide, *N*-bromotetramethylsuccinimide, and molecular bromine for the allylic halogenation of a series of diarylpropenes. This provides strong evidence that all these reagents react by way of a common chain carrier; the bromine atom. *N*-Bromotetrafluorosuccinimide was also studied but this compound showed greater relative reactivity than the other halogenating reagents for electron-rich olefins, owing to the tendency of this compound to react by an ionic pathway with such olefins.

A similarity in the relative reactivity of *N*-bromosuccinimide to that of bromine towards benzylic bromination of a series of substituted toluenes (toluene, ethylbenzene, cumene and diphenylmethane) was also taken by Walling, Rieger and Tanner (1963) as evidence for the existence of a bromine atom chain in the analogous benzylic bromination reaction.

$$C_6H_5CHRR' \xrightarrow[\text{or bromine}]{N\text{-Bromosuccinimide}} C_6H_5CBrRR'$$

2. *Allylic Rearrangement*

Allylic rearrangements must be expected during bromination with *N*-bromosuccinimide and in fact during any radical halogenation reaction owing to the delocalisation of the odd electron in the allylic radical intermediate. The reaction of *N*-bromosuccinimide with oct-1-ene (X) (Kharasch, Malec and Yang, 1957) serves to illustrate this point.

$CH_3(CH_2)_4CH_2CH{=}CH_2$ (X) $\xrightarrow[\text{(Br·)}]{N\text{-Bromosuccinimide}}$ $CH_3(CH_2)_4\dot{C}HCH{=}CH_2$ $\longleftrightarrow$ $CH_3(CH_2)_4CH{=}CH{-}CH$

$\updownarrow$

$\{CH_3(CH_2)_4CH{\cdots}CH{\cdots}CH_2\}\cdot$

Br_2 $\swarrow$ $\searrow$ Br_2

$CH_3(CH_2)_4CH(Br){-}CH{=}CH_2$ (12%) + $CH_3(CH_2)_4CH{=}CHCH_2B$ (53%)

Hexa-1,5-diene (XI) reacts with *N*-bromosuccinimide to give a monobromo-derivative containing about 90% of 1-bromohexa-2,5-diene and 10% of 3-bromohexa-1,5-diene (Bateman, Cunneen, Fabian and Koch, 1950).

$CH_2{=}CHCH_2CH_2CH{=}CH_2$ (XI) $\xrightarrow{N\text{-bromosuccinimide}}$ $BrCH_2CH{=}CH{-}CH_2CH{=}CH_2$ (90%)

$\searrow$ $CH_2{=}CHCH(Br)CH_2CH{=}CH_2$ (10%)

It is surprising that although *trans*- and *cis*-but-2-enonitriles [(XII) and (XIII) respectively] both give the same *cis-trans* mixture

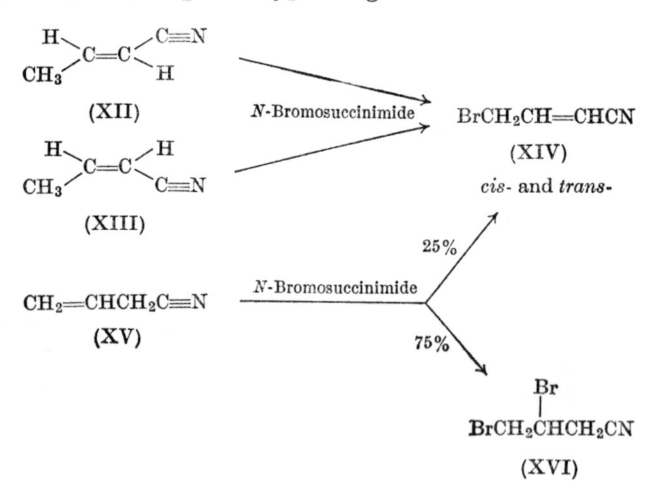

of 4-bromobut-2-enonitriles (XIV) in 70% yield, the isomeric unconjugated but-3-enonitrile (XV) gave only 25% of the same *cis-trans* mixture, the main product being the adduct 2,3-dibromobutanonitrile (XVI) (Couvreur and Bruylants, 1952).

Corey (1953) found that methyl but-3-enoate (XVII) was not brominated in the allylic position when treated with *N*-bromosuccinimide; instead the addition product (XVIII) was formed. These unexpected reactions were thought to be due to the polar nature of the substituents at the end of the chains in the substrates.

$$\underset{\text{(XVII)}}{CH_2{=}CHCH_2CO_2CH_3} \xrightarrow{N\text{-bromosuccinimide}} \underset{\text{(XVIII)}}{BrCH_2\overset{\overset{\textstyle Br}{|}}{C}HCH_2CO_2CH_3}$$

3. *Other Brominating Agents*

1,3-Dibromo-5,5-dimethylhydantoin (XIX) has been used to brominate cyclohexene in the allylic position (Orazi and Meseri, 1950) and some analogous compounds which are soluble in carbon tetrachloride such as 1,3-dibromo-5-methyl-5-ethylhydantoin have been found to be efficient allylic halogenating agents (Corral, Orazi and Bonafede, 1957). Walling and Rieger (1963) suggest that these

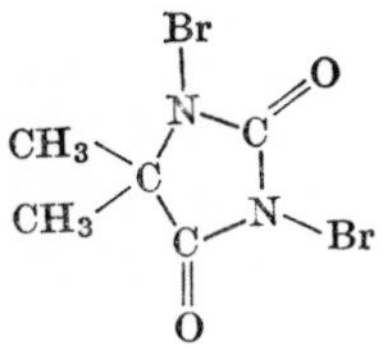

(XIX) 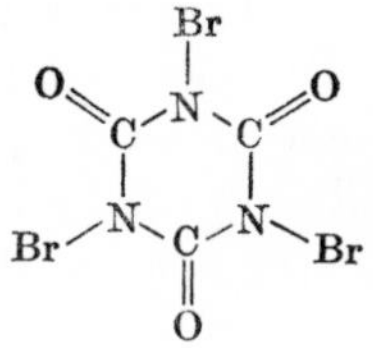 (XX)

reagents probably react in a similar manner to *N*-bromosuccinimide by way of a bromine atom chain.

Isocyanuryl tribromide (XX) has been prepared by Gottardi (1967) but so far its effectiveness as an allylic brominating agent has not been tried.

Allylic bromination has been effected with *N*-bromo-t-butylamine (XXI) (Boozer and Moncrief, 1962). Using isohexane rather than carbon tetrachloride as a solvent, 3-bromocyclohexene (59%) was obtained from cyclohexene. Azobisisobutyronitrile was used as an initiator but the nature of the chain-carrying radical was not established.

2,4,4,6-Tetrabromocyclohexa-2,5-dienone (XXII) reacts slowly with cyclohexene to form about 50 % of 3-bromocyclohexene (Messmer, Varady and Pinter, 1958). The reaction may go by way of ionic

$(CH_3)_3CNHBr$

(XXI)

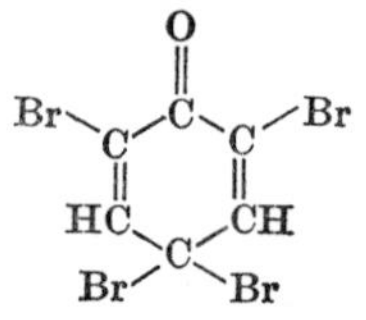

(XXII)

intermediates, since 1,2-dibromocyclohexane is also formed. The allylic bromide could be formed as the result of elimination of hydrogen bromide.

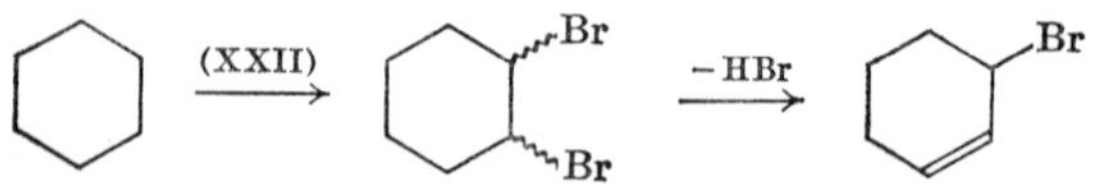

(XXII) → −HBr →

An example of a photochemically or thermally initiated allylic substitution reaction was described by Huyser and DeMott (1963) who found that cyclohexene, α-pinene, β-pinene, 2,2,4-trimethylpent-1-ene, hex-1-ene, oct-1-ene, 2-methylbut-1-ene and 2,2,4-trimethylpent-2-ene all gave allylic bromides when irradiated in the presence of 1,2-dibromotetrachloroethane.

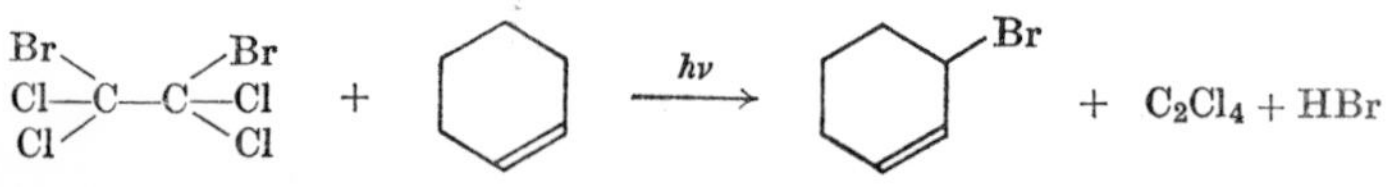

$$CBrCl_2{-}CBrCl_2 + C_6H_{10} \xrightarrow{h\nu} C_6H_9Br + C_2Cl_4 + HBr$$

Bromotrichloromethane, when irradiated, gives rise to bromine atoms and trichloromethyl radicals which usually add to carbon–carbon double bonds by way of a chain reaction (Walling and Huyser, 1963). The reaction is also initiated by peroxides which give trichloromethyl radicals by abstraction of bromine from bromotrichloromethane. Huyser (1961) found that, as well as adding to a double bond, trichloromethyl radicals simultaneously abstract

$$\cdot CCl_3 + >C{=}C< \longrightarrow >C{-}\dot{C}< \text{ (with } CCl_3 \text{ on first C)}$$

$$>C(CCl_3){-}\dot{C}< + BrCCl_3 \longrightarrow >C(CCl_3){-}C(Br)< + \cdot CCl_3$$

hydrogen from an allylic position, resulting in the production of allylic halides. In general, the ratio of the rate of allylic abstraction (k_s) to addition (k_a) is small (approximately 0·1) but with cyclohexene and with compounds containing a tertiary allylic carbon–hydrogen bond such as 4-methylpent-1-ene, the rates of abstraction and addition are equal.

$$BrCCl_3 \xrightarrow{h\nu} Br\cdot + \cdot CCl_3$$

$$\text{cyclohexene} + \cdot CCl_3 \xrightarrow{k_a} \text{2-(trichloromethyl)cyclohexyl radical}$$

$$\text{cyclohexene} + \cdot CCl_3 \xrightarrow{k_s} \text{cyclohex-2-enyl radical}$$

$$\text{2-(trichloromethyl)cyclohexyl radical} + BrCCl_3 \longrightarrow \text{1-bromo-2-(trichloromethyl)cyclohexane} + \cdot CCl_3$$

$$\text{cyclohex-2-enyl radical} + BrCCl_3 \longrightarrow \text{3-bromocyclohexene} + \cdot CCl_3$$

Tedder and Walton (1967) have studied the reaction of trichloromethyl radicals with propene in the gas phase and have found that the rates of addition and abstraction are approximately of the same order at temperatures ranging from 100 to 200°.

t-Butyl hypobromite has been shown to brominate cyclohexene and propene in the allylic position (Walling and Padwa, 1962) in a manner analogous to that of t-butyl hypochlorite (Section D.2) but the reagent is not readily available and has not been much used for allylic bromination.

D. Chlorination

1. *Chlorine*

Reaction of alkenes with chlorine in the liquid phase is rapid at room temperature, even in the dark (Poutsma, 1969). Both ionic and radical processes are responsible for this reaction, the more branched alkenes such as 2-methylpropene reacting by a predominantly ionic pathway. Initiation in the dark is suggested (Poutsma, 1965a) to be the result of the interaction of a molecule of alkene and a chlorine molecule, leading to a complex. This complex reacts with another alkene molecule to produce two chloroalkyl radicals which serve to initiate the chain reaction.

$$>C{=}C< \; + \; Cl_2 \longrightarrow >C{\cdots}C<\,(\cdots Cl{-}Cl) \xrightarrow{>C=C<} 2\; \cdot C{-}C{-}Cl$$

The radical pathway is favoured by illumination and by the absence of oxygen, and leads to substitution products as well as addition products. The allylic substitution products observed in the radical chlorination of the but-2-enes are formed without the characteristic rapid isomerisation of the starting alkene observed for bromination reactions, but with the allylic rearrangement which is expected for an intermediate methylallyl radical:

$$(CH_3CH{=\!=}CH{=\!=}CH_2)\cdot.$$

Poutsma (1965b) found that *cis*-but-2-ene gave a roughly similar ratio of rearranged and unrearranged substitution products as did *trans*-but-2-ene but in each case the geometry of the resulting 4-chlorobut-2-ene was unaltered during the chlorination, that is to say, *cis*-but-2-ene gave the *cis*-4-chlorobut-2-ene and *trans*-but-2-ene gave the *trans*-4-chlorobut-2-ene. In each case, however, the adduct 2,3-dichlorobutane (XXIII) was the main product.

It can be deduced from these observations that the abstraction step at this temperature is not reversible, i.e. k_{-7} is approximately equal

$$\underset{cis\text{-}}{CH_3CH{=}CHCH_3} \xrightarrow[\text{N}_2,\ \text{low temperature}]{Cl_2} \underset{\substack{10\% \\ cis\text{-}}}{ClCH_2CH{=}CHCH_3} + \underset{6\%}{CH_2{=}CHCHClCH_3} + \underset{\substack{(XXIII) \\ \text{(main product)}}}{CH_3CHClCHClCH_3}$$

$$\underset{trans\text{-}}{CH_3CH{=}CHCH_3} \xrightarrow[\text{N}_2,\ \text{low temperature}]{Cl_2} \underset{\substack{14\% \\ trans\text{-}}}{ClCH_2CH{=}CHCH_3} + \underset{4\%}{CH_2{=}CHCHClCH_3} + \underset{\text{(main product)}}{(XXIII)}$$

to zero, in contrast to the fast reversal of Equation (7) (p. 178) observed in allylic bromination.

$$Cl\cdot + CH_3CH{=}CHCH_3 \underset{k_{-7}}{\overset{k_7}{\rightleftharpoons}} HCl + (CH_2{\cdots}CH{\cdots}CHCH_3)\cdot$$

The allylic chloride formed as a major product of the chlorination of 2-methylpropene arises as the result of the elimination of a proton from the tertiary carbonium ion, as is shown by the chlorination of isotopically labelled 1-^{14}C-2-methylpropene (Reeve, Chambers and Prickett, 1952). This alkene reacts by way of an ionic pathway which is difficult to suppress even by illuminating the reactants under nitrogen.

$$(CH_3)_2C{=}\overset{*}{C}H_2 + Cl{-}Cl \longrightarrow (H{-}CH_2)(CH_3)\overset{+}{C}{-}\overset{*}{C}H_2Cl \xrightarrow{-H^+} CH_2{=}C(CH_3){-}\overset{*}{C}H_2Cl$$

Alkynes such as but-1-yne undergo slow radical chlorination at low temperatures in the liquid phase (Poutsma and Kartch, 1966). In the dark, there is very little reaction at low temperatures, in contrast to the behaviour of alkenes, confirming our ideas about the lower reactivity of alkynes as compared to alkenes. A mixture consisting mainly of addition products and a small proportion of substitution products is formed. The substitution products are 3-chloro-but-1-yne(XXIV) and 1-chlorobuta-1-2-diene(XXV) formed as a result of chlorine abstraction by the 3-but-1-ynyl radical. 4-Chlorobut-1-yne is also formed as the result of hydrogen abstraction from the terminal methyl group by chlorine atoms in the first stage of the reaction.

$$\underset{\text{but-1-yne}}{CH_3CH_2C{\equiv}CH} \xrightarrow[\text{propargylic abstraction}]{Cl\cdot} CH_3\dot{C}HC{\equiv}CH \longleftrightarrow CH_3CH{=}C{=}\dot{C}H$$

$$CH_3\dot{C}HC{\equiv}CH \xrightarrow{Cl_2} \underset{(4\%)}{\underset{(XXIV)}{CH_3CH(Cl)C{\equiv}CH}} \qquad CH_3CH{=}C{=}\dot{C}H \xrightarrow{Cl_2} \underset{(1\%)}{\underset{(XXV)}{CH_3C{=}C{=}CH_2Cl}}$$

Poutsma (1965b) showed that there is no activation in the allylic position towards hydrogen abstraction by chlorine atoms at 0°, and in fact the reactivity of an allylic carbon–hydrogen bond is about the same as that of the carbon–hydrogen bond in cyclohexane.

Thus the reactivities per allylic hydrogen atom of *trans*-but-2-ene, *cis*-but-2-ene, and but-1-ene are 0·69, 0·60 and 0·76 relative to cyclohexane as a standard. This means that stabilisation of the allyl radical arising from delocalisation of the unpaired electron over the three carbon atoms does not play a great part in determining the position of hydrogen abstraction by a chlorine atom in the liquid phase at this temperature. This is an indication that the activated complex for the abstraction step does not involve much extension or weakening of the carbon–hydrogen bond, and that the complex resembles (XXVI) more than (XXVII).

$$\underset{(XXVII)}{\left(>\dot{C}\text{------}H{-}Cl\right)^{\cdot}} \qquad \underset{(XXVI)}{\left(>C{-}H\text{------}Cl\right)^{\cdot}}$$

In an investigation of transannular ionic and radical addition to *cis*- and *trans*-cyclodecene [(XXVIII) and (XXIX) respectively],

Traynham and DeWitt (1970) found that chlorine reacts with *trans*-cyclodecene by a radical pathway to give dichlorides as a result of addition, and 3-chlorocyclodecene (XXX) (43 % of the products) as a result of abstraction at the allylic position. The thermodynamically more stable *cis*-isomer gave 22 % of 3-chlorocyclodecene. Iodobenzene-dichloride also gives 3-chlorocyclodecene by way of a radical abstraction step; only 6 % in the case of the *cis*- and 31 % in the case of the *trans*-cyclodecene. The formation of a higher yield of (XXX)

cis-
(XXVIII)

trans-
(XXIX)

$\xrightarrow[C_6H_5ICl_2]{Cl_2 \text{ or}}$ Adducts +

Cl

Cl

(*cis*- and *trans*-)
(XXX)

from the *trans*-cyclodecene is thought due to the shielding of one side of the double bond towards addition in the thermodynamically less stable *trans*-isomer.

The successful high temperature chlorination of propene to form allyl chloride is probably homolytic in nature, and is successful because of the reversibility of the addition step (7) (p. 178) at that temperature. The low temperature reactions discussed above do not constitute a suitable method of preparing allylic chlorides in general because of the high proportions of addition products formed. High temperature chlorination of the higher alkenes, although giving only small amounts of addition products, would be expected to give all the possible monochloro-alkenes, because firstly, of the low selectivity of chlorine atoms, and secondly, of the small difference in relative reactivity of allylic and saturated carbon–hydrogen bonds.

2. *t-Butyl Hypochlorite*

Allylic chlorination is best carried out by organic chlorinating agents, the mostly widely used example of such a reagent being t-butyl hypochlorite.

In polar solvents, this reagent adds to double bonds and behaves as a source of chlorine cations (Anbar and Ginsberg, 1954). Walling and

$$CH_2{=}CHCH_3 + (CH_3)_3COCl \xrightarrow[\text{as solvent}]{CH_3OH} ClCH_2{-}\overset{+}{C}H{-}CH_3 + (CH_3)_3CO^-$$

$$ClCH_2\overset{+}{C}H{-}CH_3 \xrightarrow{CH_3OH} ClCH_2CH(OCH_3)CH_3 + H^+$$

Jacknow (1960) used t-butyl hypochlorite to chlorinate toluene and aliphatic hydrocarbons. This chlorination was done in carbon tetrachloride solution using radical initiators such as azobisisobutyronitrile. Walling and Thaler (1961) showed that alkenes reacted with t-butyl hypochlorite, under conditions which favoured radical intermediates, to give allylic chlorides. The usual radical chain process, with the t-butoxy-radical as the chain carrier, was proposed to account for the chlorination.

Initiation steps

$$(CH_3)_2C(CN){-}N{=}N{-}C(CH_3)_2(CN) \xrightarrow[\text{heat}]{h\nu \text{ or}} 2(CH_3)_2\dot{C}CN + N_2$$

$$(CH_3)_2\dot{C}CN + (CH_3)_3COCl \longrightarrow (CH_3)_2C(Cl)CN + (CH_3)_3CO\cdot$$

Chain propagation steps

$$(CH_3)_3CO\cdot + RH \longrightarrow (CH_3)_3COH + R\cdot \quad (7)$$

$$R\cdot + (CH_3)_3COCl \longrightarrow RCl + (CH_3)_3CO\cdot \quad (8)$$

Fragmentation

$$(CH_3)_3CO\cdot \longrightarrow (CH_3)_2CO + CH_3\cdot \quad (10)$$

$$CH_3\cdot + (CH_3)_3COCl \longrightarrow CH_3Cl + (CH_3)_3CO\cdot \quad (11)$$

$$CH_3\cdot + RH \longrightarrow CH_4 + R\cdot \quad (12)$$

Termination steps

$$2R\cdot \longrightarrow R_2 \quad (4)$$

$$R\cdot + (CH_3)_3CO\cdot \longrightarrow (CH_3)_3COR \quad (5)$$

$$2(CH_3)_3CO\cdot \longrightarrow (CH_3)_3COOC(CH_3)_3 \quad (6)$$

The chlorination does not proceed by way of a chlorine atom chain analogous to that postulated for *N*-bromosuccinimide since the selectivities *RS* (relative rate of attack per hydrogen atom at the various positions in a molecule as measured by the concentration of product formed for a small extent of reaction) for a hydrocarbon such as butane are quite different when the results for chlorination by t-butyl hypochlorite and by molecular chlorine are compared (see Table 1).

TABLE 1

Relative reactivities in chlorination of butane

		Primary	Secondary	Reference
Gas phase	*RS* for Cl_2 at 70°	1	3·6	Tedder (1960)
	RS for $(CH_3)_3COCl$ at 60°	1	9	Brokenshire, Nechvatal and Tedder (1970)
Liquid phase (CCl_4 soln)	*RS* for Cl_2 at 25°	1	2·5	Walling and Mayahi (1959)
	RS for $(CH_3)_3COCl$ at 40°	1	8	Walling and Jacknow (1960)

For allylic chlorination, the allylic position is much more reactive to chlorination by t-butyl hypochlorite than it is to chlorination by molecular chlorine (Table 2).

TABLE 2

Relative reactivities in chlorination of alkenes

	Alkane CH	Allylic CH	Reference
RS for Cl_2 in CCl_4 at 0°	1	0·6	Poutsma (1965)
RS for $(CH_3)_3COCl$ in CCl_4 at 40°	1	12 to 20	Walling and Thaler (1961)

Only a small amount of addition accompanies allylic chlorination, so that t-butyl hypochlorite is more suitable than chlorine for preparing allylic chlorides. The geometry about the double bond in compounds undergoing allylic chlorination by t-butyl hypochlorite is retained, as is the case in allylic chlorination by molecular chlorine. Here again then, the abstraction step is not reversible, i.e. $k_{-7} = 0$.

Allylic rearrangement occurs, however, as is illustrated for the reaction of *trans*-but-2-ene with t-butyl hypochlorite (Walling and Thaler, 1961).

$$(CH_3)_3COCl \xrightarrow{\text{Initiation}} (CH_3)_3CO\cdot + Cl\cdot$$

$$CH_3CH{=}CHCH_3\ (trans) + (CH_3)_3CO\cdot \underset{k_{-7}}{\overset{k_7}{\rightleftharpoons}} (CH_3)_3COH + \cdot CH_2CH{=}CHCH_3\ (trans)$$

$$\cdot CH_2CH{=}CHCH_3\ (trans) + (CH_3)_3COCl \longrightarrow ClCH_2CH{=}CHCH_3\ (trans) + (CH_3)_3CO\cdot \quad (80\%)$$

$$\updownarrow$$

$$CH_2{=}CH{-}\dot{C}HCH_3 + (CH_3)_3COCl \longrightarrow CH_2{=}CH{-}CHClCH_3 + (CH_3)_3CO\cdot \quad (20\%)$$

For *cis*-but-2-ene the yields due to abstraction at the allylic position are slightly different (60 % of 1-chlorobut-2-ene and 40 % of 3-chlorobut-1-ene). The latter product is formed as a result of allylic rearrangement. Slightly more of the adduct (XXXI) was formed during the reaction with *cis*-but-2-ene than during that with *trans*-but-2-ene. With alkenes such as pent-2-ene the products are more complex since there are two possible allylic positions, but nevertheless no loss of stereochemistry is observed during the reaction of *cis*- or *trans*-pent-2-ene with t-butyl hypochlorite. This means that spreading of the unpaired electron in the allylic radical does not involve loss of stereochemistry. Thus the C2–C3 bond in the *trans*-1-but-2-enyl radical (XXXII), for example, retains sufficient double-bond character to preclude *cis-trans* isomerisation.

$$CH_3CHClCH(OC(CH_3)_3)CH_3 \quad \text{(XXXI)}$$

$$\left(CH_2{\cdots}CH{\cdots}CHCH_3\ (trans)\right)\cdot \quad \text{(XXXII)}$$

The incidence of the competing reaction (10), that of fragmentation of the t-butoxy-radical to acetone and a methyl radical, can be minimised at low temperatures. Above 80° it is likely to become important.

Other hypochlorites are known (Walling and Padwa, 1963; Bacha and Kochi, 1965) but none have supplanted t-butyl hypochlorite for convenience of bringing about allylic chlorination. Having demonstrated that the chain carrier in allylic chlorination by t-butyl hypochlorite is the t-butoxy-radical, it is appropriate to return to the analogous compound, t-butyl hypobromite. This compound is less easily available than the hypochlorite and its reactions have not been studied in great detail, but it has been shown by Walling and Padwa (1962) that the hypobromite reacts with alkenes by a mechanism in which the t-butoxy-radical acts as the chain carrier.

Bromination of saturated hydrocarbons with a mixture of t-butyl hypochlorite and bromotrichloromethane was also attempted (Walling and Padwa, 1962) but the method is effective only with hydrocarbons such as cyclohexane and cyclobutane which have strong carbon–hydrogen bonds so that reactive radicals are formed in the abstraction step. It was thought that the reaction probably proceeded by a t-butoxy-radical chain.

$$(CH_3)_3COCl \xrightarrow{h\nu} (CH_3)_3CO\cdot + Cl\cdot$$

$$(CH_3)_3CO\cdot + RH \longrightarrow (CH_3)_3COH + R\cdot$$

$$R\cdot + BrCCl_3 \longrightarrow RBr + \cdot CCl_3$$

$$(CH_3)_3COCl + \cdot CCl_3 \longrightarrow (CH_3)_3CO\cdot + CCl_4$$

t-Butyl hypoiodite, prepared by the interaction of t-butyl hypochlorite and mercuric iodide, has been used as an iodinating reagent

$$2(CH_3)_3COCl + HgI_2 \longrightarrow 2(CH_3)_3COI + HgCl_2$$

$$2(CH_3)_3COI \longrightarrow (CH_3)_3CO{-}\underset{\displaystyle I}{\underset{|}{I}}{-}OC(CH_3)_3$$

$$(CH_3)_3CO{-}\underset{\displaystyle I}{\underset{|}{I}}{-}OC(CH_3)_3 \xrightarrow{h\nu} (CH_3)_3CO{-}\underset{\bullet}{I}{-}OC(CH_3)_3$$

$$(CH_3)_3CO{-}\underset{\bullet}{I}{-}OC(CH_3)_3 + RH \longrightarrow (CH_3)_3CO{-}\underset{\displaystyle H}{\underset{|}{I}}{-}OC(CH_3)_3 + R\cdot$$

$$R\cdot + (CH_3)_3CO{-}\underset{\displaystyle I}{\underset{|}{I}}{-}OC(CH_3)_3 \longrightarrow RI + (CH_3)_3CO{-}\underset{\bullet}{I}{-}OC(CH_3)_3$$

$$2(CH_3)_3CO{-}\underset{\displaystyle I}{\underset{|}{I}}{-}OC(CH_3)_3 \longrightarrow (CH_3)_3COH + (CH_3)_3COI$$

for alkanes by Tanner and Gidley (1968), but the selectivity and relative reactivities indicate that the abstraction step is brought about by a radical more selective than the t-butoxy-radical. It was therefore suggested that the abstracting radical is a di-t-butoxy-iodonyl radical.

Propargylic chlorination of propyne by t-butyl hypochlorite has been observed by Caserio and Pratt (1967), who found that prop-1-yne is slightly more reactive than toluene.

$$(CH_3)_3COCl + CH_3C{\equiv}CH \longrightarrow ClCH_2C{\equiv}CH$$

But-1-yne, when chlorinated with t-butyl hypochlorite to low conversion, gave products analogous to those obtained by the reaction of chlorine with but-1-yne (Poutsma and Kartch, 1966). However, the relative reactivity of chlorine and of t-butyl hypochlorite towards but-1-yne was not discussed in this report. The ratio of the yields of (XXXIII) to that of (XXXIV) was approximately 10:1. A trace of the allene (XXXV) was formed.

$$(CH_3)_3COCl + CH_3CH_2C{\equiv}CH \xrightarrow{h\nu} \underset{\text{(XXXIII)}}{CH_3CH(Cl)C{\equiv}CH} + \underset{\text{(XXXIV)}}{ClCH_2CH_2C{\equiv}CH} + \underset{\text{(XXXV)}}{CH_3CH{=}C{=}CHCl}$$

Walling, Heaton and Tanner (1965) found that t-butyl hypochlorite reacted rapidly with but-2-yne in the dark to yield 1-chloro-but-2-yne (40 %) as the main product resulting from hydrogen abstraction at the terminal methyl group, together with some 2,3-dichlorobut-2-ene (10 %), the result of addition. In this reaction it was suggested that initiation involves the simultaneous breaking of the oxygen–chlorine bond and the addition of chlorine to the triple bond.

$$(CH_3)_3COCl + CH_3C{\equiv}CCH_3 \longrightarrow (CH_3)_3CO\cdot + CH_3\dot{C}{=}C(Cl)CH_3$$

Chloroallenes were not found as reaction products. The alkyne was slightly less reactive than but-2-ene. The same general observations applied to bromination of but-2-yne by t-butyl hypobromite.

An instance of the sole formation of adduct from a reaction with t-butyl hypochlorite, whereas t-butyl hypobromite effects allylic bromination, has been described by Alden and Davies (1968) in an investigation of the halogenation of 2-methylenenorbornane (XXXVI) and 2-methylnorborn-2-ene (XXXVII). t-Butyl hypochlorite gave a complex mixture of addition products with these two compounds. Bromination of either 2-methylnorborn-2-ene or 2-methylenenorbornane could be effected with either t-butyl hypobromite or *N*-bromosuccinimide. t-Butyl hypobromite gave only one product [3-exo-bromo-2-methylenenorbornane (XXXVIII)] from both alkenes, whereas *N*-bromosuccinimide gave this same bromo-compound together with 10–25% of *endo*-isomer (XXXIX) as well. Neither of these reagents effected isomerisation of either of the starting alkenes. This difference in reactivity was thought to be due to the

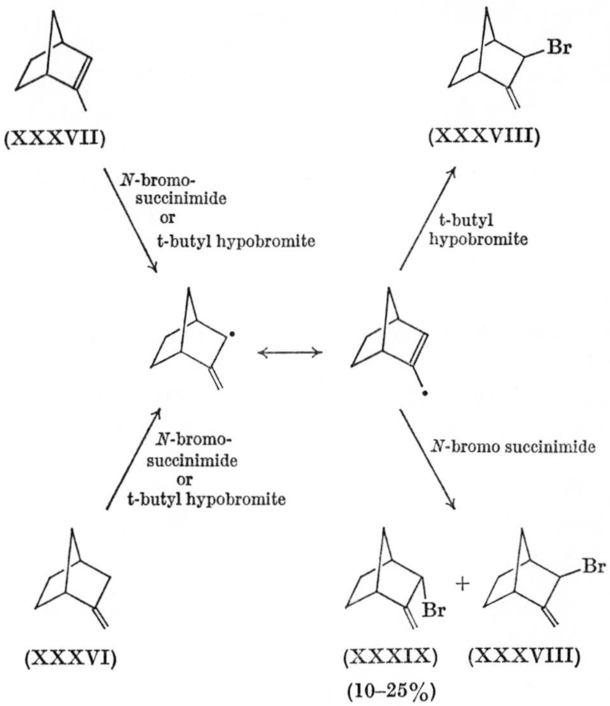

(XXXVII)

(XXXVIII)

N-bromosuccinimide or t-butyl hypobromite

t-butyl hypobromite

$\longleftrightarrow$

N-bromosuccinimide or t-butyl hypobromite

N-bromo succinimide

(XXXVI)

(XXXIX) (10–25%) + (XXXVIII)

more stringent steric requirement of the t-butyl hypobromite as compared to bromine in the chain transfer step.

In contrast, Ahmad, Gedye and Nechvatal (1968) found that *N*-bromosuccinimide, 1,3-dibromo-5,5-dimethylhydantoin and t-butyl hypochlorite reacted with methyl 3-methylbut-2-enoate (V) by means of a radical reaction to give *cis*- and *trans*-4-halo-3-methylbut-2-enoates (XL) in the proportions *cis*:*trans* = 2:3 for the bromo-compounds; and 10:11 for the chloro-compounds.

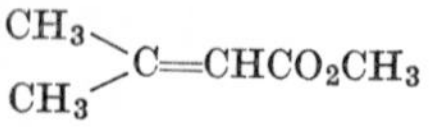

$(CH_3)_2C{=}CHCO_2CH_3$

(V)

$XCH_2(CH_3)C{=}CHCO_2CH_3$

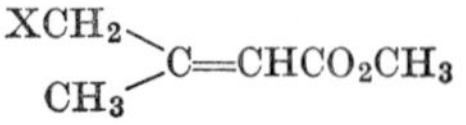

(XL)
(*cis*- and *trans*-)

(X = Cl or Br)

3. *Other Chlorinating Agents*

N-chlorosuccinimide is not an efficient allylic chlorinating agent (Ziegler *et al.*, 1942), although it chlorinates aromatic compounds (Hebbelynck and Martin, 1949) and the radical-initiated reaction with toluene served as a basis for the elucidation of the mode of action of *N*-chloro- and *N*-bromo-succinimide (Adam, Gosselain and Goldfinger, 1953). However Ucciani and Naudet (1960) report an instance of the allylic chlorination of methyl oleate. Since *N*-chloro-succinimide reacts by way of a chlorine atom chain, the selectivity would be expected to be as low as that for molecular chlorine, so that on grounds of greater reactivity, and also higher selectivity, t-butyl hypochlorite is preferred for allylic chlorination.

A number of other *N*-chloro-compounds are known, including 1,3-dichloro-5,5-dimethylhydantoin (XLI) (Bogaert-Verhogen and Martin, 1949), 1,3,5-trichloro-s-triazine-2,4,6-(1*H*,3*H*,5*H*)-trione (iso-cyanuryl chloride) (XLII) (Ziegler *et al.*, 1942), and *N*-chloro-*N*-cyclohexylbenzene sulphonamide (XLIII) (Theilacker and Wessel, 1967). In general, the main reaction of these compounds with alkenes

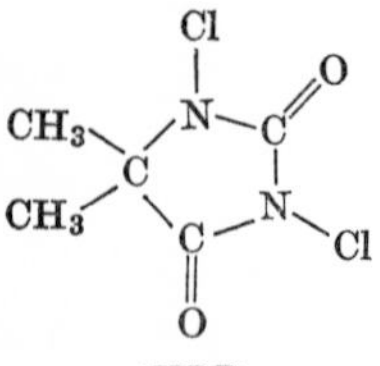

(XLI) 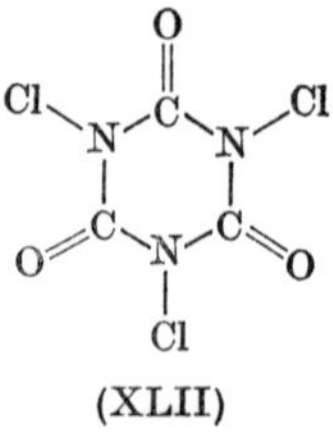 (XLII) 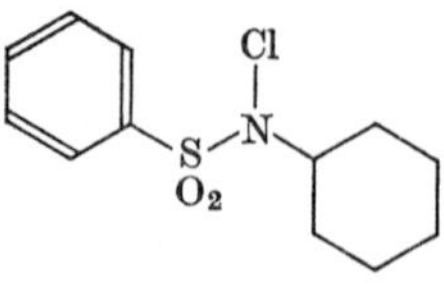 (XLIII)

is that of addition to the double bond, although this reaction is accompanied by a little allylic substitution. It is probable that these compounds react by way of chlorine atom chains, and are therefore not sufficiently selective to be useful allylic halogenating agents.

Radical chlorination has also been observed by Kochi (1962), who used mixtures of copper(II) chloride and lithium chloride in acetonitrile solution. The selectivity of these reagents in alkane chlorination is the same as for photohalogenations by molecular chlorine, and hence the following mechanism was postulated. Cyclohexene gives a

$$CuCl_2 + nCl^- \longrightarrow CuCl_{n+2}^{n-}$$

$$CuCl_{n+2}^{n-} \xrightarrow{h\nu} CuCl_{n+1}^{n-} + Cl\cdot$$

$$Cl\cdot + RH \longrightarrow HCl + R\cdot$$

$$R\cdot + CuCl_{n+2}^{n-} \longrightarrow RCl + CuCl_{n+1}^{n-}$$

mixture of addition and substitution products containing 23% of the allylic 3-chlorocyclohexene.

An example of allylic substitution of dithiocyanogen has been described by Bacon, Guy, Irwin and Robinson (1959). The reaction is analogous to those of halogens.

S=C=N—N=C=S + (1-methylcyclohexene) ⟶ (1-methylcyclohexene with N=C=S at the allylic position) + HCNS

In conclusion, it can be seen that the preparative usefulness of the allylic substitution reaction has led to attempts to understand the mechanism of the reaction and to a search for more selective reagents. Much quantitative work has been done on the relative reactivities of different alkenes, but, for a deeper understanding of the reaction, there remains a need for precise measurement of energies of activation of the abstraction step. This can probably be done best by studying gas-phase reactions.

References

Adam, J., Gosselain, P. A. and Goldfinger, P. (1953) *Nature*, **171**, 704.
Ahmad, I., Gedye, R. N. and Nechvatal, A. (1968) *J. Chem. Soc.* (*C*), 185.
Alden, C. K. and Davies, D. I. (1968) *J. Chem. Soc.* (*C*), 709.
Anbar, M. and Ginsburg, D. (1954) *Chem. Rev.*, **54**, 925.
Bacha, J. D. and Kochi, J. K. (1965) *J. Org. Chem.*, **30**, 3272.

Bacon, R. G. R., Guy, R. G., Irwin, R. S. and Robinson, T. A. (1959) *Proc. Chem. Soc.*, 304.
Bateman, L., Cunneen, J. I., Fabian, J. M. and Koch, H. P. (1950) *J. Chem. Soc.*, 936.
Benson, S. W. (1965) *J. Chem. Ed.*, **42**, 507.
Bloomfield, G. F. (1944) *J. Chem. Soc.*, 114.
Bogaert-Verhoogen, D. and Martin, R. H. (1949) *Bull. Soc. chim. belges.*, **58**, 567.
Boozer, C. E. and Moncrief, J. W. (1962) *J. Org. Chem.*, **27**, 623.
Brokenshire, J. L., Nechvatal, A. and Tedder, J. M. (1970) *Trans. Faraday Soc.*, **66**, 2029.
Caserio, M. C. and Pratt, R. E. (1967) *Tetrahedron Letters*, 91.
Charles, E. (1958) *Bull. assoc. franc. techniciens Petrole*, **130**, 385.
Corey, E. J. (1953) *J. Amer. Chem. Soc.*, **75**, 2251.
Corral, R. A., Orazi, O. O. and Bonafede, J. D. (1957) *Anales Assoc. quim. argentina*, **45**, 151.
Couvreur, P. and Bruylants, A. (1952) *Bull. Soc. chim. belges*, **61**, 253.
Dauben, H. J. and McCoy, L. L. (1959) *J. Amer. Chem. Soc.*, **81**, 4863.
Djerassi, C. (1948) *Chem. Rev.*, **43**, 271.
Djerassi, C. and Lenk, C. T. (1953) *J. Amer. Chem. Soc.*, **75**, 3493.
Gedye, R. N. and Nechvatal, A. (1964) *J. Chem. Soc.*, 5925.
Gosselain, P. A., Adam, J. and Goldfinger, P. (1956) *Bull. Soc. chim. belges*, **65**, 533.
Gottardi, W. (1967) *Monatsh.*, **98**, 1613.
Groll, H. P. A., Hearne, G., Burgin, J. and LaFrance, D. S. (1938) *U.S. Patent* 2,130,084.
Hebbelynck, M. F. and Martin, R. H. (1949) *Experientia*, **5**, 69.
Horner, L. and Winkelmann, E. H. (1959) *Angew. Chem.*, **71**, 349 (English transl. *Newer Methods of Preparative Organic Chemistry*, Ed. Foerst, W., Academic Press, 1964, **3**, 151).
Huyser, E. S. (1961) *J. Org. Chem.*, **26**, 3261.
Huyser, E. S. and DeMott, D. N. (1963) *Chem. and Ind.*, 1954.
Incremona, J. S. and Martin, J. C. (1970) *J. Amer. Chem. Soc.*, **92**, 627.
Kerr, J. A. (1966) *Chem. Rev.*, **66**, 477.
Kharasch, M. S., Malec, R. and Yang, N. C. (1957) *J. Org. Chem.*, **22**, 1443.
Kochi, J. K. (1962) *J. Amer. Chem. Soc.*, **84**, 2121.
McGrath, B. P. and Tedder, J. M. (1961) *Proc. Chem. Soc.*, 80.
Messmer, A., Varady, J. and Pinter, I. (1958) *Acta Chim. Acad. Sci. Hung.*, **15**, 183.
Orazi, O. O. and Meseri, J. (1950) *Anales Assoc. quim. argentina*, **38**, 5.
Poutsma, M. L. (1963) *J. Amer. Chem. Soc.*, **85**, 3511.
Poutsma, M. L. (1965a) *J. Amer. Chem. Soc.*, **87**, 2161.
Poutsma, M. L. (1965b) *J. Amer. Chem. Soc.*, **87**, 2172.
Poutsma, M. L. (1969) *Methods in Free-Radical Chemistry*, Ed. Huyser, E. S., Marcel Dekker, New York, **1**, 79.
Poutsma, M. L. and Kartch, J. L. (1966) *Tetrahedron*, **22**, 2167.
Reeve, W., Chambers, D. H. and Prickett, C. S. (1952) *J. Amer. Chem. Soc.*, **74**, 5369.
Rust, F. F. and Vaughan, W. E. (1940) *J. Org. Chem.*, **5**, 472.

Schaltegger, H. (1957) *U.S. Patent* 2,790,757.

Sisido, K., Kondo, K., Nozaki, H., Tuda, M. and Udo, Y. (1960) *J. Amer. Chem. Soc.*, **82**, 2286.

Sixma, F. L. J. and Riem, R. H. (1958) *Proc. k. ned. Akad. Wetenschap*, B, **61**, 183.

Sosnovsky, G. (1964) *Free Radical Reactions in Preparative Organic Chemistry*, Macmillan, New York, 337.

Stewart, T. D., Dod, K. and Stenmark, G. (1937) *J. Amer. Chem. Soc.*, **59**, 1765.

Tanner, D. D. and Gidley, G. C. (1968) *J. Amer. Chem. Soc.*, **90**, 808.

Tedder, J. M. (1960) *Quarterly Reviews*, **14**, 336.

Tedder, J. M. and Walton, J. C. (1967) *Trans. Faraday Soc.*, **63**, 2678.

Thaler, W. A. (1969) *Methods in Free-Radical Chemistry*, Ed. Huyser, E. S., Marcel Dekker, New York, **2**, 198.

Theilacker, W. and Wessel, H. (1967) *Ann.*, **703**, 34.

Traynham, J. G. and DeWitt, B. S. (Jr.) (1970) *J. Org. Chem.*, **35**, 2025.

Ucciani, E. and Naudet, M. (1960) *Bull. Soc. chim. France*, 1151.

Walling, C., Heaton, L. and Tanner, D. D. (1965) *J. Amer. Chem. Soc.*, **87**, 1715.

Walling, C. and Huyser, E. S. (1963) *Org. Reactions*, Wiley, New York, **13**, 91.

Walling, C. and Jacknow, B. B. (1960) *J. Amer. Chem. Soc.*, **82**, 6108.

Walling, C. and Mayahi, M. F. (1959) *J. Amer. Chem. Soc.*, **81**, 1485.

Walling, C. and Padwa, A. (1962) *J. Org. Chem.*, **27**, 2976.

Walling, C. and Rieger, A. L. (1963) *J. Amer. Chem. Soc.*, **85**, 3134.

Walling, C., Rieger, A. L. and Tanner, D. D. (1963) *J. Amer. Chem. Soc.* **85**, 3129.

Walling, C. and Thaler, W. (1961) *J. Amer. Chem. Soc.*, **83**, 3877.

Wendt, G., (1941) *Ber.*, **74**, 1243.

Wohl, A. (1919) *Ber.*, **52**, 51.

Ziegler, K., Späth, A., Schaaf, E., Schumann, W. and Winkelmann, E. (1942) *Ann.*, **551**, 80.

4

FREE RADICAL REACTIONS IN THE PRESENCE OF METAL IONS—REACTIONS OF NITROGEN COMPOUNDS

GEORGE SOSNOVSKY

Department of Chemistry,
University of Wisconsin-Milwaukee,
Milwaukee, Wisconsin, U.S.A.

and

DAVID J. RAWLINSON

Department of Chemistry,
Western Illinois University,
Macomb, Illinois, U.S.A.

A. INTRODUCTION

Recently, in several articles (Sosnovsky and Zaret, 1970; Sosnovsky and Rawlinson, 1970a, b, 1971a, b), we reviewed the chemistry of peroxides in the presence of metal ions which are capable of redox

We are pleased to dedicate this article to Dr. Eugen Müller, Professor of Chemistry, University of Tübingen, on the occasion of his sixty-fifth birthday, June 21, 1970.

We are grateful to the Graduate School of the University of Wisconsin-Milwaukee, and to the Research Council of Western Illinois University for grants to G.S. and D.J.R., respectively.

processes. The initial step in these reactions involves the formation of ionic and radical species in accordance with the following scheme.

$$R—O—O—R + M^{n} \rightarrow RO\cdot + RO^{-} + M^{n+1}$$

R = alkyl, aralkyl, acyl, alkoxycarbonyl; M = metal ion capable of valency change; $n = 1, 2, 3$ (usually)

Besides peroxides, various other classes of organic compounds may, in the presence of metal ions, undergo analogous types of reactions involving ionic and radical species in the initiation step.

$$YX + M^{n} \rightarrow Y\cdot + X^{-} + M^{n+1}$$

In this review, the discussion is restricted to the chemistry of nitrogen compounds. In an overwhelming number of cases the reactions involve either nitrogen radicals or nitrogen cation radicals. No special effort was made to differentiate between metal ion-catalysed reactions and reactions which involve larger than catalytic amounts of metal ions, since a strict separation was often found impossible. However, all reactions are of the homogeneous type, i.e., no surface catalysis is involved. The emphasis in the present review is on the preparative aspects. The literature has been surveyed up to January 1, 1970. However, on several occasions more recent references and unpublished results were included. Among articles unpublished at the time of writing are two reviews: one by Neale on *Nitrogen Radicals as Synthesis Intermediates* which includes only a few references concerning metal ions, and another one by Kovacic, Lowery and Field on *Chemistry of N-Bromo- and N-Chloramines* which contains in part also some of our material in an abbreviated form.

An attempt has been made to cover all topics as comprehensively as possible. However, it was impossible, because of limited space, to accommodate all radical reactions of nitrogen compounds in one chapter.

B. Halogenation of Aliphatic Compounds

Free radical halogenation reactions have been reviewed recently (Poutsma, 1969; Thaler, 1969; Huyser, 1970). However, metal ion-catalysed halogenations with *N*-haloamines were only briefly discussed in one article (Poutsma, 1969).

$$R_2NCl + SH \xrightarrow{M^{n}/M^{n+1}} SCl + HNR_2$$

The ready availability of *N*-haloamines and their high selectivity makes this method of halogenation an important new procedure for syntheses. In Table 1 results are shown of halogenation reactions of aliphatic compounds with various *N*-haloamines. The table also includes several results with cyclo-alkanes and aralkanes. In cases where several references are quoted, for the sake of simplicity, the isomer distribution is cited from one reference only. In most experiments straight chain saturated compounds containing polar moieties, such as a halogen, ester, or ether group, have been used. The halogenations occurred preferably at the ω-1 position. In many cases, for compounds containing polar groups, 70–90% of ω-1 isomer was obtained. For example, in the chlorination of methyl hexanoate with *N*-chlorodi-isobutylamine the following isomer distribution was obtained (Minisci, Galli, Bernardi and Perchinunno, 1968):

$$\underset{(7\cdot1)}{Me}-\underset{(89\cdot9)}{CH_2}-\underset{(3\cdot0)}{CH_2}-CH_2.CH_2.CO.OMe$$

and even the reactions of unsubstituted hydrocarbons gave high yields of ω-1 isomers, e.g. the reaction of *N*-chlorodi-isobutylamine with *n*-heptane resulted in the following isomer distribution (Bernardi, Galli and Minisci, 1968).

$$\underset{(1\cdot3)}{Me}-\underset{(64\cdot4)}{CH_2}-\underset{(22\cdot9)}{CH_2}-\underset{(11\cdot3)}{CH_2}-CH_2CH_2Me$$

These yields are much higher than those usually obtained in halogenation reactions with other classes of halogenating agents. The reactions of *N*-haloalkylamines with olefins usually produce adducts (I) (Section D). However, under certain conditions only hydrogen

$$RCH{=}CHR + R'_2NX \xrightarrow{M^n/M^{n+1}} \underset{(I)}{\underset{R'_2N\;\;X}{RCHCHR}}$$

abstraction at the α and β position to the double bond can be achieved, e.g. chlorination of methyl *trans*-2-hexenoate with *N*-chlorodimethylamine resulted in the following isomer distribution (Minisci, Galli, Cecere and Bernardi, 1967).

$$\underset{(8\cdot2)}{Me}-\underset{(46\cdot6)}{CH_2}-\underset{(45\cdot2)}{CH_2}-CH{=}CHCO.OMe$$

This result shows a decrease in selectivity as compared to the results of chlorination of the saturated parent compound, methyl hexanoate.

$$\underset{(5\cdot7)}{Me}-\underset{(87\cdot3)}{CH_2}-\underset{(6\cdot3)}{CH_2}-\underset{(0\cdot7)}{CH_2}-CH_2.CO.OMe$$

TABLE 1

*Halogenation of aliphatic and alicyclic compounds**

Halogenating agent	Substrate	Initiator	Product (% isomer)	Yield %	Ref.
$Me_2\overset{+}{N}HBr$	$Me(CH_2)_3Cl$	$FeSO_4$	$Me—CH_2—CH_2—CH_2—Cl$ (7·2) (78·6) (9·5) (4·7)	~50	1, 2
$Me_2\overset{+}{N}HCl$	$Me(CH_2)_3Cl$	$FeSO_4$	$Me—CH_2—CH_2—CH_2Cl$ (5·5) (88·5) (6)	—	3, 4, 2
$Me_2\overset{+}{N}HCl$	$Me(CH_2)_2CO.OMe$	$FeSO_4$	$Me—CH_2—CH_2.CO.OMe$ (13·6) (86·4)	40–50	3, 5
$Me_2\overset{+}{N}HBr$	$Me(CH_2)_3CO.OMe$	$FeSO_4$	$Me—CH_2—CH_2—CH_2—CO.OMe$ (12·2) (81·3) (6) (0·47)	—	2, 6
$Me_2\overset{+}{N}HCl$	$Me(CH_2)_3CO.OMe$	$FeSO_4$	$Me—CH_2—CH_2—CH_2.CO.OMe$ (7·3) (77) (15·7)	—	3, 2, 5, 4
$Me_2\overset{+}{N}HCl$	$Me(CH_2)_4Cl$	$FeSO_4$	$Me—CH_2—CH_2—CH_2—CH_2Cl$ (3·9) (77·2) (16·4) (2·5)	—	3
$Me_2\overset{+}{N}HBr$	$Me(CH_2)_4CO.OMe$	$FeSO_4$	$Me—CH_2—CH_2—CH_2—CH_2.CO.OMe$ (6·8) (87) (5·9) (0·4)	—	2, 6
$Me_2\overset{+}{N}HCl$	$Me(CH_2)_4CO.OMe$	$FeSO_4$ or CuCl	$Me—CH_2—CH_2—CH_2—CH_2.CO.OMe$ (5·7) (87·3) (6·3) (0·7)	70–80	3, 6, 7, 5, 2
$Me_2\overset{+}{N}HCl$	$Me(CH_2)_4O.CO.Me$	$FeSO_4$	$Me—CH_2—CH_2—CH_2CH_2.O.CO.Me$ (5·9) (88·3) (5·8)	—	8

$Me_2\overset{+}{N}HBr$	$Me(CH_2)_5Me$	$FeSO_4$	Me—CH_2—CH_2—CH_2—CH_2.CH_2Me (1) (50·4) (32·8) (15·7)	—	2
$Me_2\overset{+}{N}HCl$	$Me(CH_2)_5Me$	$FeSO_4$	Me—CH_2—CH_2—CH_2—CH_2.CH_2Me (1·1) (55·6) (29) (14·3)	—	9, 2
$Me_2\overset{+}{N}HBr$	$Me(CH_2)_5OMe$	$FeSO_4$	Me—CH_2—CH_2—CH_2—CH_2.CH_2.OMe (4) (83) (11·7) (0·9)	—	2
$Me_2\overset{+}{N}HCl$	$Me(CH_2)_5OMe$	$FeSO_4$	Me—CH_2—CH_2—CH_2—CH_2—CH_2—OMe (4) (82·7) (11·5) (0·9) – (1·3)	—	2, 8
$Me_2\overset{+}{N}HBr$	$Me(CH_2)_5Cl$	$FeSO_4$	Me—CH_2—CH_2—CH_2—CH_2—CH_2Cl (2·7) (67·6) (19·7) (7·5) (2·5)	—	2, 4, 6
$Me_2\overset{+}{N}HCl$	$Me(CH_2)_5Cl$	$FeSO_4$	Me—CH_2—CH_2—CH_2—CH_2—CH_2Cl (2·2) (73) (19·8) (3·7) (1·3)	—	3, 4, 2
$Me_2\overset{+}{N}HBr$	$Me(CH_2)_5CO.OMe$	$FeSO_4$	Me—CH_2—CH_2—CH_2—CH_2.CH_2.CO.OMe (3·5) (82·2) (13·4) (1)	—	2
$Me_2\overset{+}{N}HCl$	$Me(CH_2)_5CO.OMe$	$FeSO_4$	Me—CH_2—CH_2—CH_2—CH_2.CH_2.CO.OMe (3·7) (80·4) (14·7) (1·2)	70–80	5, 3, 4, 10, 2
$Me_2\overset{+}{N}HCl$	$Me(CH_2)_5O.CO.Me$	$FeSO_4$	Me—CH_2—CH_2—CH_2—CH_2.CH_2.O.CO.Me (2·9) (78·6) (16·3) (2·2)	—	8
$Me_2\overset{+}{N}HCl$	$MeCH_2.CH_2.CH{=}CH.CO.OMe$ (*trans*)	$FeSO_4$	Me—CH_2—CH_2—CH=CH.CO.OMe (8·2) (46·6) (45·2)	—	7
$Me_2\overset{+}{N}HCl$	$Me(CH_2)_8CO.OMe$	$FeSO_4$	Me—CH_2—CH_2—CH_2—CH_2—CH_2—CH_2—CH_2. (1·1) (44·4) (21·6) (17·5) (10·7) (4·1) (0·5) CH_2.CO.OMe	—	9

TABLE 1—*continued*

Halogenating agent	Substrate	Initiator	Product (% isomer)	Yield %	Ref.
$Me_2\overset{+}{N}HBr$	Cyclohexane	$FeSO_4$	Bromocyclohexane Dibromocyclohexane	65 5	11
$Me_2\overset{+}{N}HCl$	Cyclohexane	$FeSO_4$	Chlorocyclohexane	62	11
$Me_2\overset{+}{N}HCl$	Chlorocyclohexane	$FeSO_4$	1,1- 1,2-*cis*- 1,2-*trans*- 1,3-*cis*- 1,3-*trans*- 1,4-*cis*- 1,4-*trans*- (5·6) – (4·3) (27·4) (24·6) (21·2) (16·9)	—	3
$Bu^t_2\overset{+}{N}HCl$	$Me(CH_2)_5Me$	$FeSO_4$	$Me—CH_2—CH_2—CH_2—CH_2.CH_2.Me$ (1·3) (64·4) (22·9) (11·3)	—	9
$Bu^t_2\overset{+}{N}HCl$	$Me(CH_2)_4CO.OMe$	$FeSO_4$	$Me—CH_2—CH_2—CH_2.CH_2.CO.OMe$ (7·1) (89·9) (3·0)	—	8
$Bu^t_2\overset{+}{N}HCl$	$Me(CH_2)_5CO.OMe$	$FeSO_4$	$Me—CH_2—CH_2—CH_2—CH_2.CH_2.CO.OMe$ (6·4) (82·8) (10·2) (0·78)	—	10
$Bu^t_2\overset{+}{N}HCl$	$Me(CH_2)_5OCO.Me$	$FeSO_4$	$Me—CH_2—CH_2—CH_2—CH_2.CH_2.O.CO.Me$ (4·4) (84·6) (10·2) (0·8)	—	8
$Bu^t_2\overset{+}{N}HCl$	$Me(CH_2)_8CO.OMe$	$FeSO_4$	$Me—CH_2—CH_2—CH_2—CH_2—CH_2—CH_2.CH_2.CH_2.$ (1·4) (57·7) (19·1) (13·3) (6·7) (1·7) $CO.OMe$	—	9
$Bu^t_2\overset{+}{N}HCl$	Chlorocyclohexane	$FeSO_4$	1,1- 1,2-*cis*- 1,2-*trans*- 1,3-*cis*- 1,3-*trans*- 1,4-*cis*- (10·4) – – (29·6) (20·1) (23·0) 1,4-*trans*- (16·9)	—	3

N-Chloro-piperidine	$Me(CH_2)_3Cl$	$FeSO_4$ $FeCl_2$, $CuCl$, $Ce_2(SO_4)_3$	$Me—CH_2—CH_2—CH_2—Cl$ (6·5) (78·3) (9·7) (5·5)	~50	1
N-Chloro-piperidine	$Me(CH_2)_3Cl$	$CoCl_2$, $CeCl_3$, $Ce_2(SO_4)_3$ –NaCl	$Me—CH_2—CH_2—CH_2—Cl$ (20·9) (50·1) (23·4) (5·6)	—	1
$Me_2\overset{+}{N}HBr$	PhMe	$FeSO_4$	$PhCH_2Br$, toluidines, *o*- & *p*-bromotoluene	—	11
$Me_2\overset{+}{N}HCl$	PhMe	$FeSO_4$	$PhCH_2Cl$ & toluidines	23, 52	12
$Me_2\overset{+}{N}HCl$	PhMe	$TiCl_3$	$PhCH_2Cl$ & toluidines	13, 65	12
$Et_2\overset{+}{N}HCl$	PhMe	$FeSO_4$	$PhCH_2Cl$ & toluidines	35, 7	12
$Bu^n{}_2\overset{+}{N}HCl$	PhMe	$FeSO_4$	$PhCH_2Cl$	12	12
$Bu^i{}_2\overset{+}{N}HCl$	PhMe	$FeSO_4$	$PhCH_2Cl$	61	12
N-Chloro-piperidine	PhMe	$FeSO_4$	$PhCH_2Cl$ & toluidines	20	12

* In cases where several references are quoted, the isomer distribution is cited from only one reference.

References: (1) Spanswick and Ingold (1970). (2) Minisci, Galli, Bernardi and Perchinunno (1969). (3) Minisci, Galli and Bernardi (1967a). (4) Minisci, Gardini and Bertini (1970). (5) Minisci, Galli, Galli and Bernardi (1967b). (6) Minisci, Galli and Bernardi (1967b). (7) Minisci, Galli, Cecere and Bernardi (1967). (8) Minisci, Galli, Bernardi and Perchinunno (1968). (9) Bernardi, Galli and Minisci (1968). (10) Minisci, Galli, Perchinunno and Bernardi (1968). (11) Minisci, Galli, Galli and Bernardi (1967a). (12) Minisci, Galli and Bernardi (1966).

Alicyclic compounds, such as cyclohexane, are readily halogenated by *N*-haloamines, e.g. the reactions of *N*-chloro- and *N*-bromodimethylamine with cyclohexane resulted in chlorinated and brominated cyclohexane derivatives, respectively (Minisci, Galli, Galli and Bernardi, 1967a). Chlorination of chlorocyclohexane gave a complex mixture of 1,1-, 1,2-*cis* and *trans*, 1,3-*cis* and *trans*, and 1,4-*cis* and *trans* dichloro-derivatives (Minisci, Galli and Bernardi, 1967a).

Halogenation of aralkanes with haloamines resulted not only in side chain and nuclear halogenations but also in amination reactions, e.g. the reaction of toluene with *N*-bromodimethylamine gave a mixture of benzyl bromide, bromotoluenes, and toluidines (Minisci, Galli, Galli and Bernardi, 1967a). Similar results were obtained in the reactions of toluene with various *N*-chloroamines (Minisci, Galli and Cecere, 1965; Minisci, Galli and Bernardi, 1966). However, the chlorination with the bulky *N*-chlorodi-isobutylamine gave exclusively benzyl chloride (Minisci, Galli and Bernardi, 1966).

Ethers sometimes undergo degradation reactions during the course of halogenation or the subsequent work-up, e.g. the small amount of chlorination which occurred in the α-position of n-hexyl methyl ether resulted in caproaldehyde (Minisci, Galli, Bernardi and Perchinunno, 1968).

$$\mathrm{Me(CH_2)_4CH_2OMe + Me_2\overset{+\bullet}{N}H \longrightarrow Me(CH_2)_4\dot{C}HOMe \longrightarrow Me(CH_2)_4CHO + CH_3\cdot}$$

or

$$\mathrm{Me(CH_2)_4\dot{C}HOMe + Me_2\overset{+}{N}HCl \longrightarrow Me(CH_2)_4CHCl.OMe + Me_2\overset{+\bullet}{N}H}$$

$$\mathrm{Me(CH_2)_4CHCl.OMe + H_2O \longrightarrow Me(CH_2)_4CHO + HCl + MeOH}$$

Similarly, the chlorination of benzyl methyl ether gave some benzaldehyde (Minisci, Bernardi, Trabucchi and Grippa, 1966).

$$\mathrm{PhCH_2OMe + Me_2\overset{+}{N}HCl \longrightarrow PhCHCl.OMe \xrightarrow{H_2O} PhCHO}$$

The results in Table 1 also indicate that essentially no halogenation occurs at the α-position to the polar group, in contrast to the result usually observed with other halogenating agents (Minisci, Galli and Bernardi, 1967a, b; Minisci, Galli, Bernardi and Perchinunno, 1969).

Another remarkable feature of these halogenation processes is the fact that even very remote positions are still selectively halogenated, e.g. the halogenation of methyl decanoate with *N*-chlorodi-isobutylamine resulted in 57·7% of the ω-1 isomer (Bernardi, Galli and Minisci, 1968).

$$\underset{(1\cdot4)}{Me}—\underset{(57\cdot7)}{CH_2}—\underset{(19\cdot1)}{CH_2}—\underset{(13\cdot3)}{CH_2}—\underset{(6\cdot7)}{CH_2}—\underset{(1\cdot7)}{CH_2}—CH_2.CH_2.CH_2.CO.OMe$$

The reactions are carried out in acidic media, preferably in concentrated sulphuric acid or a mixture of sulphuric and acetic acids. The protonating power of the solvent markedly affects the product distribution (Minisci, Galli, Galli and Bernardi, 1967b), e.g. the chlorination of n-hexyl methyl ether with *N*-chlorodimethylamine in a solution of 85% sulphuric acid gave a higher amount of ω-1 isomer,

$$\underset{(2\cdot9)}{Me}—\underset{(71\cdot5)}{CH_2}—\underset{(17\cdot5)}{CH_2}—\underset{(5\cdot5)}{CH_2}—CH_2.CH_2.OMe$$

than the same reaction in a solution of 70% sulphuric acid (Minisci, Galli, Bernardi and Perchinunno, 1968).

$$\underset{(2\cdot1)}{Me}—\underset{(64)}{CH_2}—\underset{(18\cdot5)}{CH_2}—\underset{(7)}{CH_2}—CH_2.CH_2.OMe$$

To initiate the reactions, in most cases, ferrous salts have been used, although cerium, copper and titanium compounds have been also found to be effective (Minisci, Galli, Galli and Bernardi, 1967b; Tanner and Mosher, 1969; Spanswick and Ingold, 1970). In the case of cuprous ions the selectivity was even increased as compared to that obtained with ferrous ions. Thus, the reaction of *N*-chlorodimethylamine with methyl hexanoate in the presence of ferrous sulphate resulted in 87·3% of ω-1 isomer,

$$\underset{(5\cdot7)}{Me}—\underset{(87\cdot3)}{CH_2}—\underset{(6\cdot3)}{CH_2}—\underset{(0\cdot7)}{CH_2}—CH_2.CO.OMe$$

whereas the same reaction in the presence of cuprous chloride yielded 89·4% of ω-1 isomer (Minisci, Galli and Bernardi, 1967a; Minisci, Galli, Galli and Bernardi, 1967b).

$$\underset{(5\cdot6)}{Me}—\underset{(89\cdot4)}{CH_2}—\underset{(4\cdot7)}{CH_2}—\underset{(0\cdot3)}{CH_2}—CH_2.CO.OMe$$

A wide range of haloamines, such as *N*-chloromonomethyl-, -dimethyl-, -diethyl-, -mono-n-butyl-, -di-n-butyl-, -di-isobutyl-, -methylbenzyl-, and -t-butylbenzyl-amine, *N*-chloromorpholine and *N*-chloropiperidine were compared for their selectivity and resulting isomer distribution (Minisci, Galli, Perchinunno and Bernardi, 1968). From these results it appears that *N*-chlorodi-isobutylamine is the most effective halogenating agent, although the simpler *N*-halodimethylamines showed a very similar performance (Minisci, Galli, Perchinunno and Bernardi, 1968) and from the preparative point of view they might be the preferable reagents.

The selectivity has been attributed to the inductive effect of the polar groups, and to the so-called "ω-1" effect which seems to be steric in origin (Minisci, Galli, Perchinunno and Bernardi, 1968). Thus, the bulkier *N*-haloamines give increased selectivity at the ω-1 position as compared to the less bulky *N*-haloamines, e.g. the halogenation of n-heptane with *N*-chlorodimethylamine yielded 55·6% of the ω-1 isomer,

$$\underset{(1\cdot1)}{\mathrm{Me}}\text{—}\underset{(55\cdot6)}{\mathrm{CH_2}}\text{—}\underset{(29)}{\mathrm{CH_2}}\text{—}\underset{(14\cdot3)}{\mathrm{CH_2}}\text{—}\mathrm{CH_2.CH_2.Me}$$

whereas the reaction with *N*-chlorodi-isobutylamine produced 64·4% of the ω-1 isomer (Minisci, Galli, Bernardi and Perchinunno, 1968).

$$\underset{(1\cdot3)}{\mathrm{Me}}\text{—}\underset{(64\cdot4)}{\mathrm{CH_2}}\text{—}\underset{(22\cdot9)}{\mathrm{CH_2}}\text{—}\underset{(11\cdot3)}{\mathrm{CH_2}}\text{—}\mathrm{CH_2.CH_2.Me}$$

The isomer distribution is almost the same for chlorination and bromination reactions, e.g. bromination of methyl hexanoate with *N*-bromodimethylamine gave 87% of the ω-1 isomer,

$$\underset{(6\cdot88)}{\mathrm{Me}}\text{—}\underset{(87)}{\mathrm{CH_2}}\text{—}\underset{(5\cdot9)}{\mathrm{CH_2}}\text{—}\underset{(0\cdot41)}{\mathrm{CH_2}}\text{—}\mathrm{CH_2.CO.OMe}$$

and the chlorination of the same compound with *N*-chloro-dimethylamine resulted in 87·3% of the ω-1 isomer (Minisci, Galli and Bernardi, 1967b).

$$\underset{(5\cdot7)}{\mathrm{Me}}\text{—}\underset{(87\cdot3)}{\mathrm{CH_2}}\text{—}\underset{(6\cdot3)}{\mathrm{CH_2}}\text{—}\underset{(0\cdot7)}{\mathrm{CH_2}}\text{—}\mathrm{CH_2.CO.OMe}$$

On the basis of these results it must be concluded that in both cases the same hydrogen abstracting species is involved.

The reaction mechanism which has been proposed for these halogenation reactions (Minisci, Galli and Bernardi, 1967a, b; Minisci, Galli, Galli and Bernardi, 1967b; Minisci, Galli, Bernardi and Perchinunno, 1969; Spanswick and Ingold, 1970) is essentially the same as that proposed for the metal ion-catalysed Hofmann-Löffler cyclisation reaction (Section E) involving an aminium radical, i.e. a protonated amino-radical (II), with the exception that in the halogenation reaction an intermolecular process is involved, whereas in the

Hofmann-Löffler reaction intramolecular hydrogen abstraction occurs.

$$R_2NX + H^+ \longrightarrow R_2\overset{+}{N}HX$$
$$(X = Br, Cl)$$

$$R_2\overset{+}{N}HX + M^n \longrightarrow R_2\overset{+\bullet}{N}H + M^{n+1} + X^-$$
$$(II)$$
$$(M = Ce, Cu, Fe, Ti)$$

$$R'H + R_2\overset{+\bullet}{N}H \longrightarrow R'\cdot + R_2\overset{+}{N}H_2$$
$$R'\cdot + R_2\overset{+}{N}HX \longrightarrow R'X + R_2\overset{+\bullet}{N}H$$

On the basis of the preceding discussion it must be concluded that the dialkylamino-radical cations (II) are strongly electrophilic, and exhibit a very high sensitivity to steric effects, and the highest selectivity in hydrogen abstraction of any known radical.

An alternative mechanism involving a chain of halogen atoms was also considered (Tanner and Mosher, 1969).

$$R'H + Cl\cdot \longrightarrow R'\cdot + HCl$$
$$R'\cdot + Cl_2 \longrightarrow R'Cl + Cl\cdot$$
$$R_2\overset{+}{N}HCl + HCl \longrightarrow R_2\overset{+}{N}H_2 + Cl_2$$

However, this mechanism does not explain the unusually high selectivity and the observed isomer distribution. This scheme has been criticised by Minisci and co-workers (Minisci, Galli, Galli and Bernardi, 1967a, b; Minisci, Gardini and Bertini, 1970) and by Spanswick and Ingold (1970).

C. Amination of Aromatic Compounds

Metal ion-catalysed amination of aromatic compounds has been accomplished with hydroxylamine, hydroxylamine-*O*-sulphonic acid (III) and *N*-haloalkylamines. The latter category can be divided into *N*-halomonoalkylamines (IV), *N*,*N*-dihaloalkylamines (V), *N*-halo-dialkylamines (VI), and *N*-haloheterocycloalkylamines (VII).

$^+NH_3OSO_3^-$	RNHX	RNX_2	R_2NX	(ring)NX
(III)	(IV)	(V)	(VI)	(VII)

1. *Hydroxylamine and Hydroxylamine-O-sulphonic Acid*

Within the last two decades a number of publications have appeared dealing with the metal ion-catalysed decomposition of hydroxylamine and its derivatives. The most frequently mentioned metal ions were those of iron, copper or titanium, although those of cerium, cobalt, chromium, nickel, silver, manganese, and vanadium have been used also. In the overwhelming number of cases the decomposition of hydroxylamines containing an unsubstituted amino-group produced as active species either the amino-radical, $\cdot NH_2$, or the aminium radical, i.e. the protonated amino-radical, $\dot{N}H_3^{+}$. However, many studies were devoted to the physico-chemical aspects of these decompositions, such as studies of kinetics, polymerisation, and electron paramagnetic resonance (e.p.r), without isolation of products (Uri, 1952; Gutch and Waters, 1964, 1965; Ogata and Morimoto, 1965; Brikun and Kozlovskii, 1965; Anderson, 1964, 1966; Waters and Wilson, 1966; Brown and Drury, 1967; Hlasivcova, Novak and Zyka, 1967; Lingane and Christie, 1967; Wells and Salam, 1967; Davies and Kustin, 1969; Calusaru, 1968; Schmidt, Swinehart and Taube, 1968; Jijee and Santappa, 1969). In the present section the discussion is restricted to amination reactions producing well-defined products.

The first attempts to intercept amino-radicals with aromatic compounds were apparently made by Davis, Evans and Higginson (1951), Seaman, Taylor and Waters (1954), and later by Albisetti *et al.* (1959). However, difficulties were experienced in isolating pure materials from the basic product mixtures. Some biphenyl was found in these reactions (Davis, Evans and Higginson, 1951; Albisetti *et al.*, 1959; Seaman, Taylor and Waters, 1954; Frank *et al.*, 1960) indicating their free radical nature. The difficulties in isolating amino-derivatives

$$NH_2OH.HCl + M^n \rightarrow \cdot NH_2 + M^{n+1} + Cl^- + H_2O$$

$$(M = Fe, Ti)$$

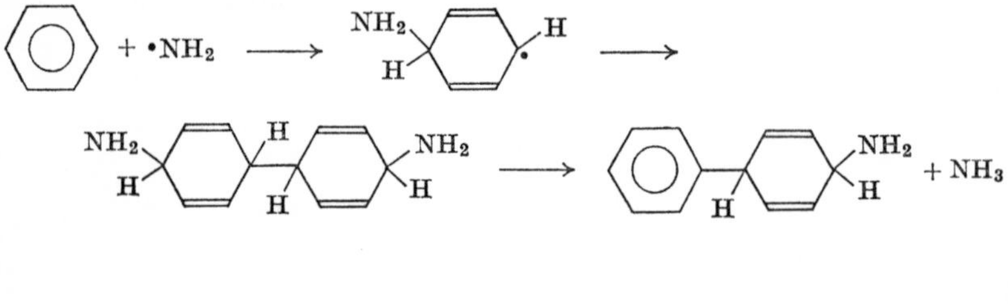

$$C_6H_6 + \cdot NH_2 \longrightarrow \text{(NH}_2\text{, H)}C_6H_6\text{(H)}\cdot \longrightarrow$$

$$\text{(NH}_2\text{, H)}C_6H_6\text{(H)}-\text{(H)}C_6H_6\text{(NH}_2\text{, H)} \longrightarrow C_6H_5-\text{(H)}C_6H_6\text{(NH}_2\text{, H)} + NH_3$$

of benzene and toluene are probably associated with the ease of formation of dimers, elimination of ammonia and only partial re-aromatisation of the systems (Seaman, Taylor and Waters, 1954; Albisetti *et al.*, 1959; Frank *et al.*, 1960; Minisci, Galli, Cecere and Mondelli, 1965).

In the case of benzene and toluene, the dimers were not directly isolated but on hydrogenation, aminocyclohexyl and aminomethyl-cyclohexyl dimers respectively were obtained (Albisetti *et al.*, 1959; Frank *et al.*, 1960). The same reaction with mesitylene produced a mixture of mesitylamine and the diaminohexamethylbicyclohexa-dienyl (VIII), which was isolated (Minisci, Galli and Cecere, 1965; Minisci, Galli, Cecere and Mondelli, 1965).

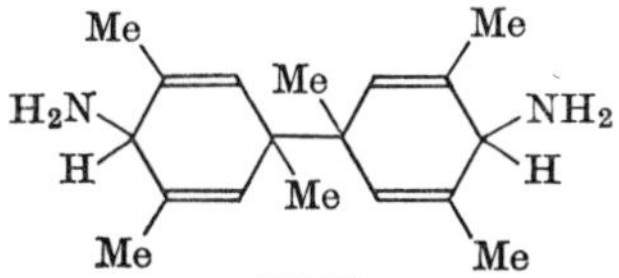

(VIII)

With an increasing number of electron-donating groups the yields of aminated product also increase. Thus, amination of toluene, *m*-xylene and mesitylene produced basic products containing 16, 33, and 50% of toluidines, xylidines and mesitylamine, respectively (Minisci, Galli, Cecere and Mondelli, 1965).

The reaction with anisole, which has also an electron-donating group, yielded a mixture of anisidines (Minisci, Galli and Cecere, 1965; Minisci, Galli, Cecere and Mondelli, 1965; Minisci and Galli, 1965b) (see equations overleaf).

The effect of oxygen on the amination of aromatic systems in the presence of titanous chloride is of interest (Minisci, Galli, Cecere and Mondelli, 1965). The reaction with toluene under nitrogen gave the following mixture of products: $o, m, p = 37$, 20·5, 42·5%; whereas in the presence of oxygen a different composition was obtained: $o, m, p =$ 25·4, 12·6, 62%. Similarly, amination of *m*-xylene under nitrogen yielded xylidines containing 73% of 2,4-dimethylaniline, whereas in the presence of oxygen the proportion rose to 83% (Minisci, Galli, Cecere and Mondelli, 1965). A very striking effect was observed with mesitylene. The reactions under nitrogen in the presence of ferrous and of titanous chloride resulted in mixtures of mesitylamine and the diaminohexamethylbicyclohexadienyl (VIII), whereas in the presence of oxygen the dimer (VIII) was absent. These results indicate

$$NH_2OH.HCl + M^n \rightarrow \cdot NH_2 + M^{n+1} + H_2O + Cl^-$$
$$(M = Fe, Ti)$$

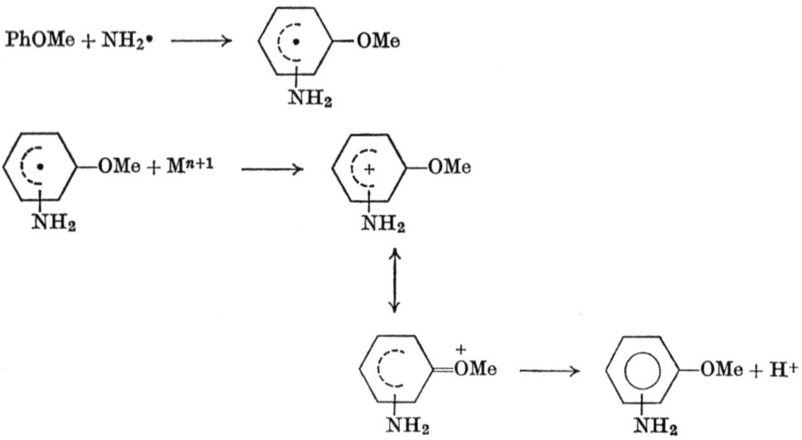

PhOMe + $NH_2\cdot$ ⟶ (·)—OMe, NH_2

(·)—OMe + M^{n+1}, NH_2 ⟶ (+)—OMe, NH_2

↕

=$\overset{+}{O}$Me, NH_2 ⟶ —OMe + H^+, NH_2

that oxygen enhances the re-aromatisation reaction (Minisci, Galli, Cecere and Mondelli, 1965).

The amination of aromatic compounds containing halogens resulted in lower yields of products than those obtained with benzene or benzene derivatives containing electron-donating groups (Minisci, Bernardi, Grippa and Trabucchi, 1966). The *para*-position is the most reactive, and less than 10% of the substitution occurs at the *meta*-positions. Thus, the reactions of chloro- and of bromo-benzene in the presence of ferrous chloride yielded product mixtures containing 3·5–4·5% *o*-isomer, 8% *m*-isomer and 86–87% *p*-isomer. In addition, a small amount (1·3–1·5%) of aniline was obtained, presumably by the following route (Minisci, Bernardi, Grippa and Trabucchi, 1966).

X + $\cdot NH_2$ ⟶ X, NH_2 (·) $\xrightarrow{Fe^{3+}}$

(X = Br, Cl)

X, NH_2 (+) ⟶ NH_2 + X^+

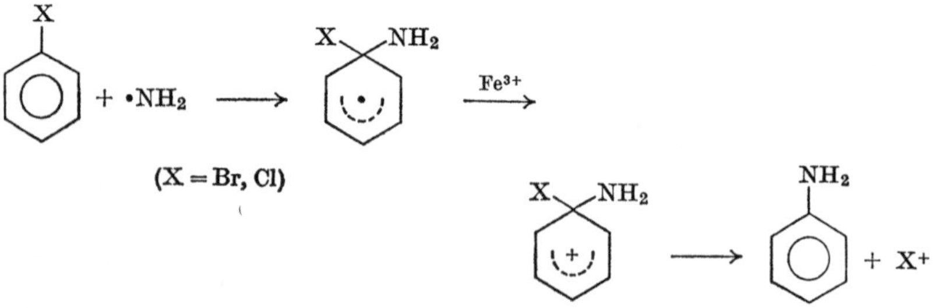

Only limited information is available about the amination of polycyclic compounds with hydroxylamine. Thus, the reaction of anthraquinone with hydroxylamine in sulphuric acid at 130° in the presence of ferrous sulphate gave a mixture of products consisting of α- and β-aminoanthraquinone, and compounds (IX), (X), (XI) and (XII) (Yoshida, Matsumoto and Oda, 1964c).

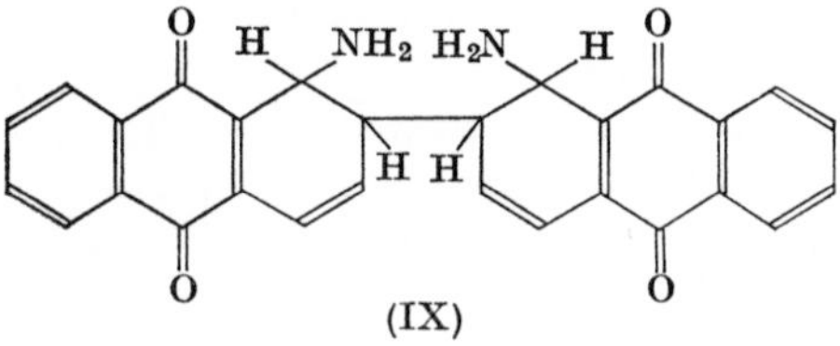

(IX)

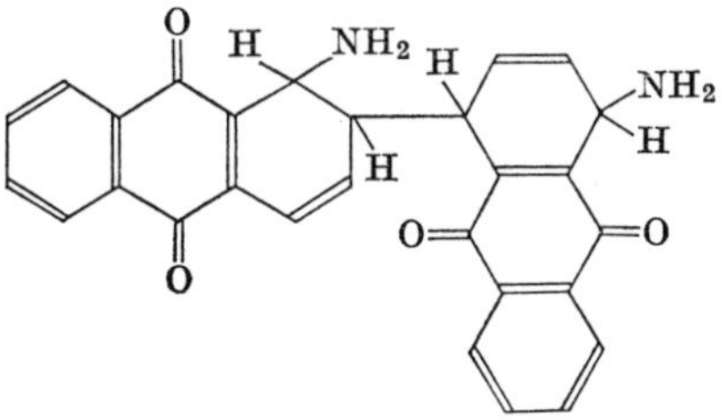

(X)

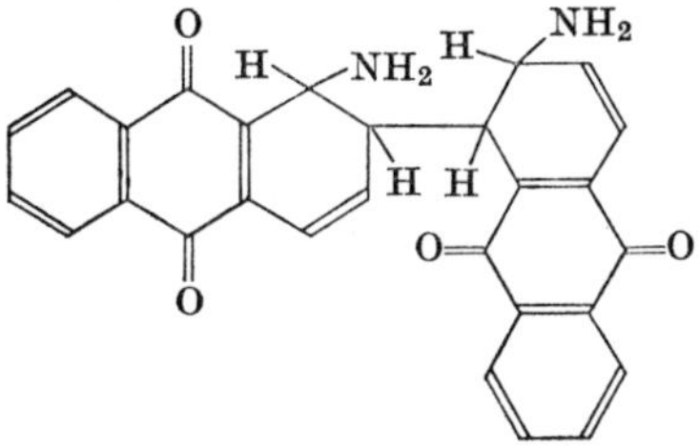

(XI)

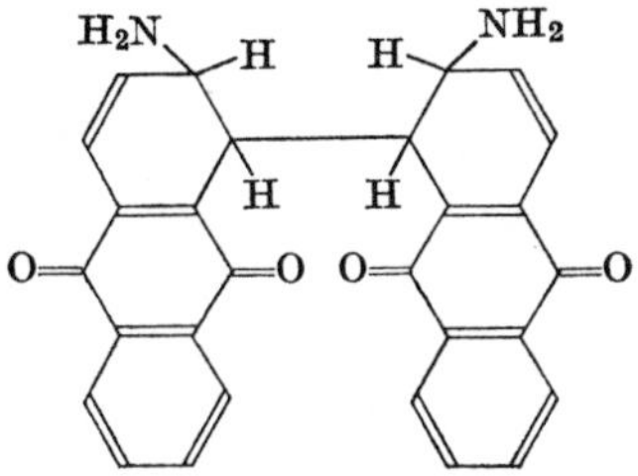

(XII)

A similar reaction of anthraquinone-β-sulphonic acid gave a mixture of 1-aminoanthraquinone-2-sulphonic acid, 1,4-diamino-anthraquinone-2-sulphonic acid, 2-aminoanthraquinone, and 2-aminoanthraquinone-3-sulphonic acid (Yoshida, Matsumoto and Oda, 1964b).

Results of various aminations are shown in Table 2. In general it must be concluded that the amination with hydroxylamine results in low yields (5–15%) of products and, therefore, is of limited synthetic value.

The aminations of aromatic compounds with hydroxylamine-*O*-sulphonic acid in the presence of ferrous ions proceed readily at 18–20° to give aromatic amines in 10–40% yields (Table 2) (Minisci, Galli and Cecere, 1965).

$$^{+}NH_3OSO_3^{-} + Fe^{2+} \rightarrow \overset{+\bullet}{N}H_3 + SO_4^{2-} + Fe^{3+}$$

$$C_6H_5R + \overset{+\bullet}{N}H_3 \longrightarrow [R\text{-}C_6H_5\text{-}\overset{+}{N}H_3]^{\bullet}$$

(R = Br, Cl, I, Me, OMe)

$$[R\text{-}C_6H_5\text{-}\overset{+}{N}H_3]^{\bullet} + Fe^{3+} \longrightarrow [R\text{-}C_6H_5\text{-}\overset{+}{N}H_3]^{+} \longrightarrow R\text{-}C_6H_4\text{-}\overset{+}{N}H_3 + H^{+}$$

Although the aminations with hydroxylamine–metal ion systems seem to be analogous to hydroxylamine-*O*-sulphonic acid systems, the reactivity and orientation are substantially different, in that the attacking amino-radical cation produced from hydroxylamine-*O*-sulphonic acid shows a much more pronounced electrophilic character than the amino-radical generated from hydroxylamine. Thus, the reaction of anisole with the hydroxylamine—titanous chloride system formed a mixture consisting of 63% of the *o*-isomer and 37% of the *p*-isomer, whereas the amination with hydroxylamine-*O*-sulphonic acid in the presence of ferrous ions yielded a mixture of 34% *o*- and 66% *p*-isomer (Minisci and Galli, 1965b).

The reactivity of anisole relative to benzene in competitive reactions gave a $K_{anisole}/K_{benzene}$ ratio of 1·7 for the reaction with hydroxylamine, whereas with hydroxylamine-*O*-sulphonic acid a value of 7·9 was obtained (Minisci, Galli and Cecere, 1965). Amination of deactivated aromatic compounds such as chloro-, bromo-, and iodo-benzene with hydroxylamine-*O*-sulphonic acid resulted in lower yields than those obtained for benzene or benzene derivatives activated by electron-donating substituents. Reactions of all three halobenzenes resulted in a high *p*-isomer content, ranging between 75–77%. The *o*- and *m*-isomer content was between 7·7–10%. In addition, 6·3–10% aniline was obtained, by a mechanism analogous to that described for the reaction of hydroxylamine with halobenzenes (Minisci, Bernardi, Grippa and Trabucchi, 1966).

2. N-*Chloromonoalkylamines and* N,N-*Dichloroalkylamines*

Monoalkylamination of aromatic compounds was achieved at 0° with *N*-chloromonoalkylamines (Minisci, Galli and Cecere, 1966c).

$$R\overset{+}{N}H_2Cl + Fe^{2+} \rightarrow R\overset{+\bullet}{N}H_2 + Fe^{3+} + Cl^-$$

$$R\overset{+\bullet}{N}H_2 + C_6H_5{-}R' \longrightarrow [R\overset{+}{N}H_2{-}C_6H_5{-}R']^{\bullet}$$

$$[R\overset{+}{N}H_2{-}C_6H_5{-}R']^{\bullet} + Fe^{3+} \longrightarrow RNH{-}C_6H_4{-}R' + Fe^{2+} + H^+$$

(R, R′ = alkyl)

However, the synthetic usefulness of this reaction seems to be limited, owing to the instability of *N*-chloromonoalkylamines in acidic solutions and to the fact that aromatic amines are not formed exclusively. The undistillable products (15–30%) always formed probably arise by incomplete re-aromatisation of intermediates.

Nevertheless, benzene, toluene, and *p*-xylene were converted with various *N*-chloromonoalkylamines to aromatic monoalkylamines in 21–45% yields. Although these yields are somewhat lower than those obtained in dialkylamination reactions using *N*-halodialkylamines, there might be certain interest in the monoalkylamination reaction because of a higher *o*-isomer content as compared to the dialkylamination. Thus, the reaction of toluene with *N*-halomethylamines produced the following isomer mixtures: *o*, *m*, *p* = 27·5, 25, 47·5%, whereas with *N*-chlorodimethylamine, because of greater steric hindrance, a decrease in the yield of the *o*-isomer was found: *o*, *m*, *p* = 9, 53, 38% (Minisci, Galli and Cecere, 1966c).

Monoalkylamination was also achieved with *N*,*N*-dichloromonoalkylamines, e.g. the reaction of benzene with *N*,*N*-dichloro-n-butylamine and *N*,*N*-dichlorocyclohexylamine yielded 10 and 7% of monoalkylamino-derivatives, respectively. The following reaction paths were suggested for this monoalkylamination. In all cases it is assumed that dichloroamine is converted into *N*-chloromonoalkylamine, which then acts as the actual aminating agent (Minisci, Galli and Cecere, 1966c).

$$RNCl_2 + 2Fe^{2+} + H_2O \longrightarrow RNHCl + Fe(OH)^{2+} + FeCl^{2+}$$

or

$$RNCl_2 + HCl \longrightarrow RNHCl + Cl_2$$

or

$$RNCl_2 + ArH \longrightarrow RNHCl + ArCl$$

3. N-*Halodialkylamines and* N-*Haloheterocycloalkylamines*

Free radical dialkyl- and heterocycloalkyl-aminations of aromatic compounds have been performed at room temperature using a variety of metal salts capable of redox reactions. Thus, iron, copper, titanium, chromium, vanadium, manganese, nickel, cobalt, and silver salts have been used (Bock and Kompa, 1965, 1966a), although in most studies iron, copper, or titanium compounds have been preferred. The reactions in the presence of copper ions are more sensitive to temperature changes than those in the presence of ferrous ions (Minisci, Galli, Bernardi and Galli, 1967). The copper ion-catalysed reactions produce larger amounts of *o*- and *p*-isomers and lesser amounts of *m*-isomers than the ferrous and titanous ion-catalysed reactions (Minisci, Galli, Bernardi and Galli, 1967). At elevated temperatures, the preferential formation of *p*-isomer is lost, e.g. the reaction of *N*-chlorodimethylamine with toluene in the presence of copper chloride at 0–30° gave a mixture of 18% *o*-, 13% *m*-, and 69% *p*-dimethyl-toluidines, whereas at 75–80° a mixture consisting of 17·4% *o*-, 41·8% *m*-, and 40·8% *p*-isomers was obtained (Minisci, Galli, Bernardi and Galli, 1967).

Although the reactions do proceed in non-acidic media, a strongly acidic medium has been preferred, particularly with activated aromatic compounds, because in these cases a more reactive aminium radical, $>\overset{+\bullet}{N}H$, is generated, and because the product of amination being an aromatic amine, and therefore susceptible to further attack, is deactivated by protonation immediately on formation to give an unreactive derivative containing $-NHR_2^+$ group. The reactions are usually conducted in sulphuric acid or in a mixture of sulphuric and acetic acids. Occasionally, trifluoroacetic acid and various organic solvents such as methanol, ether and carbon tetrachloride have been used (Minisci, Galli and Cecere, 1965; Minisci, Galli, Galli and Bernardi, 1967a). Comparative studies of amination of toluene with *N*-chlorodimethylamine using sulphuric, acetic, and trifluoroacetic acids with additions of benzenesulphonic acid, nitroethane, and ether have been carried out. The isomer ratios varied with the solvent composition ranging from 7·8 to 11·3% for the *o*-, 43·6 to 61·8% for the *m*-, and 27·9 to 48·6% for the *p*-isomers (Minisci, Galli, Galli and Bernardi, 1967a).

The reactions of benzene with *N*-halodialkyl- and *N*-halohetero-cycloalkyl-amines, e.g. *N*-chlorodimethylamine and *N*-chloropiperidine, proceed readily to give high yields of product (Bock and Kompa, 1965, 1966a, c; Montecatini Edison, 1967).

$$C_6H_6 + R_2\overset{+}{N}HCl + Fe^{2+}/Fe^{3+} \longrightarrow C_6H_5\overset{+}{N}HR_2$$

To avoid electrophilic chlorination by the *N*-haloamines and further amination of products, short reaction times and strongly acidic solutions are essential. Compounds containing activating *o*- and *p*- directing substituents, such as hydroxy-, amino-, methoxy-, and alkylamino-groups, readily undergo amination. However, these

$$RC_6H_5 + R_2\overset{+}{N}HX \longrightarrow RC_6H_4{-}X + R_2\overset{+}{N}H_2$$

(R = OH, NH_2, alkoxy, alkylamino, etc)

compounds are also rather susceptible to electrophilic halogenation (Minisci, Galli and Pollina, 1965).

Therefore the reaction conditions must be chosen so as to favour amination relative to chlorination, i.e., in the acidic media a high metal ion concentration is added to favour the redox process. Unfortunately, these conditions also favour reduction of the haloamines (Minisci, Galli and Cecere, 1966e; Minisci and Galli, 1965a), at the expense of nuclear amination. Consequently, the best conditions

$$R_2NCl + Fe^{2+} \longrightarrow R_2N\cdot + FeCl^{2+}$$

$$R_2N\cdot + Fe^{2+} + H_2O \longrightarrow R_2NH + FeOH^{2+}$$

involve the lowest possible concentration of metal ions consistent with the avoidance of electrophilic chlorination, but at the same time permitting an effective amination process (Minisci, Galli and Cecere, 1966e; Minisci, Galli, Cecere and Trabucchi, 1966; Minisci, 1967). Under these conditions, the reaction of phenol with *N*-chloropiperidine yielded 87% of *N*-(hydroxyphenyl)piperidine, consisting of 9% *o*- and 91% *p*-isomer (Minisci, Galli and Perchinunno, 1969b). The high amount of *p*-isomer has been explained by a strong steric effect. Similarly, dialkylamination of anisole with *N*-chloropiperidine resulted in 11% of product consisting of 90·5% *p*-, no *m*-, and only 4·5% *o*-isomer (Minisci, Galli and Pollina, 1965; Minisci, Galli, Cecere and Trabucchi, 1966; Minisci and Galli, 1965a). In addition, a small amount of the chloro-derivative (XIII) was formed (Minisci and Galli, 1965a).

(XIII)

Unprotonated aromatic amines are very reactive and susceptible to electrophilic chlorination (Minisci, Galli and Pollina, 1965; Minisci, Galli and Cecere, 1966e). However, in acidic media they are completely deactivated and unreactive because of protonation of the amino-group. In order to aminate aromatic amines they are converted to acetyl derivatives which are then aminated, e.g. the reaction of *N*-chloropiperidine with acetanilide gave a high yield of product consisting mainly of *p*-isomer (Minisci, Galli and Cecere, 1966e; Minisci, Galli, Cecere and Quilico, 1969).

Dialkylamination of benzene derivatives containing deactivating, but *o*- and *p*- directing substituents, such as halogens or olefinic groups with electron-attracting substituents, results in lower yields of products than those obtained with benzene (Minisci, Bernardi, Grippa and Trabucchi, 1966; Bock and Kompa, 1965, 1966a, c). The reactions of *N*-chlorodimethylamine with chloro-, bromo-, and iodobenzene gave high proportions of *p*-isomer (Minisci, Bernardi, Grippa and Trabucchi, 1966): PhCl—$o, m, p = 17, < 5, 68\%$; PhBr—$o, m, p = 19, 3, 67\%$; PhI—$o, m, p = 5, 8, 80\%$. In addition, dimethylaniline was obtained from the three halobenzenes in 9, 11, and 2% yields, respectively. This result was explained by the following scheme (Minisci, Bernardi, Grippa and Trabucchi).

$$R_2\overset{+}{N}HCl + Fe^{2+} \rightarrow R_2\overset{+\bullet}{N}H + Fe^{3+}$$

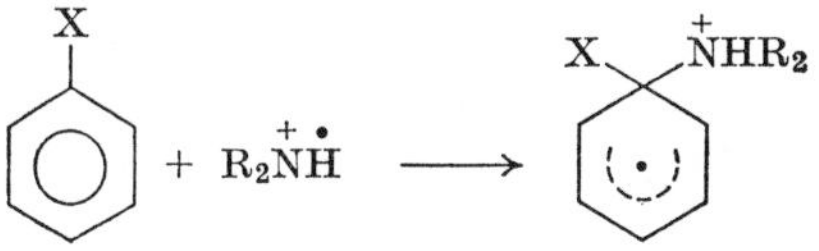

$$C_6H_5X + R_2\overset{+\bullet}{N}H \longrightarrow [C_6H_5(X)(\overset{+}{N}HR_2)]^{\bullet}$$

(X = Br, Cl, I)

$$[C_6H_5(X)(\overset{+}{N}HR_2)]^{\bullet} + Fe^{3+} \longrightarrow [C_6H_5(X)(\overset{+}{N}HR_2)]^{+} \longrightarrow C_6H_5\overset{+}{N}HR_2 + X^{+}$$

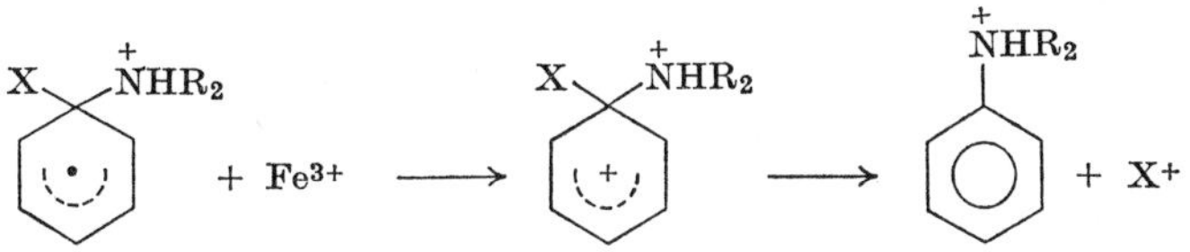

The formation of 5% of *p*-chlorodimethylaniline (XIV) from iodobenzene is significant mechanistically (Minisci, Bernardi, Grippa and Trabucchi, 1966; Minisci, 1967).

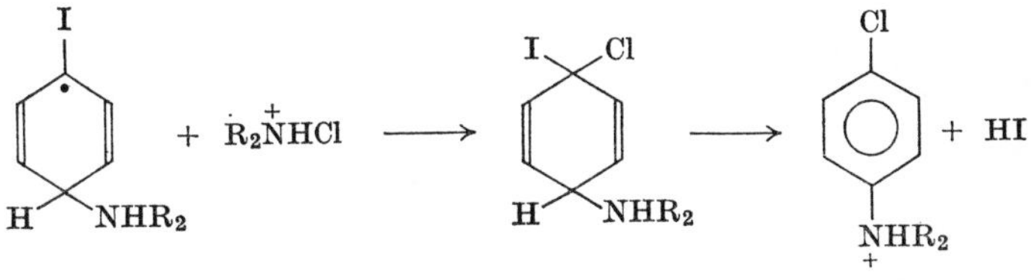

$$\text{I-C}_6\text{H}_5^{\bullet}(\text{H})(\overset{+}{N}HR_2) + R_2\overset{+}{N}HCl \longrightarrow \text{C}_6\text{H}_5(\text{I})(\text{Cl})(\text{H})(\overset{+}{N}HR_2) \longrightarrow p\text{-Cl-C}_6\text{H}_4\text{-}\overset{+}{N}HR_2 + HI$$

(XIV)

Amination of styrene derivatives containing an electron-attracting group, such as methyl cinnamate, occurs simultaneously at the *p*-position and at the double bond of the side chain (Minisci, Galli, Galli and Bernardi, 1967a). Thus, the reaction with *N*-chlorodimethylamine produced a mixture of methyl *p*-*N*,*N*-dimethylaminocinnamate and an adduct. The adduct was characterised by conversion to phenylpyruvic acid (XV) (Minisci, Galli, Galli and Bernardi, 1967a).

$$PhCH{=}CHCO.OMe + Me_2\overset{+\,\bullet}{NH} \longrightarrow Ph\overset{\bullet}{C}H{-}\underset{\underset{+}{NHMe_2}}{CH}{-}CO.OMe$$

$$Ph\overset{\bullet}{C}H\underset{\underset{+}{NHMe_2}}{C}HCO.OMe + Me_2\overset{+}{N}HCl \longrightarrow PhCHCl\underset{\underset{+}{NHMe_2}}{C}HCO.OMe \xrightarrow{-HCl}$$

$$PhCH{=}\underset{\underset{+}{NHMe_2}}{C}CO.OMe \xrightarrow{H_2O} \underset{(XV)}{PhCH_2CO.CO.OH} + Me_2NH$$

The ratio of the amounts of nuclear and side chain attack depends on the reaction conditions. For example, in sulphuric acid the reaction with *N*-chloropiperidine resulted in 60% of side chain and 40% of nuclear attack, whereas in a mixture of 60% sulphuric acid and 40% acetic acid, the reaction with *N*-chlorodimethylamine gave 93·5% side chain and only 6·5% nuclear attack (Minisci, Galli, Galli and Bernardi, 1967a).

The dialkylamination of aralkanes is very characteristic and different from reactions involving ionic electrophilic substitution. Occurrence of side chain reactions indicates free radical intermediates (Minisci, Galli and Cecere, 1965). The observed orientation indicates the involvement of an unselective electrophilic species.

In the series constituted by toluene, cumene, and t-butylbenzene the reactions with *N*-chlorodimethylamine show an increase in *p*-isomer content and a decrease in the *m*-isomer content of the product mixture: toluene, $\frac{1}{2}m/p = 0{\cdot}77$; cumene, $\frac{1}{2}m/p = 0{\cdot}38$; t-butylbenzene, $\frac{1}{2}m/p = 0{\cdot}15$. This trend is in the opposite direction to that found in ionic electrophilic substitution which follows the Baker-Nathan order of hyperconjugation, but it is consistent with the sequence for an inductive effect. The declining yield of *o*-isomer in the series toluene, 9·6%, cumene, 6·2%, and t-butylbenzene, 0%, indicates a steric effect (Minisci, Galli, Galli and Bernardi, 1967a).

A similar steric effect was observed with *m*-xylene which was dialkylaminated in position 5. This position, although electronically less favoured, is the only position not adjacent to the methyl group

(Minisci, Galli and Cecere, 1965). In dialkylaminations of benzylic compounds, halogenations of the side chains have been also observed (Minisci, Galli and Cecere, 1965; Minisci, Galli and Bernardi, 1966; Minisci, Bernardi, Trabucchi and Grippa, 1966). Thus, the reactions of *N*-chlorodialkylamines and *N*-chloroheterocycloalkylamines with toluene produced benzyl chloride by the following sequence.

$$R_2\overset{+}{N}HCl + M^{n} \longrightarrow R_2\overset{+\bullet}{N}H + M^{n+1} + Cl^-$$

$$PhMe + R_2\overset{+\bullet}{N}H \longrightarrow Ph\dot{C}H_2 + R\overset{+}{N}H_2$$

$$Ph\dot{C}H_2 + FeCl^{2+} \longrightarrow PhCH_2Cl + Fe^{2+}$$

or

$$Ph\dot{C}H_2 + R_2\overset{+}{N}HCl \longrightarrow PhCH_2Cl + R_2\overset{+\bullet}{N}H$$

In these reactions the steric effect plays an important role, since benzylic chlorination increases with the bulkiness of the alkyl group. Thus, the reaction of toluene with *N*-chlorodimethylamine gave 23% of benzyl chloride and 52% of toluidines, whereas with *N*-chlorodi-isobutylamine, benzyl chloride was formed exclusively (Minisci, Galli and Bernardi, 1966). Also solvents have a strong influence on the relative extents of nuclear and side chain attack. For example, in 100% sulphuric acid the amination of toluene with *N*-chlorodimethyl-amine gave 4·6% of side chain and 95·4% of nuclear attack whereas in 100% acetic acid, attack occurred only at the side chain (Minisci, Galli, Galli and Bernardi, 1967a).

The reaction of benzyl methyl ether with *N*-chlorodimethylamine showed a strong steric effect, since the dimethylamination product contained only 1% of the *o*-isomer. However, the proportion of *p*-isomer was also low, and the main (79%) component was the *m*-isomer. The preferential *m*-orientation was ascribed to the $-I$ effect of the ether group (Minisci, Bernardi, Trabucchi and Grippa, 1966). The occurrence of some halogenation in the side chain gave benzaldehyde as well as the above products.

$$PhCH_2OMe + Me_2\overset{+}{N}HCl + Fe^{2+}/Fe^{3+} \longrightarrow PhCHCl.OMe \xrightarrow{H_2O} PhCHO + HCl + H_2O$$

Similarly, the reaction of benzyl chloride resulted in a mixture of products containing 3·9% *o*-, 17·5% *p*-, and 60·3% *m*-isomers. Side chain halogenation also produced benzaldehyde in this case (Minisci, Bernardi, Trabucchi and Grippa, 1966).

TABLE 2

Amination of aromatic compounds

Source of amino-radical	Aromatic compound	Initiator	Products	Yield %	Ref.
NH_2OH	PhH	$TiCl_3$	$PhNH_2$	—	1, 2
NH_2OH	PhH	$TiCl_3$	H_2N, H, H, H, NH_2, H	35	3
NH_2OH	PhH	$TiCl_4$	H_2N, H, H, H, NH_2, H	23	3
NH_2OH	PhMe	$TiCl_3$	H_2N, Me	15–20	4, 1
NH_2OH	PhMe	$TiCl_3$	H_2N, Me (*o*, *m*, *p* = 37, 20·5, 42·5%)	4	5, 2
NH_2OH	PhMe	$FeCl_2$	H_2N, Me (*o*, *m*, *p* = 41·2, 19·5, 39·3%)	5–15	6
NH_2OH	*m*-Xylene	$FeCl_2$ & $TiCl_3$	2,6-Dimethylaniline (21·6%) 2,4-Dimethylaniline (68·2%) 2,5-Dimethylaniline (10·2%)	5–15	6
NH_2OH	*m*-Xylene	$TiCl_3$	*m*-Xylidines (73%)	18	5

NH_2OH	Mesitylene	$FeCl_2$ & $TiCl_3$	2,4,6-Trimethylaniline	16	5
			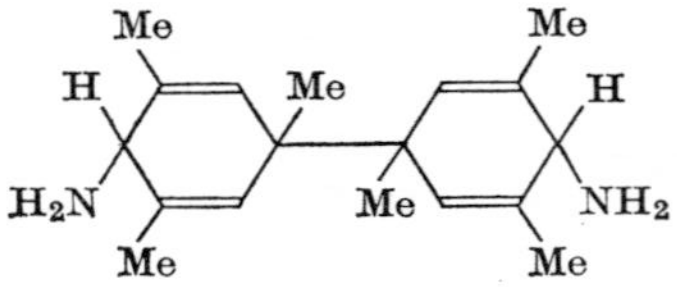 Me, Me, H, Me, H, H_2N, Me, NH_2, Me, Me	11	
NH_2OH	PhOMe	$FeCl_2$	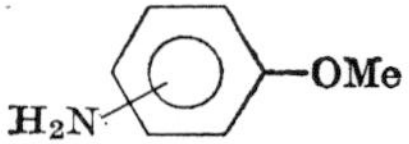 H_2N, OMe ($o, m, p = 52, 0, 48\%$)	5–15	6
NH_2OH	PhOMe	$TiCl_3$	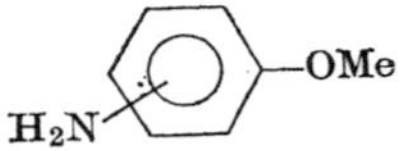 H_2N, OMe ($o, m, p = 63, 0, 37\%$)	18	7
NH_2OH	*p*-Methylanisole	$FeCl_2$	2-Amino-4-methylanisole (51·9%) 3-Amino-4-methylanisole (48·1%)	5–15	6
NH_2OH	PhCl	$FeCl_2$	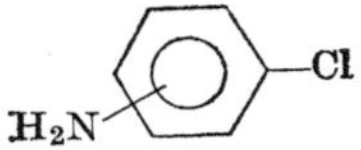 H_2N, Cl ($o, m, p = 4{\cdot}5, < 8, > 86\%$) $PhNH_2$ (1·5%)	—	8
NH_2OH	PhBr	$FeCl_2$	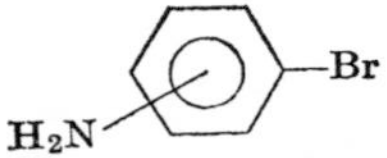 H_2N, Br ($o, m, p = 3{\cdot}5, < 8, > 87\%$) $PhNH_2$ (1·3%)	< 45	8

TABLE 2—*continued*

Source of amino-radical	Aromatic compound	Initiator	Products	Yield %	Ref.
NH_2OH	$PhCO_2H$	$FeSO_4$	H_2N–C_6H_4–CO_2H (*o*, *m*, *p* = 2·9, 4·1, 3·8%)	11	9, 10
NH_2OH	Naphthalene	$FeCl_2$	α-Aminonaphthalene (40·4%) β-Aminonaphthalene (59·6%)	5–15	6
$^+NH_3OSO_3^-$	PhMe	$FeCl_2$ or $FeSO_4$	H_2N–C_6H_4–Me (*o*, *m*, *p* = 37, 21·5, 41·5%)	15	7, 6
$^+NH_3OSO_3^-$	*m*-Xylene	$FeCl_2$	2,6-Dimethylaminobenzene (10.8%) 2,4-Dimethylaminobenzene (78.3%) 2,5-Dimethylaminobenzene (10.9%)	10–40	6
$^+NH_3OSO_3^-$	PhOMe	$FeCl_2$ or $FeSO_4$	H_2N–C_6H_4–OMe (*o*, *m*, *p* = 34, 0, 66%)	38	7, 6
$^+NH_3OSO_3^-$	*p*-Methylanisole	$FeCl_2$	2-Amino-4-methylanisole (88·5%) 3-Amino-4-methylanisole (11·5%)	10–40	6
$^+NH_3OSO_3^-$	*m*-Dimethoxy-benzene	$FeSO_4$	2,4-Dimethoxyaniline (87%) 2,6-Dimethoxyaniline (?) (13%)	—	7

$^{+}NH_3OSO_3^{-}$	PhCl	$FeCl_2$	$H_2N-C_6H_4-Cl$ ($o, m, p = 10, < 8, > 75\%$) $PhNH_2$ (10%)	—	8
$^{+}NH_3OSO_3^{-}$	PhBr	$FeCl_2$	$H_2N-C_6H_4-Br$ ($o, m, p = 7, < 8, > 77\%$) $PhNH_2$ (8%)	—	8
$^{+}NH_3OSO_3^{-}$	PhI	$FeCl_2$	$H_2N-C_6H_4-I$ ($o, m, p = 7{\cdot}7, < 8, > 76\%$) $PhNH_2$ (6·3%)	—	8
$^{+}NH_3OSO_3^{-}$	Naphthalene	$FeCl_2$	α-Aminonaphthalene (76·8%) β-Aminonaphthalene (23·2%)	—	6
MeNHCl	PhH	$FeSO_4$	PhNHMe	45	11
EtNHCl	PhH	$FeSO_4$	PhNHEt	43	11
$Pr^{i}NHCl$	PhH	$FeSO_4$	$PhNHPr^{i}$	27	11
$Bu^{n}NHCl$	PhH	$FeSO_4$	$PhNHBu^{n}$	27	11
N-chlorocyclo-hexylamine	PhH	$FeSO_4$	$PhNHC_6H_{11}$	45	11
MeNHCl	PhMe	$FeSO_4$	$MeNH-C_6H_4-Me$ ($o, m, p = 27{\cdot}5, 25, 47{\cdot}5\%$)	21	11

TABLE 2—*continued*

Source of amino-radical	Aromatic compound	Initiator	Products	Yield %	Ref.
MeNHCl	*p*-Xylene	$FeSO_4$	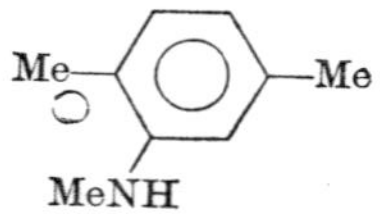 Me —Me MeNH	24	11
Me_2NCl	PhH	$FeSO_4$	Me_2NPh	61	12, 13, 14, 15
Me_2NCl	PhH	CuCl, $h\nu$	Me_2NPh	81	16
Et_2NCl	PhH	$FeSO_4$	Et_2NPh	35	13, 12
Me_2NBr	PhMe	$FeSO_4$	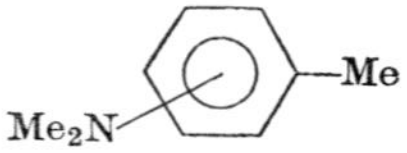 —Me Me_2N (o, m, $p = 12{\cdot}3$, $34{\cdot}5$, $53{\cdot}2\%$)	53	17
Me_2NCl	PhMe	$FeSO_4$	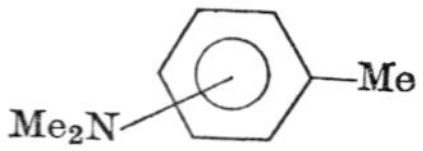 —Me Me_2N (o, m, $p = 9{\cdot}6$, $54{\cdot}2$, $36{\cdot}2\%$)	82	17, 12, 6, 11
Me_2NCl	PhMe	CuCl	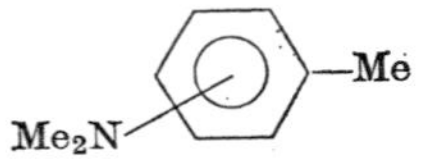 —Me Me_2N (o, m, $p = 21$, 46, 33%) (at 0–30°) (o, m, $p = 18$, 13, 69%)	80	13, 12 18

Me_2NCl	PhMe	$FeSO_4$	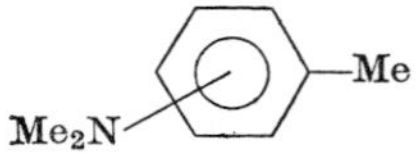 $Me_2NC_6H_4Me$ ($o, m, p = 9{\cdot}6, 43{\cdot}6, 46{\cdot}8\%$)	—	18
Me_2NCl	PhMe	$TiCl_3$	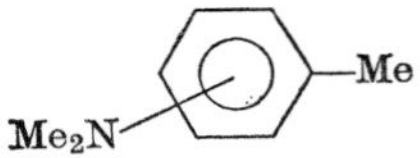 $Me_2NC_6H_4Me$ ($o, m, p = 8{\cdot}3, 36{\cdot}9, 54{\cdot}7\%$)	—	18
Me_2NCl	$PhCHMe_2$	$FeSO_4$	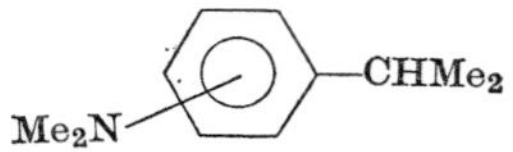 $Me_2NC_6H_4CHMe_2$ ($o, m, p = 6{\cdot}2, 40{\cdot}6, 53{\cdot}2\%$)	39	17
Me_2NCl	$PhCMe_3$	$FeSO_4$	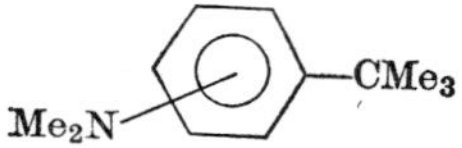 $Me_2NC_6H_4CMe_3$ ($o, m, p = 0, 22{\cdot}8, 77{\cdot}2\%$)	72	17
Me_2NCl	*m*-Xylene	$FeSO_4$	2,6-Dimethyl-*N*,*N*-dimethylaniline (0·5%) 2,4-Dimethyl-*N*,*N*-dimethylaniline (28%) 3,5-Dimethyl-*N*,*N*-dimethylaniline (71·5%)	—	19, 6
Me_2NCl	*o*-Xylene	$FeSO_4$	2,3-Dimethyl-*N*,*N*-dimethylaniline (4·8%) 3,4-Dimethyl-*N*,*N*-dimethylaniline (95·2%)	96	19
Me_2NCl	PhOH	$FeSO_4$	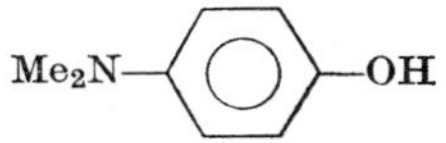 p-$Me_2NC_6H_4OH$	59	20
Me_2NCl	PhOMe	$FeSO_4$	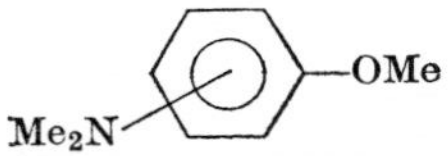 $Me_2NC_6H_4OMe$ ($o, m, p = 37, 0, 63\%$)	54	21

TABLE 2—*continued*

Source of amino-radical	Aromatic compound	Initiator	Products	Yield %	Ref.
Me_2NCl	PhCl	$FeSO_4$	$Me_2N-C_6H_4-Cl$ (*o*, *m*, *p* = 17, < 5, > 68%) $PhNMe_2$ (9%)	45	8, 12, 13
Me_2NCl	PhBr	$FeSO_4$	$Me_2N-C_6H_4-Br$ (*o*, *m*, *p* = 19, 3, 67%) $PhNMe_2$ (11%)	45	8
Me_2NCl	PhI	$FeSO_4$	$Me_2N-C_6H_4-I$ (*o*, *m*, *p* = 5, 8, 80%) *p*-Chlorodimethylaniline (5%) $PhNMe_2$ (2%)	45	8
Me_2NCl	Acetanilide	$FeSO_4$	*p*-*N*,*N*-Dimethylaminoacetanilide	93	22, 23
$PhCH_2NMeCl$	Acetanilide	$FeSO_4$	*p*-*N*-Methyl-*N*-benzylaminoacetanilide	88	22, 23
Me_2NCl	2-Methyl-acetanilide	$FeSO_4$	2-Methyl-4-*N*,*N*-dimethylaminoacetanilide	60	22
Me_2NCl	$PhCH_2.CH_2.NMe_2$	$FeSO_4$	2-(*N*,*N*-dimethylaminophenyl)ethyl-*N*,*N*-dimethylamine (*o*, *m*, *p* = 1.4, 66.3, 31·7%)	90	17

Me_2NCl	Tetralin	$FeSO_4$	α-Dimethylaminotetralin (30%) β-Dimethylaminotetralin (70%)	24	19
Me_2NCl	Naphthalene	$FeSO_4$	1-Dimethylaminonaphthalene (97%) 2-Dimethylaminonaphthalene (3%)	65	24, 6, 12, 13
Me_2NCl	1-Methylnaphthalene	$FeSO_4$	1,4-$MeC_{10}H_6NMe_2$ (95%)	74	24
Me_2NCl	1-Chloronaphthalene	$FeSO_4$	1,5-$ClC_{10}H_6NMe_2$ (75·7%) 1,8-$ClC_{10}H_6NMe_2$ (20·8%)	95	24
Me_2NCl	1-Bromonaphthalene	$FeSO_4$	1,5-$BrC_{10}H_6NMe_2$ (92·6%)	97	24
Me_2NCl	1-Nitronaphthalene	$FeSO_4$	1,5-$NO_2C_{10}H_6NMe_2$ (75·7%) 1,8-$NO_2C_{10}H_6NMe_2$ (24·3%)	20	24
Me_2NCl	8-Methoxyquinoline	$FeSO_4$	5-Dimethylamino-8-methoxyquinoline	64	21
Me_2NCl	Biphenyl	$FeSO_4$	*N*,*N*-Dimethyl-4-aminobiphenyl (18%) *N*,*N*,*N*′,*N*′-tetramethylbenzidine (78%)	—	14
Me_2NCl	4-Methoxybiphenyl	$FeSO_4$	3-*N*,*N*-Dimethylamino-4-methoxybiphenyl (65%) 4-Methoxy-4′-*N*,*N*-dimethylaminobiphenyl (35%)	—	25
Me_2NCl	4-*N*,*N*-Dimethyl-aminobiphenyl	$FeSO_4$	*N*,*N*,*N*′,*N*′-Tetramethylbenzidine	90	25
Me_2NCl	4-Chlorobiphenyl	$FeSO_4$	4-Chloro-4′-*N*,*N*-dimethylaminobiphenyl	84	25
Me_2NCl	4-Nitrobiphenyl	$FeSO_4$	4-Nitro-4′-*N*,*N*-dimethylaminobiphenyl	99	25
Me_2NCl	4-Biphenyl-sulphonic acid	$FeSO_4$	4-*N*,*N*-Dimethylamino-4′-biphenyl-sulphonic acid	86	25
Me_2NCl	Biphenyl-2-carboxylic acid	$FeSO_4$	2-*N*,*N*-Dimethylaminofluorenone	—	25
Me_2NCl	Oxindole	$FeSO_4$	5-*N*,*N*-Dimethylamino-oxindole	—	22
Me_2NCl	Fluorene	$FeSO_4$	*N*,*N*-Dimethyl-2-aminofluorene (16%) *N*,*N*,*N*′,*N*′-Tetramethyl-2,7-diamino-fluorene (72%)	—	14
Me_2NCl	Fluorenone	$FeSO_4$	2-*N*,*N*-Dimethylaminofluorenone	98	25
N-Chloro-piperidine	PhH	$FeSO_4$	N-Phenyl-piperidine	65	13 ,12, 15

TABLE 2—*continued*

Source of amino-radical	Aromatic compound	Initiator	Products	Yield %	Ref.
N-Chloro-piperidine	PhMe	$FeSO_4$	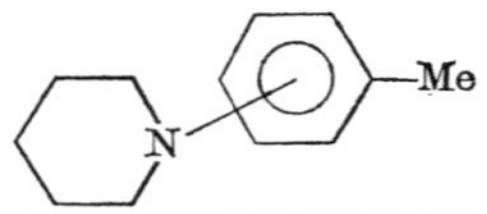 N —Me (*o*, *m*, *p* = 4·6, 39·2, 56·2%)	41	26, 17, 6
N-Chloro-piperidine	PhOH	$FeSO_4$	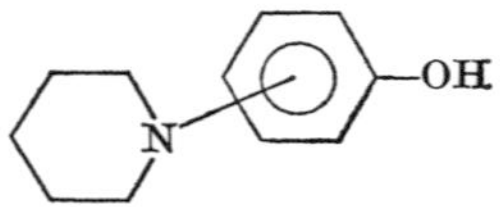 N —OH (*o*, *m*, *p* = 9, 0, 91%)	87	20
N-Chloro-piperidine	*o*-Chlorophenol	$FeSO_4$	*N*-(3-Chloro-4-hydroxyphenyl)-piperidine	92	20
N-Chloro-piperidine	PhOMe	$FeSO_4$	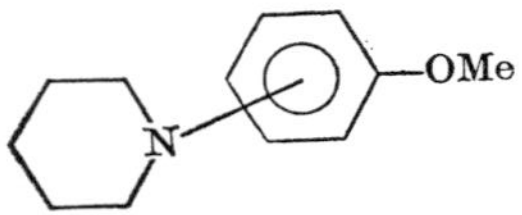 N —OMe (*o*, *m*, *p* = 4·5, 0, 90·5%)	11	27, 21, 28, 6
			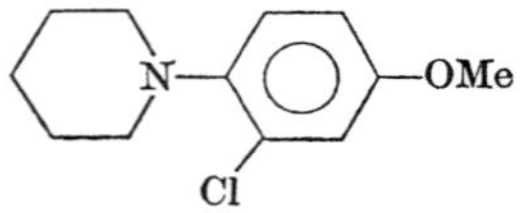 N— —OMe (5%) Cl	—	
N-Chloro-piperidine	PhOMe	iron (II)-lactate	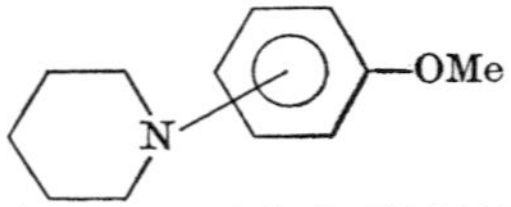 N —OMe (*o*, *m*, *p* = 4·4, 0, 37·6%)	19·5	28, 27

			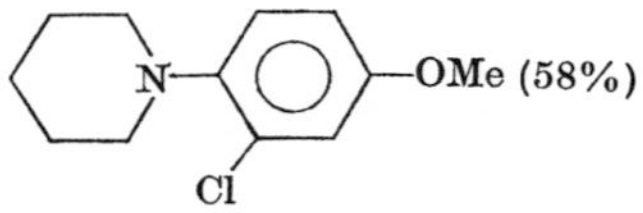 N, Cl, OMe (58%)	—	
N-Chloro-piperidine	1,3-Dimethoxy-benzene	$FeSO_4$	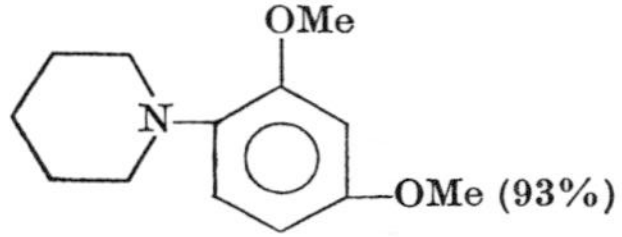 N, OMe, OMe (93%)	25·5	28, 27
N-Chloro-piperidine	Acetanilide	$FeSO_4$	*N*-(4-Acetamidophenyl)-piperidine	98·5	22, 23
N-Chloro-piperidine	8-Methoxyquinoline	$FeSO_4$	5-Piperidino-8-methoxyquinoline	—	21
N-Chloro-piperidine	4-Nitrobiphenyl	$FeSO_4$	4-Nitro-4′-piperidinobiphenyl	85	25
N-Chloro-piperidine	Biphenyl-4-sulphonic acid	$FeSO_4$	4-Piperidino-4′-biphenylsulphonic acid	76	25
N-Chloro-piperidine	Oxindole	$FeSO_4$	5-Piperidino-oxindole	—	22

References: (1) Seaman, Taylor and Waters (1954). (2) Albisetti, Coffman, Hoover, Jenner and Mochel (1959). (3) Frank, Mador, Rekers and Tonne (1960). (4) Hoover (1961). (5) Minisci, Galli, Cecere and Mondelli (1965). (6) Minisci, Galli and Cecere (1965). (7) Minisci and Galli (1965b). (8) Minisci, Bernardi, Grippa and Trabucchi (1966). (9) Yoshida, Matsumoto and Oda (1962). (10) Yoshida, Matsumoto and Oda (1964a). (11) Minisci, Galli and Cecere (1966c). (12) Bock and Kompa (1966a). (13) Bock and Kompa (1965). (14) Minisci, Trabucchi and Galli (1966). (15) Montecatini Edison (1967). (16) Bock and Kompa (1966b). (17) Minisci, Galli and Bernardi (1967a). (18) Minisci, Galli, Bernardi and Galli (1967). (19) Minisci, Bernardi, Trabucchi and Grippa (1966). (20) Minisci, Galli and Perchinunno (1969b). (21) Minisci, Galli, Cecere and Trabucchi (1966). (22) Minisci, Galli and Cecere (1966e). (23) Minisci, Galli, Cecere and Quilico (1969). (24) Minisci, Trabucchi, Galli and Bernardi (1966). (25) Minisci and Cecere (1967). (26) Minisci, Galli and Bernardi (1966). (27) Minisci, Galli and Pollina (1965). (28) Minisci and Galli (1965a).

The reactions of *N*,*N*-dimethyl-2-phenylethylamine and *N*,*N*-dimethylbenzylamine with *N*-chlorodimethylamine also produced mainly *m*-isomers, and showed a strong inductive effect. The difference in reactivity of the former over the latter was more than 1000-fold (Minisci, Galli, Galli and Bernardi, 1967). The dialkylamination of unsubstituted polycyclic aromatic compounds, such as naphthalene, is highly selective, producing 97% of α-isomers (Minisci, Galli and Cecere, 1965; Minisci, Trabucchi, Galli and Bernardi, 1966). In polynuclear aromatic compounds containing electron-donating groups, the dialkylamination is preferentially directed into the substituted ring. Thus, the reaction of α-methylnaphthalene with *N*-chlorodimethylamine yielded 74% of dimethylaminomethylnaphthalene containing 95% of the 1,4-isomer (Minisci, Trabucchi, Galli and Bernardi, 1966).

Compounds containing deactivating groups, such as $—\overset{+}{N}HR_2$, $—NO_2$, —CO.OH, etc., undergo no substantial reactions. However, introduction into such compounds of electron-donating substituents can make these compounds reactive towards dialkylamination. For example, whereas the decomposition of *N*-chlorodimethylamine in quinoline produced no reaction, the reaction with 8-methoxyquinoline yielded 5-dimethylamino-8-methoxyquinoline (Minisci, Galli, Cecere and Trabucchi, 1966; Montecatini Edison, 1967).

In polycyclic aromatic systems with electron-attracting groups, the dialkylamination is directed towards the unsubstituted ring (Minisci, Trabucchi, Galli and Bernardi, 1966). Thus, reaction of α-nitronaphthalene with *N*-chlorodimethylamine gave a mixture of 1-nitro-5- and 1-nitro-8-dimethylaminonaphthalene derivatives (Minisci, Trabucchi, Galli and Bernardi, 1966; Montecatini Edison, 1967). Similarly, various 4-substituted biphenyl derivatives were dialkylaminated with *N*-chlorodimethylamine and *N*-chloropiperidine to give the corresponding 4′-dialkylaminated derivatives in high yields (Table 2) (Minisci, Trabucchi and Galli, 1966; Minisci and Cecere, 1967; Montecatini Edison, 1967). In an analogous way, the reaction of fluorene yielded a mixture of 16% *N*,*N*-dimethyl-2-aminofluorene and 72% of *N*,*N*,*N*′,*N*′-tetramethyl-2,7-diaminofluorene (Minisci, Trabucchi and Galli, 1966). However, although the reaction of biphenyl-2-carboxylic acid resulted in the cyclized product (XVI) (Minisci and Cecere, 1967), the orientation of the substitution reaction remained the same as in preceding reactions, i.e. *para* (Minisci, Trabucchi and Galli, 1966).

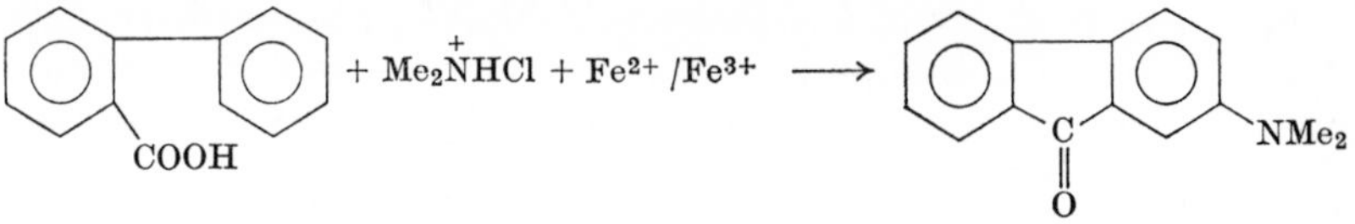

$$\text{(biphenyl-2-carboxylic acid)} + Me_2\overset{+}{N}HCl + Fe^{2+}/Fe^{3+} \longrightarrow \text{(XVI)}$$

COOH

NMe$_2$

(XVI)

4. *Mechanism of Dialkylamination Reactions*

In summary, on the basis of the preceding discussion, and the fact that the reactions are inhibited by oxygen, nitrogen oxides and acrylic esters (Minisci, Trabucchi and Galli, 1966; Minisci, 1967) it is concluded that the dialkylamination reactions are chain processes involving an electrophilic radical species (Minisci, 1967). The initiating step in acid solution is as follows.

$$R_2\overset{+}{N}HCl + M^n \rightarrow R_2\overset{+\bullet}{N}H + M^{n+1} + Cl^-$$

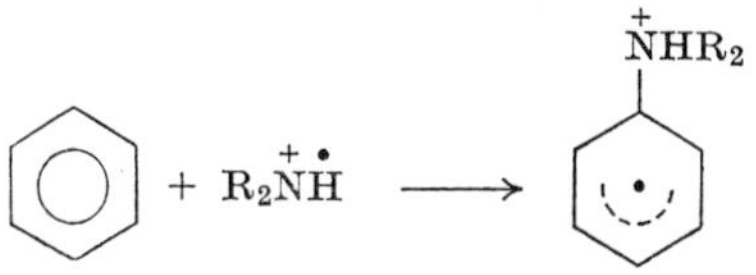

$$C_6H_6 + R_2\overset{+\bullet}{N}H \longrightarrow \overset{+}{N}HR_2\text{-cyclohexadienyl radical}$$

The re-aromatisation step can proceed either by path (*a*),

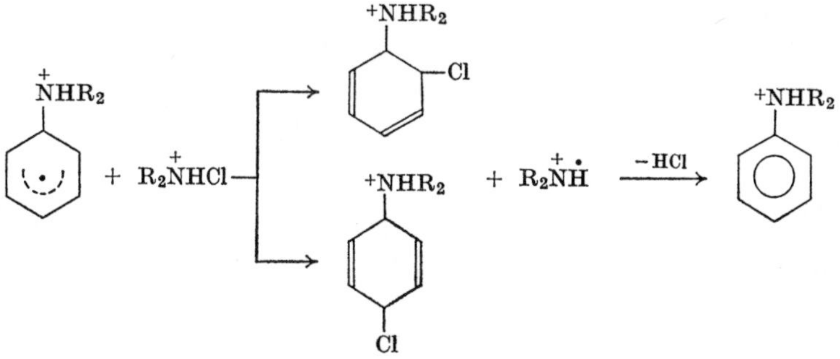

$$\overset{+}{N}HR_2\text{-cyclohexadienyl radical} + R_2\overset{+}{N}HCl \longrightarrow \text{(}{}^{+}NHR_2\text{, Cl-cyclohexadienes)} + R_2\overset{+\bullet}{N}H \xrightarrow{-HCl} C_6H_5\overset{+}{N}HR_2 \qquad (a)$$

or by path (*b*).

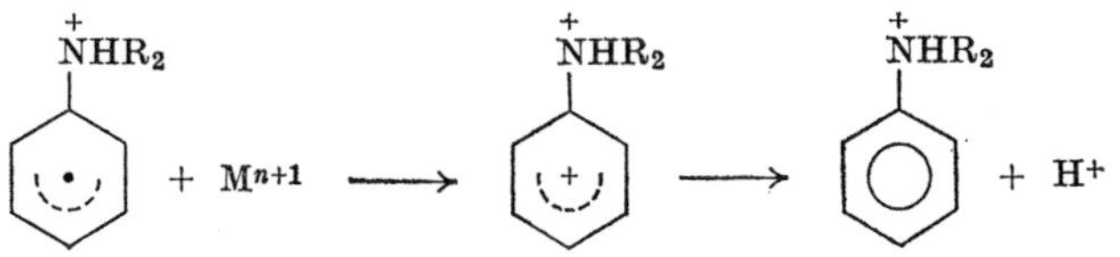

$$\overset{+}{N}HR_2\text{-cyclohexadienyl radical} + M^{n+1} \longrightarrow \overset{+}{N}HR_2\text{-cyclohexadienyl cation} \longrightarrow C_6H_5\overset{+}{N}HR_2 + H^+ \qquad (b)$$

The aromatisation process (*a*) is favoured by low metal ion concentrations and high chloroamine concentrations, whereas the aromatisation process (*b*) is favoured by high metal ion concentrations and low chloroamine concentrations.

D. Amination and Aminohalogenation of Unsaturated Compounds

During the last two decades a number of publications have appeared dealing with the formation of amino-radicals from hydroxylamine and derivatives in the presence of chromium, copper, iron, titanium, and vanadium compounds. The amino-radicals were used to initiate polymerisation (Howard, 1951, 1954; Schuster, Gehm and Völkl, 1958; Fischer and Pilgram, 1961; Bro and Scheyer, 1962), to study kinetics and electron paramagnetic resonance (Corjava, Fischer and Giacometti, 1965; Dewing, Longster, Myatt and Todd, 1965; Griffiths, Longster, Myatt and Todd, 1967) and in organic syntheses. Only the last topic will be discussed in this section (Albisetti *et al.*, 1959; Minisci, 1967, 1968; Neale, 1964, 1967). The decomposition of hydroxylamine and derivatives in non-acidic media produces the unprotonated amino-radical (XVII) (Albisetti *et al.*, 1959; Minisci, 1967).

$$R_2NOH + M^n \longrightarrow \underset{\text{(XVII)}}{R_2N\cdot} + M^{n+1} + OH^-$$

(R = H, alkyl; M = Cu, Cr, Fe, V, Ti; n = 1, 2, 3)

In strongly acidic media or with hydroxylamine-*O*-sulphonic acid, the aminium radical, i.e., the protonated amino radical (XVIII) is probably formed. The amino-radicals can undergo the following

$$^+NH_3OSO_3^- + M^n \longrightarrow \underset{\text{(XVIII)}}{\overset{+\ \bullet}{N}H_3} + M^{n+1} + SO_4^{2-}$$

series of reactions, depending on the prevailing reaction conditions. In the presence of large quantities of reducing ions amino-radicals may be converted to the corresponding amine (Minisci, 1967).

$$R_2N\cdot + M^n + H_2O \longrightarrow RN_2H + M^{n+1} + OH^-$$

In the presence of mono-olefins, addition of amino-radicals to the olefinic bond can occur to give the corresponding carbon radical (XIX).

$$R_2N\cdot + CH_2{=}CHR' \longrightarrow \underset{\text{(XIX)}}{R_2NCH_2\dot{C}HR'}$$

(R′ = alkyl, aryl)

These radicals (XIX) may then undergo either reduction to give saturated amines (Seaman, Taylor and Waters, 1954; Minisci, 1967; Albisetti *et al.*, 1959),

$$R_2NCH_2\dot{C}HR' + M^n + H_2O \longrightarrow RCH_2CH_2R' + M^{n+1} + OH^-$$
$$\text{(XIX)}$$

or, they can be involved in an electron-ligand transfer process to give chloroamine derivatives (XX). For example, the reaction of ethylene

$$R_2NCH_2\dot{C}HR' + M^{n+1} + X^- \longrightarrow \underset{\text{(XX)}}{R_2NCH_2CHXR'} + M^n$$
$$(X = OH, \text{halogens}, N_3, NR_2, \text{etc.})$$

with hydroxylamine in the presence of vanadium-(III)-sulphate gave a mixture consisting of ethylamine, butylamine, ethanolamine, 4-hydroxybutylamine, plus higher boiling amines (Albisetti *et al.*, 1959). These products may be explained by the following scheme.

$$\overset{+}{N}H_3OH + V^{3+} \longrightarrow \cdot NH_2 + V^{4+} + H_2O$$
$$nCH_2{=}CH_2 + \cdot NH_2 \longrightarrow NH_2(CH_2CH_2)_n\cdot$$
$$NH_2(CH_2CH_2)_n\cdot + V^{3+} + H_2O \longrightarrow NH_2(CH_2CH_2)_nH + V^{4+} + OH^-$$
$$NH_2(CH_2CH_2)_n\cdot + V^{4+} + H_2O \longrightarrow NH_2(CH_2CH_2)_nOH + V^{3+} + OH^-$$

The reaction of styrene with hydroxylamine-*O*-sulphonic acid in methanol in the presence of ferrous sulphate formed the amino-ether (XXI) (Minisci and Galli, 1965b),

$$PhCH{=}CH_2 + \dot{\overset{+}{N}}H_3 \longrightarrow Ph\dot{C}HCH_2\overset{+}{N}H_3$$
$$Ph\dot{C}HCH_2\overset{+}{N}H_3 + Fe^{3+} + MeOH \longrightarrow \underset{\underset{\text{(XXI)}}{OMe}}{PhCHCH_2\overset{+}{N}H_3} + Fe^{2+} + H^+$$

whereas this reaction in the presence of ferrous chloride gave the chloroamine derivative (XXII) (Minisci and Galli, 1965b).

$$Ph\dot{C}HCH_2\overset{+}{N}H_3 + Fe^{3+} + Cl^- \longrightarrow \underset{\text{(XXII)}}{PhCHCl.CH_2\overset{+}{N}H_3} + Fe^{2+}$$

The addition of a base to this type of chloroamine results in formation of aziridines. Thus, compound (XXII) and 2-chloro-2-methyl-2-phenylethylamine prepared from hydroxylamine or hydroxylamine-*O*-sulphonic acid and ferrous chloride, were converted to 2-phenyl- and 2-methyl-2-phenyl-aziridine (XXIII), respectively (Minisci, Galli and Cecere, 1966b).

$$\underset{\substack{| \\ R}}{PhCClCH_2NH_2} \xrightarrow{\text{base}} \text{PhC(R)—CH}_2\text{ (aziridine ring via NH)}$$

(R = H, Me) (XXIII)

In the presence of large quantities of sodium azide, aminozidation of styrene occurred producing 2-azido-2-phenylethylamine (XXIV) (Minisci, Galli and Cecere, 1966a).

$$Ph\dot{C}HCH_2\overset{+}{N}H_3 + Fe^{3+} + N_3^- \longrightarrow PhCH(N_3)CH_2\overset{+}{N}H_3$$

(XXIV)

In the case of conjugated dienes, such as butadiene and derivatives, the addition of amino-radicals to the diene yields resonance-stabilised allyl radical (XXV).

$$R_2N\cdot + CH_2{=}CH{-}CH{=}CH_2 \longrightarrow R_2NCH_2\dot{C}HCH{=}CH_2 \longleftrightarrow R_2NCH_2CH{=}CHCH_2\cdot$$

(XXV)

As in the case of the amino-radical adducts to mono-olefins, radical (XXV) may undergo, depending on the reaction conditions, either reduction to give alkeneamines (XXVIa, b), or a concomitant

$$R_2NC_4H_6\cdot \longrightarrow R_2NCH_2CH{=}CHCH_2NR_2 + R_2NCH_2\underset{\substack{| \\ NR_2}}{C}HCH{=}CH_2$$

(XXVIa) (XXVIb)

electron and ligand transfer process to give derivatives (XXVII), or

$$R_2NC_4H_6\cdot + M^{n+1} + X^- \longrightarrow R_2NC_4H_6X + M^n$$

(XXVII)

dimerisation to give a mixture of unsaturated diamines (XXVIIIa, b).

$$R_2NC_4H_6\cdot \longrightarrow R_2N(CH_2CH{=}CHCH_2)_2NR_2 +$$

(XXVIIIa)

$$R_2NCH_2CH{=}CHCH_2CH(CH{=}CH_2)CH_2NR_2$$

(XXVIIIb)

Because of lack in selectivity these reactions are in most cases not particularly useful from the synthetic point of view since mixtures of products are obtained in poor to moderate yields.

For example, the reaction of butadiene in acidic methanolic solution at $-7°$ in the presence of titanous chloride gave a 42% yield of isomeric diamino-octadienes which, on hydrogenation, resulted in a mixture of diamino-octanes, consisting of more than 70% of straight chain isomer (Coffman and Jenner, 1957; National Distillers Prod. Corp., 1957; Albisetti *et al.*, 1959; Mador and Rekers, 1962).

N-Haloalkylamines and *N*-haloheterocycloalkylamines can also undergo the above series of reactions (Neale, 1964, 1967; Minisci and Galli, 1964; Minisci, 1967 and references therein) (Table 3). The initiating step in these reactions is as follows (Minisci and Galli, 1963; Minisci, 1967).

$$R_2NX \text{ or } \bigcirc NX + M^n \rightarrow R_2N\bullet \text{ or } \bigcirc N\bullet + M^{n+1} + X^-$$

$$R = \text{alkyl}$$

$$\bigcirc N = \text{piperidino}, \text{ morpholino}$$

For example, the reaction of *N*-chloropiperidine with styrene or *p*-chlorostyrene in the presence of ferric chloride resulted in adducts (XXIX) which were transformed into amino-ethers (XXX) (Minisci, Galli and Pollina, 1965).

$$RCH{=}CH_2 + \bullet N\bigcirc + FeCl^{2+} \longrightarrow$$

$$RCHClCH_2N\bigcirc + Fe^{2+} \xrightarrow{MeO^-} RCH(OMe)CH_2N\bigcirc + Cl^-$$

(XXIX) (XXX)

$(R = Ph, p\text{-Cl-Ph})$

The reactions of haloamines with butadiene and derivatives give mainly two types of products (Minisci, Galli and Pollina, 1965; Minisci, 1967). In the presence of cuprous or titanous chlorides, or ferrous sulphate, plus large amounts of halide ions, 1-chloro-4-dialkylamino-but-2-ene derivatives are formed (Minisci and Galli, 1963), whereas in the presence of ferrous sulphate alone mainly 1,8-dialkylamino-octa-2,6-diene, accompanied by some branched chain isomers and 1-chloro-4-alkylaminobut-2-ene derivatives is obtained (Minisci and Galli, 1964a) (Table 3). For example, the

TABLE 3

Amination and aminochlorination of olefinic hydrocarbons

Source of amino radicals	Olefinic hydrocarbon	Initiator	Products	Yield %	Ref.
NH_2OH	$CH_2{=}CH_2$	$V_2(SO_4)_3$	$EtNH_2$	5	1
			$BuNH_2$	15	
			$HOCH_2CH_2NH_2$	38	
			4-Hydroxybutylamine	18	
NH_2OH	*Trans*-but-2-ene	$TiCl_3$	1,2,3,4-Tetramethyltetramethylenediamine	12	1
			1-Methylpropylamine	8·5	
NH_2OH	Trimethylethylene	$FeCl_2$	2-Amino-3-chloro-3-methyl butane	12	2
NH_2OH	$CH_2{=}CH.CH{=}CH_2$	$TiCl_3$	Diaminooctane (after hydrogenation of adduct dimer)	73	3, 1
NH_2OH	$CH_2{=}CH.CH{=}CHCH_3$	$TiCl_3$	Diaminodecadiene	54	4
NH_2OH	Hex-1-ene	$FeCl_2$	1-Amino-2-chlorohexane	32	2
NH_2OH	Hex-1-ene	$TiCl_3$	2,3-Dibutyltetramethylenediamine	10	1
NH_2OH	Cyclohexene	$FeCl_2$ & $FeCl_3$	1-Amino-2-chlorocyclohexane (*cis*-72%, *trans*-28%)	24	2, 5, 9
NH_2OH	Cyclohexadiene	$TiCl_3$	Diaminobicyclohexyl	12	1
NH_2OH	Styrene	$FeCl_2$	2-Chloro-2-phenylethylamine	13·5	2
MeNHOH	$CH_2{=}CH.CH{=}CH_2$	$TiCl_3$	*N*,*N'*-Dimethylaminooctadienes	39	1
$^+NH_3OSO_3^-$	Cyclohexene	$FeCl_3$ or $FeCl_2$	1-Amino-2-chlorocyclohexane (*cis* 30%, *trans* 70%)	28	2, 5
$^+NH_3OSO_3^-$	Hex-1-ene	$FeCl_2$	1-Amino-2-chlorohexane	24	5
$^+NH_3OSO_3^-$	Styrene	$FeCl_2$	2-Chloro-2-phenylethylamine	38	2, 5
Et_2NCl	$CH_2{=}C{=}CH_2$	$Fe(NH_4)_2(SO)_4)_2$	$Et_2NCH_2CCl{=}CH_2$	35	6, 7
Et_2NCl	$CH_2{=}CHMe$	$Fe(NH_4)_2(SO_4)_2$	$Et_2NCH_2.CHClMe$	42	6, 7
Et_2NCl	$CH_2{=}CH.CH{=}CH_2$	$FeSO_4$ or CuCl	$Et_2NCH_2.CH{=}CH.CH_2Cl$ (*trans*)	—	8

Et_2NCl	$CH_2{=}CH.CH_2Me$	$Fe(NH_4)_2(SO_4)_2$	$Et_2NCH_2.CHCl.CH_2Me$	33	6, 7
Et_2NCl	$(CH_2{=}CH.CH_2CH_2)_2$	$Fe(NH_4)_2(SO_4)_2$	$(Et_2NCH_2.CHCl.CH_2.CH_2)_2$	49	7
$Bu^n{}_2NCl$	$CH_2{=}CH.CH{=}CH_2$	$FeCl_3$ or $CuCl_2$	$Bu^n{}_2NCH_2.CH{=}CH.CH_2Cl$ (*trans*)	—	8
$Bu^n{}_2NCl$	$CH_2{=}CH.CH_2Me$	$Fe(NH_4)_2(SO_4)_2$	$Bu^n{}_2NCH_2.CHCl.CH_2Me$	16	7
$Bu^n{}_2NCl$	$CH_2{=}CH(CH_2)_3Me$	$FeCl_3$	Adduct	—	8
$Bu^i{}_2NCl$	Cyclohexene	$TiCl_3$	*N*-2-Chlorocyclohexyldi-isobutylamine (*cis* & *trans*)	—	9
N-Chloromorpholine	$CH_2{=}CH.CH{=}CH_2$	$TiCl_3$	$O(CH_2CH_2)_2N{-}CH_2CH{=}CHCH_2Cl$ (*trans*)	—	8
N-Chloropiperidine	Hex-1-ene	$FeCl_3$ & $FeSO_4$	*N*-Chlorohexylpiperidine	62	8, 10, 11
N-Chloropiperidine	But-2-ene	$FeCl_3$	Adduct	—	8
N-Chloropiperidine	Cyclohexene	$FeSO_4$ (acidic)	*N*-2-Chlorocyclohexylpiperidine (*cis* & *trans*)	54	11
N-Chloropiperidine	Cyclohexene	$FeCl_3$ & $FeSO_4$	*N*-2-Chlorocyclohexylpiperidine	80	8, 9
N-Chloropiperidine	Cyclohexene	$TiCl_3$	*N*-2-Chlorocyclohexylpiperidine	68	8
N-Chloropiperidine	Cyclohexene	CuCl	*N*-2-Chlorocyclohexylpiperidine	—	8
N-Chloromorpholine	Cyclohexene	$FeCl_3$	*N*-2-Chlorocyclohexylmorpholine	—	8
N-Chloropiperidine	Styrene	$FeSO_4$ & $FeCl_3$	*N*-(2-Chloro-2-phenylethyl)-piperidine	58	11, 8

References: (1) Albisetti, Coffman, Hoover, Jenner and Mochel (1959). (2) Minisci, Galli and Cecere (1966a). (3) Mador and Rekers (1962). (4) Gadzhiev, Mamedov and Gabelaya (1969). (5) Minisci and Galli (1965b). (6) Neale (1964). (7) Neale (1967). (8) Minisci, Galli and Pollina (1965). (9) Minisci, Galli and Cecere (1966b). (10) Minisci and Galli (1964c). (11) Minisci, Galli and Cecere (1966d).

reaction of *N*-chloromorpholine in the presence of titanous chloride gave adduct (XXXI) (Minisci and Galli, 1964a; Minisci, 1967), and

$$\text{O}\langle\rangle\text{NCl} + \text{Ti}^{3+} \longrightarrow \text{O}\langle\rangle\text{N}\cdot + \text{TiCl}^{3+}$$

$$\text{O}\langle\rangle\text{N}\cdot + \text{CH}_2{=}\text{CH}{-}\text{CH}{=}\text{CH}_2 \longrightarrow \text{O}\langle\rangle\text{NCH}_2\text{CH}{=}\text{CHCH}_2\cdot$$

$$\text{O}\langle\rangle\text{NCH}_2\text{CH}{=}\text{CHCH}_2\cdot + \text{TiCl}^{3+} \longrightarrow \text{O}\langle\rangle\text{NCH}_2\text{CH}{=}\text{CHCH}_2\text{Cl}$$

(XXXI)

the reaction of *N*-chloropiperidine in the presence of ferrous sulphate yielded a substantial amount of a mixture of products (XXXIIa, b) (Minisci, Galli and Pollina, 1965).

$$\langle\rangle\text{N}\cdot + \text{CH}_2{=}\text{CH}{-}\text{CH}{=}\text{CH}_2 + \text{Fe}^{2+} \longrightarrow \langle\rangle\text{N(CH}_2\text{CH}{=}\text{CHCH}_2)_2\text{N}\langle\rangle$$

(XXXIIa)

$$+ \langle\rangle\text{NCH}_2\text{CH}{=}\text{CHCH}_2\underset{\underset{\text{CH}{=}\text{CH}_2}{|}}{\text{C}}\text{HCH}_2\text{N}\langle\rangle$$

(XXXIIb)

In general the aminochlorination of unsaturated compounds with hydroxylamine or hydroxylamine-*O*-sulphonic acid in the presence of large quantities of halide ions gives low yields of products, usually less than 40% (Minisci, Galli and Cecere, 1966a; Minisci and Galli, 1965b). From the synthetic point of view, hydroxylamine-*O*-sulphonic acid is less useful than other aminochlorination methods because of its instability, although the products are rather pure and free from by-products.

Aminohalogenation using *N*-haloalkylamines seems to produce better results than those obtained with hydroxylamine derivatives. To avoid ionic reactions, conditions similar to those employed in aromatic substitution are used, i.e., in highly acidic media, large amounts of metal ions, usually a mole per mole of haloamine. As previously mentioned in the case of hydroxylamine systems, under these conditions, reductive side processes may occur at the expense of aminochlorination and electrophilic chlorination (Minisci, Galli and Pollina, 1965; Minisci, Galli and Cecere, 1966d; Minisci, 1967).

$$R_2\overset{+}{N}H + M^n + H + H^+ \longrightarrow R_2\overset{+}{N}H_2 + M^{n+1}$$

The orientation in addition reactions of haloalkylamines to straight chain olefins indicates the radical nature of the aminohalogenation reactions, namely, the alkylamino-radical attacks the terminal carbon atom of the double bond.

$$R_2NCl + CH_2{=}CHR' + M^n \longrightarrow R_2NCH_2CHClR' + M^{n+1}$$
$$(R' = \text{alkyl, aryl})$$

Further evidence for a radical mechanism is obtained by carrying out the reaction of an olefin with *N*-haloalkylamines in the presence of nitric oxide and ferrous sulphate to give a mixture of nitrosoamine (XXXIII) and the amino-oxime derivative (XXXIV).

$$R_2NCl + Fe^{2+} \longrightarrow R_2N\cdot + FeCl^{2+}$$

$$R_2N\cdot + NO \longrightarrow \underset{\text{(XXXIII)}}{R_2NNO}$$

$$R_2N\cdot + CH_2{=}CHR' \longrightarrow R_2NCH_2\dot{C}HR'$$

$$R_2NCH_2\dot{C}HR' + NO \longrightarrow R_2NCH_2\underset{\substack{|\\ NO}}{C}HR' \longrightarrow \underset{\text{(XXXIV)}}{R_2NCH_2\underset{\substack{\|\\ NOH}}{C}R'}$$

Thus, the reaction of styrene with *N*-chloropiperidine resulted in formation of *N*-nitrosopiperidine and the amino-oxime derivative (XXXIV; $R_2N =$ (piperidino ring)N, $R' = Ph$) (Minisci and Galli, 1966b).

Similarly the corresponding nitroso-derivatives were prepared from *N*-chloromorpholine, *N*-chlorodiethylamine and *N*-chlorodi-n-butylamine in the absence of olefin (Minisci and Galli, 1964b).

The reactions of *N*-chlorodialkylamines and *N*-chloroheterocycloalkylamines with olefins in the presence of ferrous or titanous ion and large quantities of oxygen result in formation of α-aminoketones (XXXV) (Minisci and Galli, 1964d; Minisci, Galli and Pollina, 1965).

Thus, these reactions of styrene, *trans*-stilbene and isosafrole with *N*-chloro-dibutylamine, -morpholine, and -piperidine produced the corresponding α-aminoketones, ranging in yield from 21·7 to 76%

$$R_2N\cdot + CH_2{=}CHR' \longrightarrow R_2NCH_2\dot{C}HR'$$

$$R_2NCH_2\dot{C}HR' + O_2 \longrightarrow R_2NCH_2CHR'(O{-}O\cdot)$$

$$R_2NCH_2CHR'(O{-}O\cdot) + Fe^{2+} \longrightarrow R_2NCH_2C({=}O)R' + FeOH^{2+}$$

(XXXV)

(R = aliphatic, alicyclic, heterocyclic; R′ = alkyl, aryl)

(Minisci and Galli, 1964d; Minisci, Galli and Pollina, 1965). Reactions with aliphatic and cyclic olefins produced less reliable results. An analogous reaction of diethylhydroxylamine, ferrous sulphate, titanous chloride, and styrene gave the corresponding aminoketone (Minisci, Galli, Cecere and Mondelli, 1965).

The reactions of *N*-chloropiperidine and *N*-chloromorpholine with the same olefin gave markedly different yields. Although from the steric point of view both molecules are comparable, the morpholino-radical has a more pronounced electrophilic character, and because of this fact, it is more susceptible to reduction than the piperidino-radical at the expense of the product.

The use of olefins containing strongly electron-withdrawing groups, e.g., vinylpyridine, results in very low yields of product (Minisci and Galli, 1964d; Minisci, Galli and Pollina, 1965). The electrophilic character of amino-radicals was further demonstrated by the following experiments. The reaction of *N*-chloropiperidine with hex-1-ene in the presence of ferric chloride produced a 62% yield of chloramine (XXXVI).

$$\text{(cyclic)}N\cdot + CH_2{=}CH(CH_2)_3Me + FeCl^{2+} \longrightarrow \text{(cyclic)}NCH_2CHCl(CH_2)_3Me + Fe^{2+}$$

(XXXVI)

Addition of acrylonitrile or methyl acrylate to this reaction, however, did not change the yields, and there was no polymer formation. It was concluded that because amino-radicals do not attack such deactivated double bonds they must be strongly electrophilic (Minisci and Galli, 1964c). The electrophilic nature of radicals formed in metal ion catalysed reactions of *N*-haloalkylamines is further demonstrated by competitive experiments. Thus, the reaction of

N-chloropiperidine with an equimolar mixture of cyclohexene and 1-chlorocyclohexene in acidic medium produced a 4:1 mixture of products consisting of *N*-2-chlorocyclohexylpiperidine and *N*-2,2-dichlorocyclohexylpiperidine. This result shows a higher reactivity with cyclohexene, which is compatible with the electrophilic nature of aminium radicals (Minisci, Galli and Cecere, 1966d). With unprotonated *N*-chloroamine the reaction yielded 95% of *N*-2-cyclohexylpiperidine.

The reactions of cyclic olefins with protonated and unprotonated *N*-haloalkylamines also result in different *cis* to *trans* ratios of adducts. Thus, the addition of *N*-chloropiperidine to cyclohexene in acidic medium gave a mixture of *cis* and *trans* products, whereas in a non-acidic solution with unprotonated *N*-chloropiperidine mainly *cis* isomer was obtained. This stereoselectivity was explained by a coordination of unprotonated amino-group with ferric salt which is mainly responsible for the transfer of the chlorine atom.

The protonated amino-nitrogen would be expected not to form a complex with the iron salt, and as a result, a less stereospecific reaction would occur (Minisci, Galli and Cecere, 1966b, d) as observed.

Similar results were also obtained in the aminochlorination reaction using hydroxylamine and hydroxylamine-*O*-sulphonic acid in the presence of ferrous and ferric chlorides. In the former reaction a mixture of 30% *trans* and 70% *cis* and in the latter reaction a mixture of 70% *trans* and 30% *cis* chloroamines were obtained. It is evident that the two processes cannot follow the same mechanism. As in the case of *N*-chloroalkylamines, the preferential formation of *cis* isomer from hydroxylamine was interpreted to proceed through a coordination step involving an unprotonated amino-radical, whereas from hydroxylamine-*O*-sulphonic acid a protonated amino-radical is formed, $\overset{+\,\bullet}{N}H_3$, which is incapable of complex formation (Minisci and Galli, 1965b; Minisci, Galli and Cecere, 1966a).

Although a great variety of olefins have been aminochlorinated comparatively little is known about reactions of acetylenic compounds. The reaction of *N*-chloropiperidine with phenylacetylene in the presence of ferric chloride and ferrous sulphate at 0–8° resulted in the formation of α-chlorophenylacetaldehyde (XXXVII) (Minisci, Galli and Pollina, 1965).

$$PhC{\equiv}CH + {\bullet}NC_5H_{10} \longrightarrow Ph\dot{C}{=}CHNC_5H_{10}$$

$$Ph\dot{C}{=}CHNC_5H_{10} + FeCl^{2+} \longrightarrow PhC(Cl){=}CHNC_5H_{10}$$

$$PhC(Cl){=}CHNC_5H_{10} + H_2O \longrightarrow \underset{\text{(XXXVII)}}{PhCHClCHO} + C_5H_{10}NH$$

An analogous result was obtained from the reaction of phenylacetylene with hydroxylamine-*O*-sulphonic acid in the presence of ferrous chloride to give (XXXVII) (Minisci and Galli, 1965b).

$$PhCCl{=}CHNH_2 + H_2O \longrightarrow \underset{\text{(XXXVII)}}{PhCHCl.CHO} + NH_3$$

An unusual reaction was also obtained in the dialkylamination of methyl cinnamate with *N*-chlorodimethylamine in a mixture of 60% sulphuric and 40% acetic acid to give 6·5% nuclear attack and 93·5% side chain attack producing methyl ester of phenylpyruvic acid (Minisci, Galli, Galli and Bernardi, 1967a; this review C3).

$$\text{cyclohexene} + \underset{\text{(XXXVIIIa)}}{Me_3CN(Cl){-}C{\equiv}N} \rightleftharpoons \underset{\text{(XXXVIIIb)}}{Me_3CN{=}C{=}N{-}Cl} \xrightarrow[\text{FeCl}_3,\ \text{MeOH}]{\text{FeSO}_4,}$$

$$\underset{\text{(XXXIX)}}{\text{2-chloro-1-methoxycyclohexane}} + [Me_3CN{=}C{=}NH] \xrightarrow[H_2O]{H^+} Me_3CNH.CO.NH_2$$

An interesting reaction was observed between cyclohexene and *N*-*t*-butyl-*N*-chlorocyanamide (XXXVIIIa, b) in the presence of a mixture of ferrous sulphate and ferric chloride in methanol. The products were 40% 1-chloro-2-methoxycyclohexane (XXXIX), 32% t-butylurea, and a few percent 1,2-dichlorocyclohexane (Neale and Marcus, 1969). It is thought that t-butylcarbodi-imide is an intermediate in the formation of t-butylurea.

E. Cyclisation Reactions

The thermal and photochemical reactions for cyclisations of *N*-chloroamines are known under the names of Hofmann-Löffler or Hofmann-Löffler-Freytag Reaction (Hofmann, 1883; Löffler and Freytag, 1909, 1910; Wolff, 1963). These reactions can also be initiated with metal ions, usually ferrous ions. Unlike the thermal and photochemical reactions (Wolff, 1963), comparatively few metal ion catalysed reactions have been investigated to date (Table 4). The products are in most cases five and six membered ring structures, i.e., pyrrolidine and piperidine derivatives (Table 4).

For example, the reaction of *N*-chlorobutylamine (XL) with ferrous ammonium sulphate in 85% sulphuric acid gave a 72% yield of pyrrolidine (XLI) (Schmitz and Murawski, 1966). Similarly, the

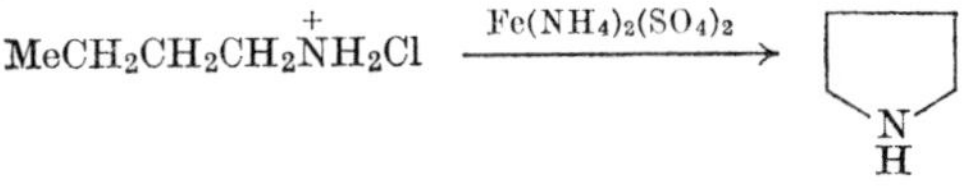

$$MeCH_2CH_2CH_2\overset{+}{N}H_2Cl \xrightarrow{Fe(NH_4)_2(SO_4)_2} \text{pyrrolidine (NH)}$$

(XL) (XLI)

reactions of substituted *N*-chlorohexylamines (XLII) with ferrous sulphate in sulphuric acid resulted in various piperidine derivatives (XLIII) in 60–65% yield (Minisci, Galli and Rossetti, 1967).

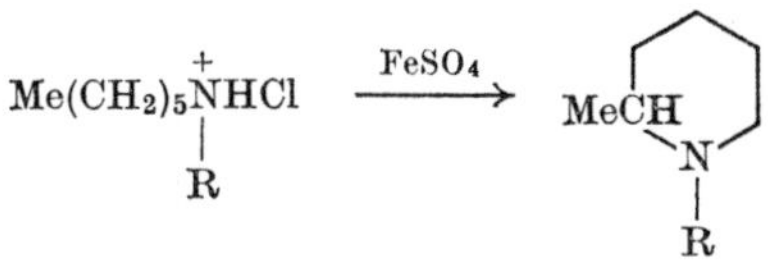

$$Me(CH_2)_5\overset{+}{N}HCl(R) \xrightarrow{FeSO_4} \text{2-MeCH-piperidine, N-R}$$

(XLII) (XLIII)

($R = H, Pr^n, Pr^i$, Ph, cyclohexyl)

TABLE 4

Cyclisation reactions

Haloamine	Initiator	Products	Yield %	Ref.
$Me(CH_2)_3\overset{+}{N}H_2Cl$	$Fe(NH_4)_2(SO_4)_2$	Pyrrolidine	72	1
$Me(CH_2)_4\overset{+}{N}H_2Cl$	$Fe(NH_4)_2(SO_4)_2$	2-Methylpyrrolidine	60	1
$Me(CH_2)_6\overset{+}{N}H_2Cl$	$Fe(NH_4)_2(SO_4)_2$	2-Propylpyrrolidine	61	1
$Pr_2CH\overset{+}{N}H_2Cl$	$Fe(NH_4)_2(SO_4)_2$	2-Propylpyrrolidine	80	1
$Bu^n{}_2\overset{+}{N}HCl$	$Fe(NH_4)_2(SO_4)_2$	1-Butylpyrrolidine	69	2
$CH_2{=}CH(CH_2)_3\overset{+}{N}HPrCl$	$CuCl + O_2$	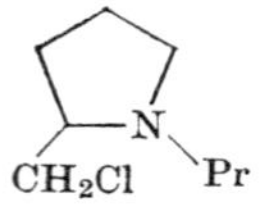 N, Pr, CH_2Cl	24·5	3
$CH_2{=}CH(CH_2)_3\overset{+}{N}HPrCl$	CuCl, CuBr, $FeSO_4$, or $TiCl_3$	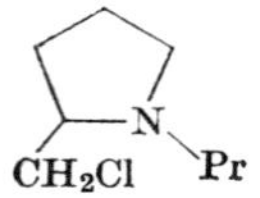 N, Pr, CH_2Cl	38–81	3
$[Me(CH_2)_4]_2\overset{+}{N}HCl$	$Fe(NH_4)_2(SO_4)_2$	*N*-Amyl-2-methylpyrrolidine	80	4
$Me(CH_2)_5\overset{+}{N}H_2Cl$	$FeSO_4$	2-Methylpiperidine	60–65	5

$Me(CH_2)_5\overset{+}{N}HPr^nCl$	$FeSO_4$	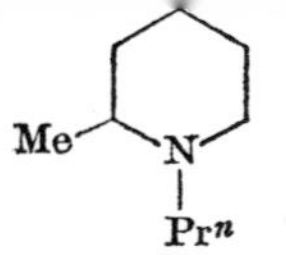 Me, N, Pr^n	60–65	5
$Me(CH_2)_5\overset{+}{N}HPr^iCl$	$FeSO_4$	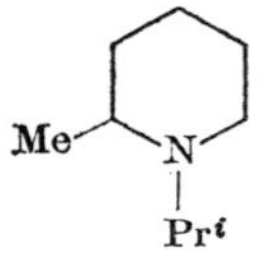 Me, N, Pr^i	60–65	5
$Me(CH_2)_5\overset{+}{N}HPhCl$	$FeSO_4$	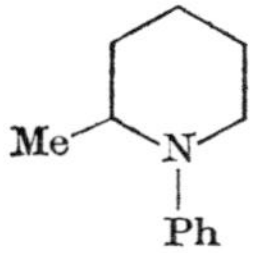 Me, N, Ph	60–65	5
$Me(CH_2)_5\overset{+}{N}H$-cyclohexyl-Cl	$FeSO_4$	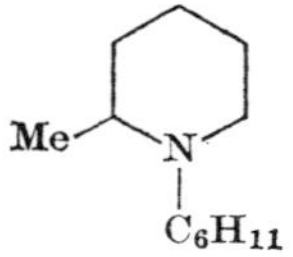 Me, N, C_6H_{11}	60–65	5
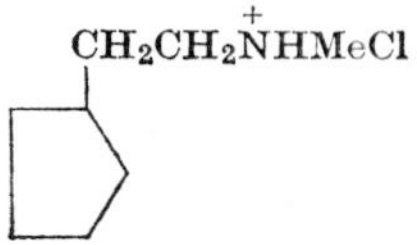 $CH_2CH_2\overset{+}{N}HMeCl$	$FeSO_4$	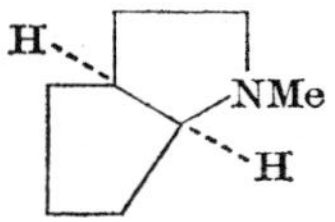 H, NMe, H	20	6
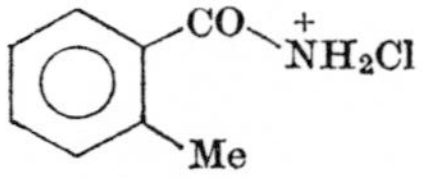 $CO\overset{+}{N}H_2Cl$, Me	CuCl	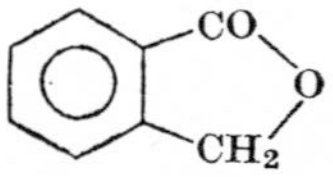 CO, O, CH_2	31	7

TABLE 4—*continued*

Haloamine	Initiator	Products	Yield %	Ref.
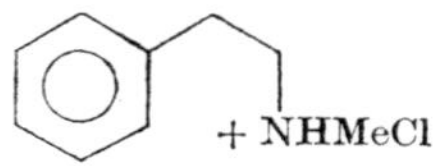 + NHMeCl	$FeSO_4$	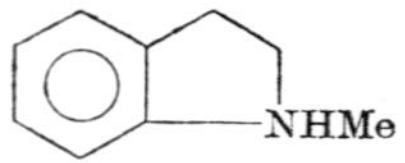 NHMe	27	8, 9, 10
		+ $PhCH_2Cl$	44	
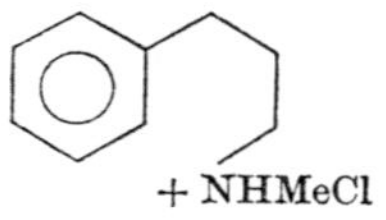 + NHMeCl	$FeSO_4$	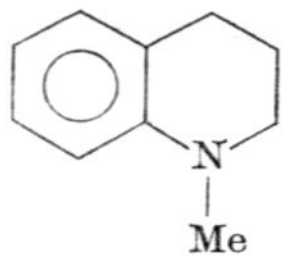 N Me	81	8, 9, 10
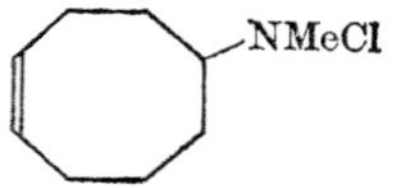 NMeCl	$AgClO_4$ or $FeSO_4$	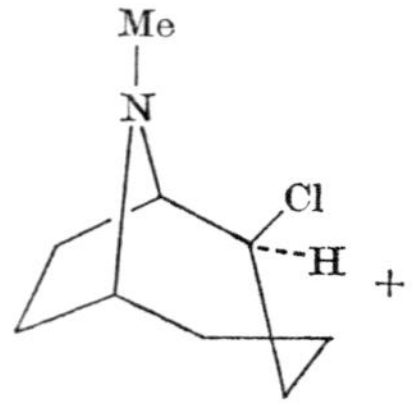 Me N Cl H +	11	11
		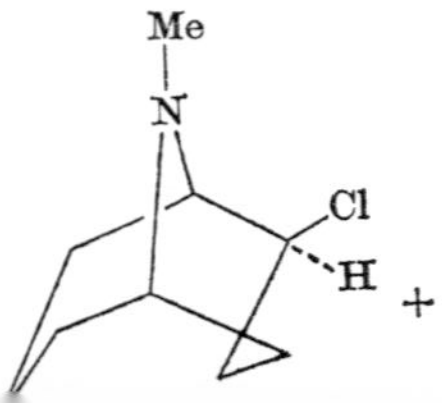 Me N Cl H +	4	

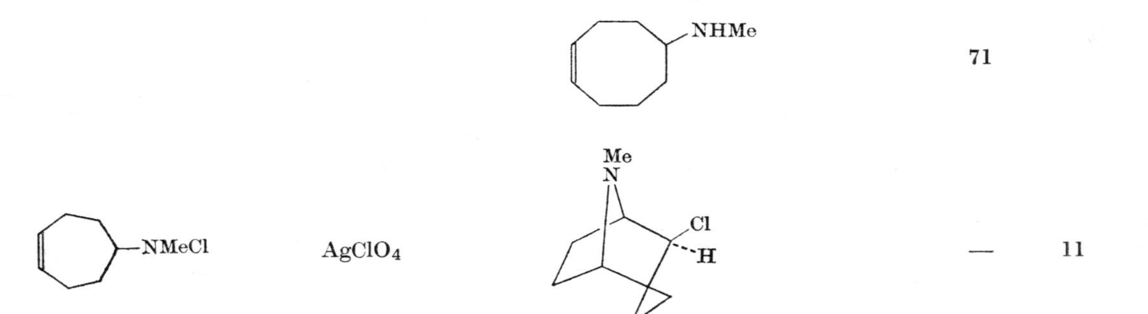

References: (1) Schmitz and Murawski (1966). (2) Corey and Hertler (1960). (3) Surzur, Stella and Tordo (1970). (4) Neale (1964). (5) Minisci, Galli and Rossetti (1967). (6) Kessar, Rampal and Mahajan (1962). (7) Beckwith and Goodrich (1965). (8) Minisci and Galli (1966a). (9) Minisci (1967). (10) Minisci, Galli and Perchinunno (1969a). (11) Hobson and Riddell (1968).

A mechanism explaining these observations has been proposed, based on the radical chain mechanism involving aminium radical cation (XLIV) as suggested by Wawzonek and Thelen (1950) (Corey and Hertler, 1960) for the thermal and photochemical Hofmann-Löffler reaction (Kessar, Rampal and Mahajan, 1963; Minisci, 1964; Minisci, Galli and Rossetti, 1967).

$$\underset{\text{(XLII)}}{MeCH_2(CH_2)_4\overset{+}{N}(R)HCl} + Fe^{2+} \longrightarrow \underset{\text{(XLIV)}}{MeCH_2(CH_2)_4\overset{+\bullet}{N}(R)H} + Fe^{3+} + Cl^-$$

or

$$\underset{\text{(XLII)}}{MeCH_2(CH_2)_4\overset{+}{N}(R)HCl} + Fe^{2+} \longrightarrow MeCH_2(CH_2)_4N(R)H + Fe^{3+} + Cl\cdot$$

$$R'H + Cl\cdot \longrightarrow R'\cdot + HCl$$

$$R'\cdot + MeCH_2(CH_2)_4\overset{+}{N}(R)HCl \longrightarrow R'Cl + \underset{\text{(XLIV)}}{MeCH_2(CH_2)_4\overset{+\bullet}{N}(R)H}$$

$$MeCH_2(CH_2)_4\overset{+\bullet}{N}(R)H \longrightarrow Me\overset{\bullet}{C}H(CH_2)_4\overset{+}{N}(R)H_2$$

$$Me\overset{\bullet}{C}H(CH_2)_4\overset{+}{N}(R)H_2 + Me(CH_2)_5\overset{+}{N}(R)HCl \longrightarrow MeCH(Cl)(CH_2)_4\overset{+}{N}(R)H_2 + \underset{\text{(XLIV)}}{Me(CH_2)_5\overset{+\bullet}{N}(R)H}$$

$$MeCH(Cl)(CH_2)_4\overset{+}{N}(R)H_2 \xrightarrow{OH^-} \underset{\text{(XLIII)}}{\text{MeCH–N(R) (piperidine ring)}}$$

The cyclisation can also be carried out by mixing an appropriate amine (XLV) with *N*-chlorodimethylamine (XLVI), concentrated sulphuric acid, and ferrous sulphate, followed by the addition of sodium hydroxide. In this case, a somewhat modified reaction mechanism was proposed (Minisci, 1964; Minisci, Galli and Rossetti, 1967).

$$\mathrm{Me_2\overset{+}{N}HCl + Fe^{2+} \longrightarrow Me_2\overset{+\bullet}{N}H + Fe^{3+} + Cl^-}$$
(XLVI)

or

$$\mathrm{Me_2\overset{+}{N}HCl + Fe^{2+} \longrightarrow Me_2NH + Fe^{3+} + Cl\cdot}$$
$$\mathrm{R'H + Cl\cdot \longrightarrow HCl + R'\cdot}$$
$$\mathrm{R'\cdot + Me_2\overset{+}{N}HCl \longrightarrow R'Cl + Me_2\overset{+\bullet}{N}H}$$
(XLVI)

$$\mathrm{MeCH_2(CH_2)_4\overset{+}{N}H_2R + Me_2\overset{+\bullet}{N}H \longrightarrow Me\dot{C}H(CH_2)_4\overset{+}{N}H_2R + Me_2\overset{+}{N}H_2}$$
(XLV)

$$\mathrm{Me\dot{C}H(CH_2)_4\overset{+}{N}H_2R + Me_2\overset{+}{N}HCl \longrightarrow MeCHCl(CH_2)_4\overset{+}{N}H_2R + Me_2\overset{+}{N}H}$$
(XLVI)

$$\mathrm{MeCH(Cl)(CH_2)_4\overset{+}{N}H_2R \xrightarrow{OH^-} \text{2-Me-piperidine-}N\text{-R}}$$

Intramolecular aromatic amination reactions can occur when aromatic nuclei are present in appropriate positions in the *N*-chloroamine. Thus, the reaction of methyl-2-phenylethyl-*N*-chloroamine (XLVII) with ferrous sulphate in sulphuric acid produced *N*-methylindoline (XLVIII) in 27% yield, together with 44% of benzyl

$$\mathrm{PhCH_2CH_2N(Cl)Me \xrightarrow{FeSO_4} PhCH_2CH_2\overset{+}{N}\dot{H}Me \longrightarrow}$$
(XLVII)

$$\text{(cyclohexadienyl radical of } N\text{-methylindoline)} \xrightarrow{Fe^{3+}} N\text{-methylindoline} + \mathrm{Fe^{2+} + H^+}$$
(XLVIII)

chloride (Minisci, 1967; Minisci and Galli, 1966a; Minisci, Galli and Perchinunno, 1969a).

$$PhCH_2CH_2NClMe \xrightarrow{FeSO_4} PhCH_2CH_2\overset{+\bullet}{N}HMe \longrightarrow PhCH_2\cdot + CH_2{=}\overset{+}{N}HMe$$

$$PhCH_2\cdot + PhCH_2CH_2\overset{+}{N}HClMe \longrightarrow PhCH_2Cl + PhCH_2CH_2\overset{+\bullet}{N}HMe$$

Similarly, the reaction of methyl-3-phenylpropyl-*N*-chloroamine (XLIX) produced kairoline (L) in 81% yield (Minisci, 1967; Minisci and Galli, 1966a; Minisci, Galli and Perchinunno, 1969a).

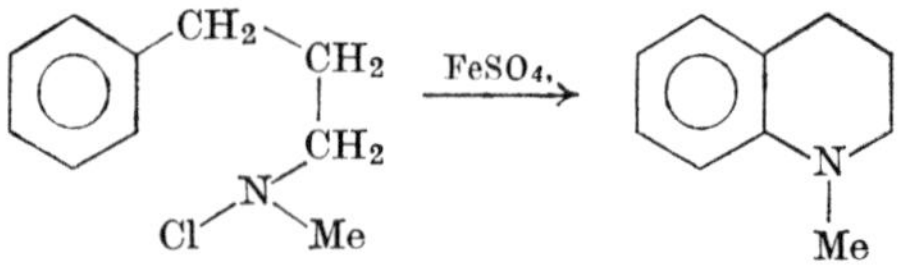

$$\text{(XLIX)} \xrightarrow{FeSO_4} \text{(L)}$$

The metal ion catalysed modification of Hofmann-Löffler reaction was applied to steroidal chemistry (Schaffner, Arigoni and Jeger, 1960). Thus, the cyclisation of compound (LI; R = H) with ferrous sulphate formed conanine (LII; R = H). Similarly, the reaction of *N*-chloroderivative of (20S)-3β-acetoxy-20-methylamino-5-α-pregnan (LI; R = AcO) yielded 3β-acetoxyconanine (LII; R = AcO).

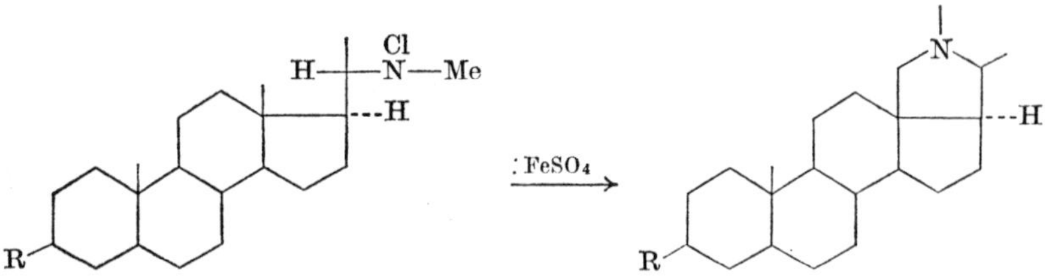

$$\text{(LI)} \xrightarrow{FeSO_4} \text{(LII)}$$

(R = H, AcO)

Ring closure to a pyrrolidine derivative (LIV) was achieved in 38–81% yield using an unsaturated *N*-chloroamine (LIII) and copper, iron or titanium ions (Surzur, Tordo and Stella, 1970; Surzur, Stella and Tordo, 1970). Oxygen greatly inhibits the reaction.

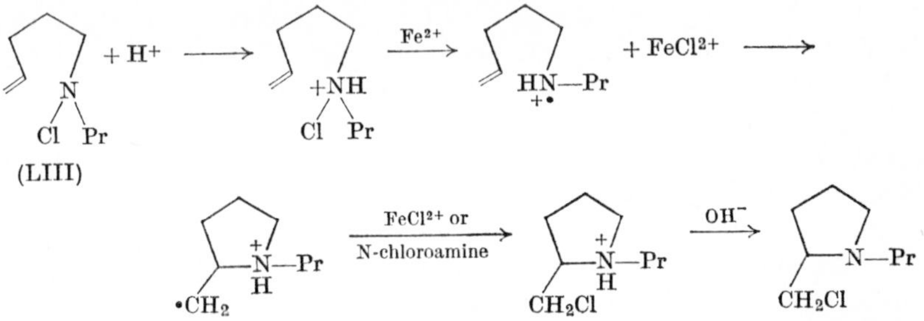

(LIII) $+ H^+ \longrightarrow$ $\xrightarrow{Fe^{2+}}$ $+ FeCl^{2+} \longrightarrow$ $\xrightarrow[\text{N-chloroamine}]{FeCl^{2+}\text{ or}}$ $\xrightarrow{OH^-}$ (LIV)

Perhaps a related intramolecular cyclisation reaction of *N*-chloro-cycloalkenes occurs in the presence of catalytic amounts of silver and ferrous ions (Hobson and Riddell, 1968). Thus, the reaction of *N*-chloro-*N*-methyl-4-cyclo-octeneamine (LV) in the presence of silver perchlorate yielded 71% of *N*-methyl-4-cyclo-octeneamine (LVI), 11% of (LVII), and 4% (LVIII). A ten-fold increase in concentration of silver perchlorate resulted in 25% of (LVII) and 11% of (LVIII). With ferrous sulphate in a sulphuric-acetic acid mixture the reaction produced 26% of (LVIII).

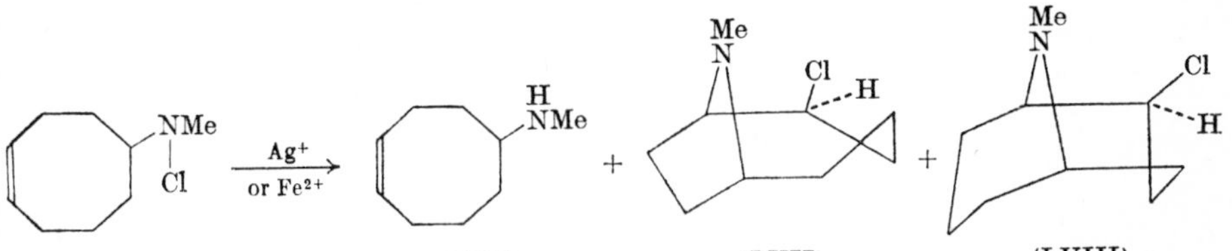

(LV) $\xrightarrow[\text{or } Fe^{2+}]{Ag^+}$ (LVI) + (LVII) + (LVIII)

The reaction of *N*-chloro-*N*-methyl-4-cyclo-hepteneamine (LIX) with silver perchlorate yielded a single product (LX).

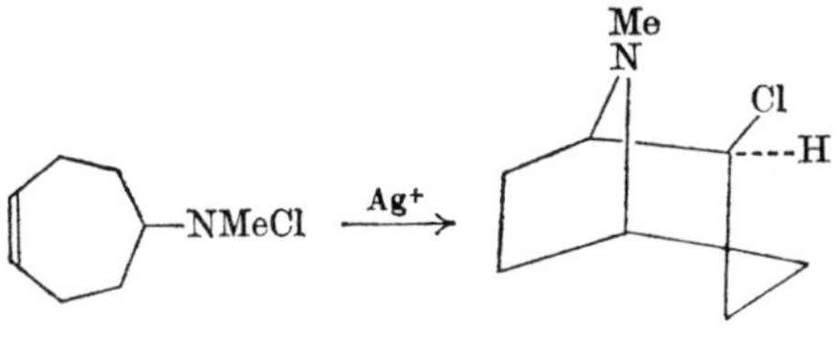

(LIX) $\xrightarrow{Ag^+}$ (LX)

In view of the stereospecific *cis*-addition of chlorine atoms, and the experimental conditions, it was suggested that coordination of catalyst ions to the nitrogen atom occurs, followed by homolysis of the N—Cl bond to give aminium radicals which in the presence of an olefinic bond in the vicinity undergo a radical chain addition process to give β-chlorodialkylamine (Hobson and Riddell, 1968).

Silver ions also catalyse cyclisation reactions of saturated alicyclic *N*-chloroamines. For example, the reaction of *N*-chloroazacyclononane (LXI) yielded 68% of indolizidine (LXII) (Edwards, Vocelle,

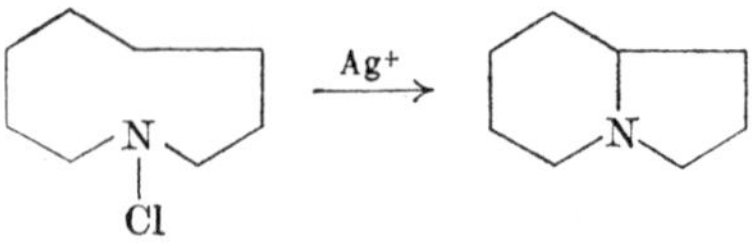

$\xrightarrow{Ag^+}$

(LXI) (LXII)

ApSimon and Haque, 1965), and the reaction of *N*-chloro-*N*-methylcyclo-octylamine (LXIII) formed 5% of *N*-methylgranatamine

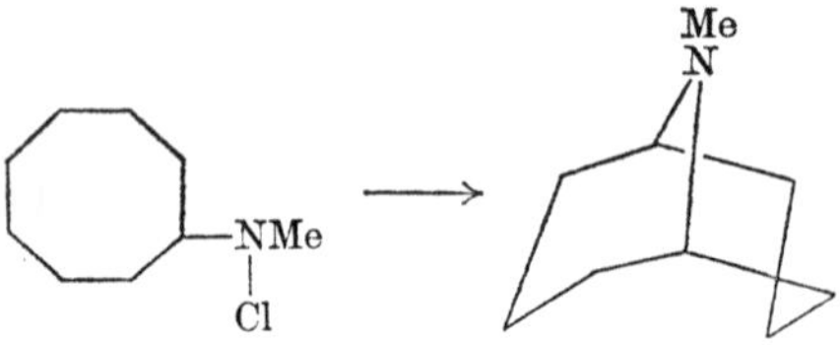

(LXIII) (LXIV)

(LXIV) (Edwards, Vocelle, ApSimon and Haque, 1965). However, for these reactions non-radical intermediates were suggested (Edwards, Vocelle, ApSimon and Haque, 1965).

An interesting cyclisation reaction was observed with *N*-chloroamides leading to lactones. For example, cyclisation of *N*-chloro-*o*-toluamide (LXV) in boiling carbon tetrachloride in the presence of cuprous ions yielded 31% of phthalide (LXVI) (Beckwith and Goodrich, 1965).

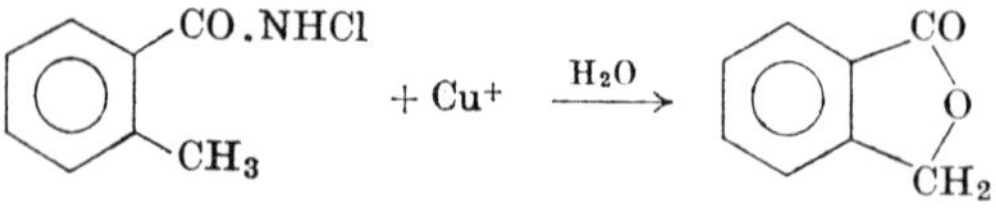

CO.NHCl, CH_3 $+ Cu^+ \xrightarrow{H_2O}$ CO, O, CH_2

(LXV) (LXVI)

Similarly, the reaction of *N*-chlorotetradecanamide gave a mixture of γ and δ-lactones in 30% yield (Beckwith and Goodrich, 1965). These reactions were also achieved with light. The metal ion-catalysed reaction might be interpreted to proceed by the following mechanism (Beckwith and Goodrich, 1965).

$$o\text{-}C_6H_4(CO.NHCl)(CH_3) + CuCl \longrightarrow o\text{-}C_6H_4(CO\dot{N}H)(CH_3) + CuCl_2$$

(LXVII)

$$o\text{-}C_6H_4(CO\dot{N}H)(CH_2\text{—}H) \longrightarrow o\text{-}C_6H_4(CONH_2)(CH_2\cdot)$$

$$o\text{-}C_6H_4(CONH_2)(CH_2\cdot) + CuCl_2 \text{ or (LXV)} \longrightarrow o\text{-}C_6H_4(CONH_2)(CH_2Cl) + CuCl \text{ or (LXVII)}$$

$$o\text{-}C_6H_4(CONH_2)(CH_2Cl) + H_2O \longrightarrow o\text{-}C_6H_4\text{(CO—O—}CH_2\text{)}$$

F. Ring-Opening Reactions of Oxaziridines

Oxaziridines, which are readily obtained, for example, by the reactions of ketones, amines and peroxy-acids or ketones and halo-amines (Schmitz, 1963; Schmitz, Ohme and Murawski, 1965), decompose in the presence of metal ions to give useful products (Table 5). Thus, the reaction of 2-t-butyl-3-phenyloxaziridine (LXVIII) gives a 98% yield of *N*-t-butylbenzamide (LXIX) (Emmons, 1957). This

$$\underset{\text{(LXVIII)}}{PhCH\overset{O}{\text{——}}NCMe_3} \xrightarrow{Fe(NH_4)_2(SO_4)_2} \underset{\text{(LXIX)}}{PhCONHCMe_3}$$

reaction has been extensively investigated by Emmons (1957), Schmitz and Murawski (1965), and Minisci and co-workers (Minisci, Galli, Malatesta and Caronna, 1970).

TABLE 5

Ring-opening reactions of oxaziridines

Oxaziridine	Initiator	Products	Yield %	Ref.
2,3,3-Triethyloxaziridine	$FeSO_4$	Diethyl ketone *N*-Ethylpropionamide, Ammonia Ethane, Ethylene, Butane	50 32 55	1
2,3,3-Triethyloxaziridine	$FeCl_2$	Diethyl ketone *N*-Ethylpropionamide, Ammonia Ethane, Ethylene, Butane, Ethyl chloride	50 32 55	1
2-Isobutyl-3-isopropyloxaziridine	$Fe(NH_4)_2(SO_4)_2$	*N*-Isobutylformamide Propane, Propene	83	1
2-t-Butyl-3-isopropyloxaziridine	$Fe(NH_4)_2(SO_4)_2$	*N*-t-Butylformamide Propane, Propene	82	1
2-t-Octyloxaziridine	$Fe(NH_4)_2(SO_4)_2$	*N*-t-Octylformamide	87	1
2-(α-Phenylethyl)-3,3-diethyloxaziridine	$FeSO_4$	*N*-(α-Phenylethyl)-propionamide Ammonia, Ethane, Ethylene, Butane, Acetophenone (trace)	81	1
2-(α-Phenylethyl)-3,3-diethyloxaziridine	$FeCl_2$	*N*-(α-Phenylethyl)-propionamide, Ammonia, Ethane, Ethylene, Butane, Acetophenone (trace), Ethyl chloride	—	1

O, PhCH——$NCMe_3$	$Fe(NH_4)_2(SO_4)_2$	$PhCO.NH.CMe_3$	98	1
O, NMe	$FeSO_4$	$[(CH_2)_4CO.NHMe]_2$	44	2
O, NMe	$FeSO_4$	$[(CH_2)_5CO.NHMe]_2$	85	2
O, NMe	$FeSO_4$ & $CuSO_4$	Methylamine, Cyclohexanone, Hex-5-enoic acid *N*-methylamide	46	3
O, NMe	$Ti_2(SO_4)_3$	Cyclohexanone Caproic acid *N*-methylamide	 26	3
O, NMe	$TiCl_3$	Cyclohexanone ε-Chlorocaproic acid *N*-methylamide, Caproic acid *N*-methylamide	 6 29	3

TABLE 5—*continued*

Oxaziridine	Initiator	Products	Yield %	Ref.
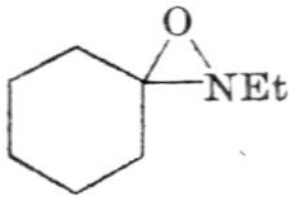 O, NEt	$FeSO_4$	$\frac{1}{}$$[(CH_2)_5CO.NHEt]_2$	28	2
O, $N–Pr^n$	$FeSO_4$	$\frac{1}{}$$[(CH_2)_4CO.NHPr]_2$	36	2
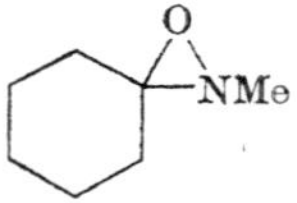 O, NMe	$FeSO_4$ & NO	Cyclohexanone ε-Nitrosohydroxylaminocaproic acid *N*-methylamide	 41	3
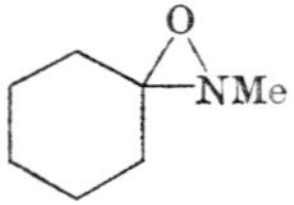 O, NMe	$FeSO_4$ & $FeCl_3$ or CuCl & $CuCl_2$	ε-Chlorocaproic acid *N*-methylamide	98 11	4
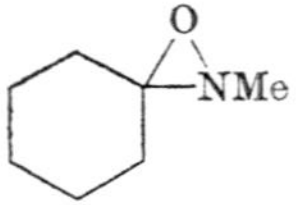 O, NMe	$FeSO_4$ & $FeBr_3$	ε-Bromocaproic acid *N*-methylamide	83	4
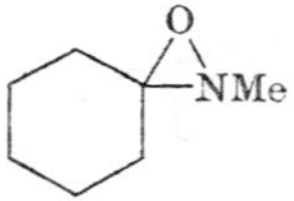 O, NMe	$FeSO_4$ & $Fe(SCN)_3$ or CuSCN & NH_4SCN	ε-Thiocyanocaproic acid *N*-methylamide	46 19	4

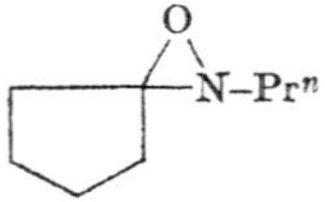

2-methyl-1-oxa-2-azaspiro[2.5]octane (NMe)	$FeSO_4$ & NaN_3	Cyclohexanone		4
		ε-Azidocaproic acid *N*-methylamide	41	
2-methyl-1-oxa-2-azaspiro[2.5]octane (NMe)	[a]$FeSO_4$	Pyridyl–$(CH_2)_5CO.NHMe$ 2-isomer (34%) 4-isomer (66%) $[$–pyridyl-2–$(CH_2)CO.NHMe]_2$	79	5, 6
2-methyl-1-oxa-2-azaspiro[2.5]octane (NMe)	[b]$FeSO_4$	Quinolyl–$(CH_2)_5CO.NHMe$ 2-isomer (53%) 4-isomer (47%)	80	5, 6
2-methyl-1-oxa-2-azaspiro[2.5]octane (NMe)	[c]$FeSO_4$	9-Acridinyl–$(CH_2)_5CO.NHMe$	76	5, 6

There is no substrate except for: [a]pyridine; [b]quinoline; [c]acridine.

References: (1) Emmons (1957). (2) Schmitz and Murawski (1965). (3) Minisci, Cecere and Galli (1968b). (4) Minisci, Cecere and Galli (1968a). (5) Minisci, Cecere, Malatesta and Caronna (1968). (6) Minisci, Galli, Malatesta and Caronna (1970).

Emmons (1957) initially proposed the following mechanism involving oxygen radicals (LXIX). However, in a footnote Emmons

$$\underset{\text{RCH}\text{—}\text{NR}'}{\overset{\text{O}}{\triangle}} + Fe^{2+} + H^{+} \longrightarrow \underset{(\text{LXIX})}{\text{RCH(O}\cdot\text{).NHR}'} + Fe^{3+}$$

$$\text{RCH(O}\cdot\text{).NHR}' + \underset{\text{RCH}\text{—}\text{NR}'}{\overset{\text{O}}{\triangle}} \longrightarrow \text{RCO.NHR}' + \text{RCH(O}\cdot\text{).NHR}'$$

(1957) has also suggested a radical scheme involving nitrogen radicals (LXX).

$$\underset{\text{RCH}\text{—}\text{NR}'}{\overset{\text{O}}{\triangle}} + Fe^{2+} + H^{+} \longrightarrow \text{RCH(O}\cdot\text{).NHR}' + Fe^{3+}$$

$$\text{RCH(O}\cdot\text{).NHR}' + \underset{\text{RCH}\text{—}\text{NR}'}{\overset{\text{O}}{\triangle}} \longrightarrow \text{RCH(OH).NHR} + \underset{(\text{LXX})}{\text{RCO.}\dot{\text{N}}\text{R}'}$$

$$\underset{(\text{LXX})}{\text{RCO.}\dot{\text{N}}\text{R}'} + \underset{\text{RCH}\text{—}\text{NR}'}{\overset{\text{O}}{\triangle}} \longrightarrow \text{RCO.NHR}' + \text{RCO.}\dot{\text{N}}\text{R}'$$

A more complex mixture of products is obtained in cases where R = alkyl (Emmons, 1957). For example, the reaction of 2,3,3-triethyloxaziridine (LXXI) with ferrous ions gives a mixture of 50% diethyl ketone, 32% of *N*-ethylpropionamide (LXXII), 55% ammonia, with ethane, ethylene and butane (Emmons, 1957). To account for these products Emmons (1957) proposed a mechanism which was accepted with minor modifications by Schmitz (1963) in his review article. It was suggested that the carbon radical (LXXIII) rearranges to the oxygen radical (LXXIV). Acetaldehyde imine (LXXV) is readily hydrolysed to give ammonia and acetaldehyde

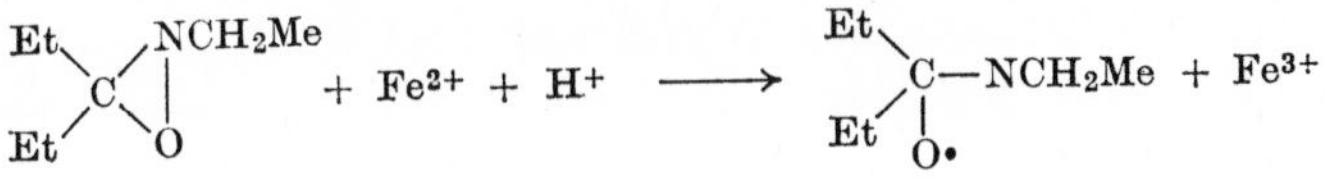

$$\mathrm{Et_2C(-O-)NCH_2Me} + Fe^{2+} + H^+ \longrightarrow \mathrm{Et_2C(O^{\bullet})-NCH_2Me} + Fe^{3+}$$

(LXXI)

$$\mathrm{Et_2C(O^{\bullet})-NCH_2Me} \longrightarrow \mathrm{EtCO.NHCH_2Me} + \mathrm{Et^{\bullet}}$$

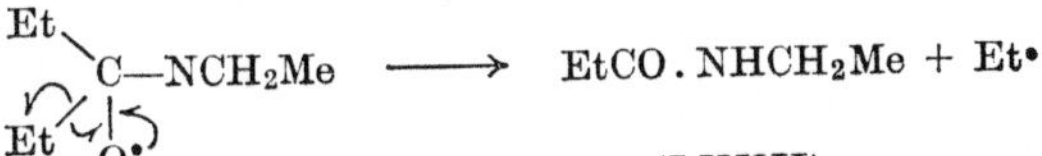

(LXXII)

$$\mathrm{Et^{\bullet}} \rightarrow CH_2{=}CH_2 + C_2H_6 + C_4H_{10}$$

$$\mathrm{Et_2C(-O-)NCH_2Me} + \mathrm{Et^{\bullet}} \longrightarrow \mathrm{Et_2C(-O-)\dot{N}CHMe} + C_2H_6$$

(LXXI) (LXXIII)

$$\mathrm{Et_2C(-O-)\dot{N}CHMe} \longrightarrow \mathrm{Et_2C(O^{\bullet})-N{=}CHMe} \xrightarrow{RH} \mathrm{Et_2C{=}O} + \mathrm{HN{=}CHMe} + \mathrm{R^{\bullet}}$$

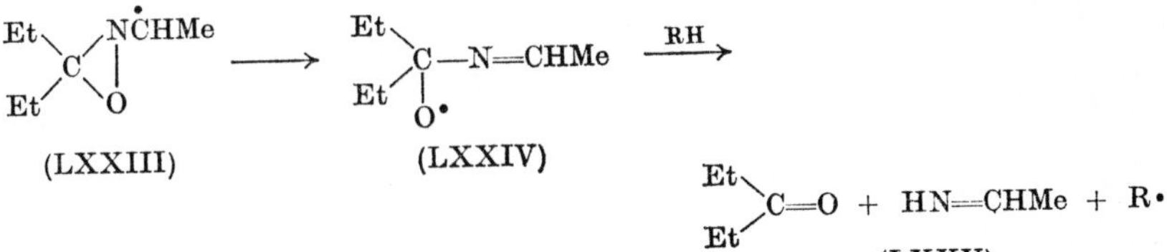

(LXXIII) (LXXIV) (LXXV)

(Emmons, 1957). The same products are obtained by the following reaction.

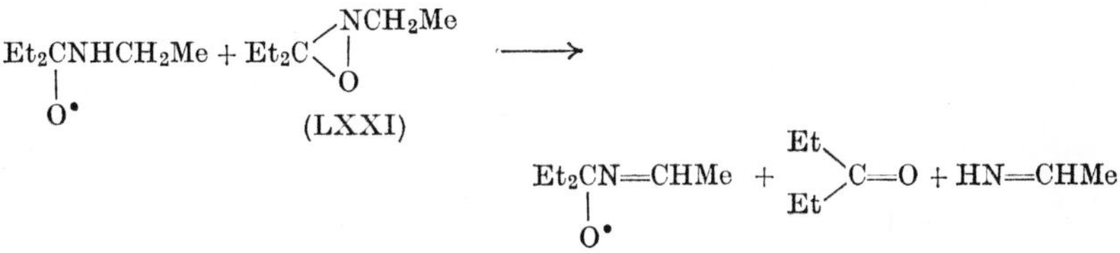

$$\mathrm{Et_2C(O^{\bullet})NHCH_2Me} + \mathrm{Et_2C(-O-)NCH_2Me} \longrightarrow \mathrm{Et_2C(O^{\bullet})N{=}CHMe} + \mathrm{Et_2C{=}O} + \mathrm{HN{=}CHMe}$$

(LXXI)

In those cases where R is a cyclic system, e.g., compound (LXXVI), ring opening of the cyclic system also occurs resulting in carbon radical (LXXVII) which can dimerise or undergo other types of reactions common to carbon radicals. Thus, such reactions occur with

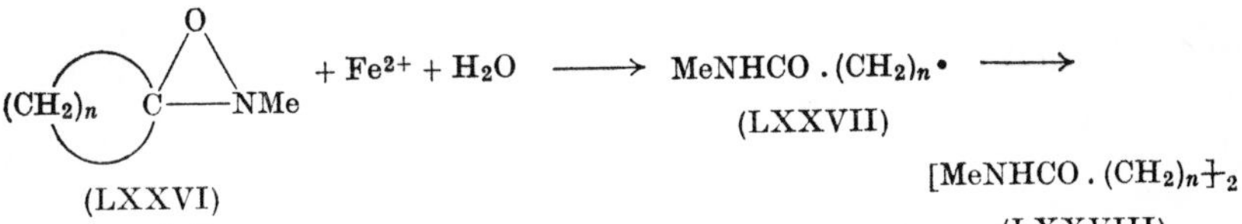

$$\mathrm{(CH_2)_n C(-O-)NMe} + Fe^{2+} + H_2O \longrightarrow \mathrm{MeNHCO.(CH_2)_n^{\bullet}} \longrightarrow [\mathrm{MeNHCO.(CH_2)_n}]_2$$

(LXXVI) ($n = 4, 5$) (LXXVII) (LXXVIII)

2-methyl-3,3-tetramethyleneoxaziridine (LXXVI, $n = 4$) and 2-methyl-3,3-pentamethyleneoxaziridine (LXXVI, $n = 5$) to yield dimers LXXVIII ($n = 4$) and LXXVIII ($n = 5$), respectively (Schmitz and Murawski, 1965).

In the presence of large amounts of anions, ligand and electron transfer processes occur. Thus, the reaction of radical (LXXVII) and an anion forms products (LXXIX) (Minisci, Cecere and Galli, 1968a).

$$\underset{\text{(LXXVII)}}{MeNHCO.(CH_2)_n\cdot} + X^- + Fe^{3+} \longrightarrow \underset{\text{(LXXIX)}}{MeNHCO.(CH_2)_nX}$$

$$(X = Cl, Br, SCN, N_3)$$

In the presence of cupric ions, an electron transfer process occurs to give also unsaturated products, e.g., 5-hexenoic acid *N*-methylamide (LXXX) is formed (Minisci, Cecere and Galli, 1968b) by the oxidation of radical (LXXVII).

$$MeNHCO.(CH_2)_4CH_2\cdot + Cu^{2+} \longrightarrow MeNHCO.(CH_2)_4CH_2^+ \longrightarrow \underset{\text{(LXXX)}}{MeNHCO.(CH_2)_3CH{=}CH_2} + H^+$$

Reaction of (LXXVI) ($n = 5$) with titanous sulphate yields caproic acid N-methylamide by reduction of radical (LXXVII) (Minisci, Cecere and Galli, 1968b).

$$\underset{\substack{\text{(LXXVII)}\\(n=5)}}{MeNHCO.(CH_2)_4CH_2\cdot} + Ti^{3+} + H_2O \longrightarrow \underset{\text{(LXXXI)}}{MeNHCO.(CH_2)_4Me} + Ti(OH)^{3+}$$

Reaction of (LXXVI) ($n = 5$) with titanous chloride forms ε-chlorocaproic acid *N*-methylamide (LXXXII) by an electron and ligand

$$MeNHCO.(CH_2)_4CH_2\cdot + TiCl^{3+} \longrightarrow \underset{\text{(LXXXII)}}{MeNHCO.(CH_2)_4CH_2Cl} + Ti^{3+}$$

transfer process (Minisci, Cecere and Galli, 1968b). When the formation of radical (LXXVII) occurs in the presence of nitric oxide, the carbon radical is intercepted by nitric oxide to give ε-nitrosohydroxylaminocaproic acid *N*-methylamide (LXXXIII) (Minisci, Cecere and Galli, 1968b).

$$MeNHCO.(CH_2)_4CH_2\cdot + Fe(NO)^{2+} \longrightarrow MeNHCO.(CH_2)_4CH_2\dot{N}OFe^{2+} \xrightarrow{NO} \underset{\text{(LXXXIII)}}{MeNHCO.(CH_2)_5\underset{\displaystyle NO}{\underset{|}{N}}OFe^{2+}}$$

In addition the original ketone and amine are frequently formed by a reductive process (Minisci, Cecere and Galli, 1968b).

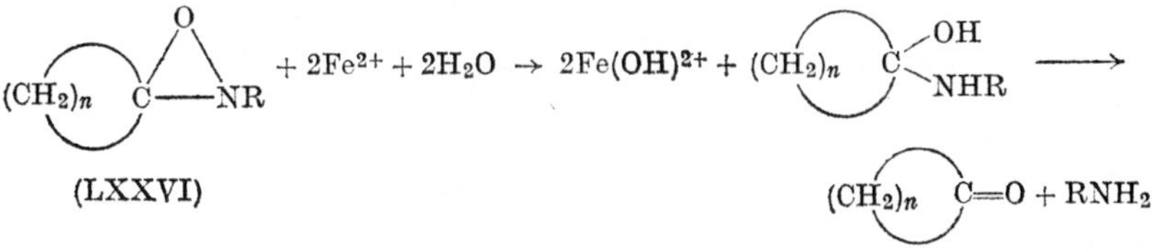

$$(CH_2)_n\,C(O)NR + 2Fe^{2+} + 2H_2O \rightarrow 2Fe(OH)^{2+} + (CH_2)_n\,C(OH)(NHR) \longrightarrow (CH_2)_n\,C{=}O + RNH_2$$

(LXXVI)

The decomposition of oxaziridine (LXXVI, $n = 5$) in the presence of butadiene and an olefin (LXXXIV) containing electron-withdrawing substituents gives additive dimers (LXXXV) (Minisci, Galli, Cecere, Malatesta and Caronna, 1968).

$$(CH_2)_5\,C(O)NMe + Fe^{2+} \longrightarrow MeNHCO.(CH_2)_5\bullet + Fe^{3+}$$

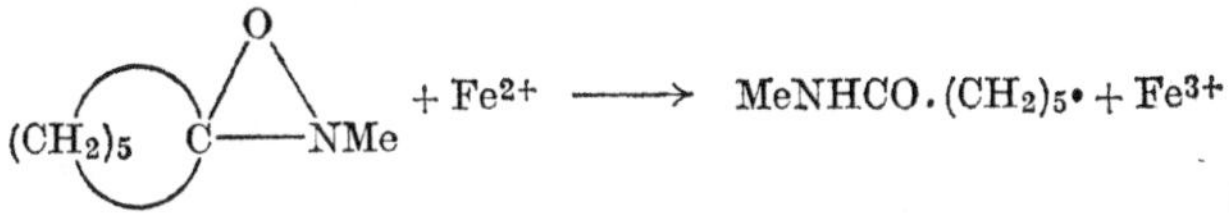

$$R\bullet + CH_2{=}CHX \longrightarrow RCH_2\dot{C}HX \xrightarrow{CH_2=CHCH=CH_2} RCH_2.CHX.C_4H_6\bullet \longrightarrow (RCH_2.CHX.C_4H_6)_2$$

(LXXXIV) (LXXXV)

($R = MeNHCO.(CH_2)_5$; $X = CN, CO.OH, CO.OMe, CO.Me$)

In the presence of cupric sulphate and methanol, an oxidative process occurs to give (LXXXVI) (Minisci, Galli, Cecere, Malatesta

$$RCH_2.CHX.C_4H_6\bullet + Cu^{2+} + MeOH \longrightarrow RCH_2.CHX.C_4H_6OMe + Cu^{+} + H^{+}$$

(LXXXVI)

and Caronna, 1968). When X is a carbonyl moiety an intramolecular reaction gives products (LXXXVII) (Minisci, Galli, Cecere, Mala-

$$RCH_2CH(CO_2H)C_4H_6^{+} \longrightarrow RCH_2.\underset{|}{CH}.CH_2.\underset{|}{CH}.CH{=}CH_2 + H^{+}$$

(with CO——O lactone bridge between the two CH groups)

(LXXXVII)

testa and Caronna, 1968). Similarly, the reactions involving styrene or α-methylstyrene instead of butadiene give rise to products (LXXXVIII) and (LXXXIX) in 70–80% yields (Minisci, (Galli,) Cecere, Malatesta and Caronna, 1968).

$$RCH_2.CHX.CH_2CR'Ph(OMe) \quad (LXXXVIII; R'=H, Me)$$

$$RCH_2.CHCH_2.CR'Ph \text{ (CO—O lactone)} \quad (LXXXIX; R'=H, Me)$$

Alkylation of benzoquinone results in a dialkylated derivative (XC) (Minisci, Galli, Cecere, Malatesta and Caronna, 1968). Addition of

$$O{=}C_6H_4{=}O + 2R\cdot \longrightarrow O{=}C_6H_2R_2{=}O \quad (XC)$$

butadiene does not alter this result, indicating a high reactivity of radical R· towards benzoquinone (Minisci, Galli, Cecere, Malatesta and Caronna, 1968).

The reaction of R· with biacetyl indicates the nucleophilic nature of R· radicals (Minisci, Galli, Cecere, Malatesta and Caronna, 1968).

$$R\cdot + MeCO.CO.Me \longrightarrow RMeC(O\cdot).CO.Me \longrightarrow RCO.Me + Me\dot{C}O$$

In the presence of aromatic substrates, such as pyridine, pyrazine, quinoline, isoquinoline, 2-aminopyrimidine, benzothiazole, and acridine, homolytic aromatic substitution occurs, often in good yield. However, benzene, chlorobenzene, dimethylaniline sulphate, and naphthalene showed no significant reactions (Minisci, Galli, Cecere, Malatesta and Caronna, 1968; Minisci, Galli, Malatesta and Caronna, 1970).

The following mechanism was proposed for this substitution reaction (Minisci, Galli, Cecere, Malatesta and Caronna, 1968; Minisci, Galli, Malatesta and Caronna, 1970).

$$C_5H_5\overset{+}{N}H + R\cdot \longrightarrow [R\text{-}C_5H_5\overset{+}{N}H]^{\cdot} \xrightarrow{Fe^{3+}} R\text{-}C_5H_4\overset{+}{N}H + Fe^{2+} + H^+$$

The pattern of selectivity and orientation seems to suggest a nucleophilic character in the reagent responsible for this radical alkylation (Minisci, Galli, Cecere, Malatesta and Caronna, 1968; Minisci, Galli, Malatesta and Caronna, 1970). Specifically, the reaction of pyridine produced 2- and 4-isomers, i.e., (XCI) and (XCII), and a dimer (XCIII) (Minisci, Galli, Cecere, Malatesta and Caronna, 1968; Minisci, Galli, Malatesta and Caronna, 1970).

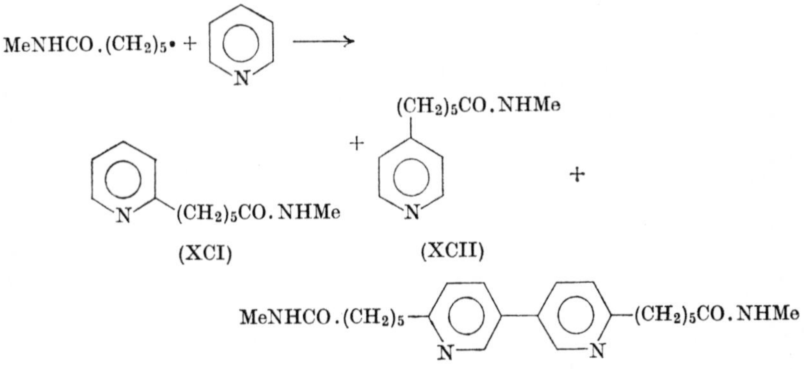

$MeNHCO.(CH_2)_5\cdot$ + N ⟶

N $(CH_2)_5CO.NHMe$ (XCI) + $(CH_2)_5CO.NHMe$ N (XCII) +

$MeNHCO.(CH_2)_5$— N N —$(CH_2)_5CO.NHMe$

(XCIII)

Similarly, the reaction with quinoline yielded 2-(XCIV) and 4-(XCV) isomers (Minisci, Galli, Cecere, Malatesta and Caronna, 1968; Minisci, Galli, Malatesta and Caronna, 1970).

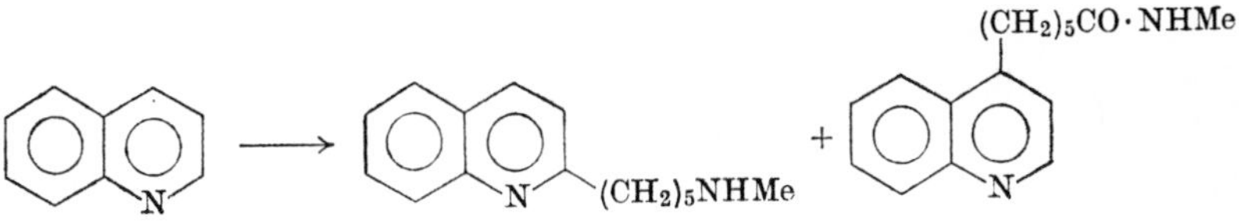

N ⟶ N $(CH_2)_5NHMe$ (XCIV) + $(CH_2)_5CO\cdot NHMe$ N (XCV)

Pyrazine gave the 2-isomer and a dimer (XCVI) (Minisci, Galli, Cecere, Malatesta and Caronna, 1968), while isoquinoline gave the 1-isomer and the dimer (XCVII) (Minisci, Galli, Cecere, Malatesta

R— N N N N —R

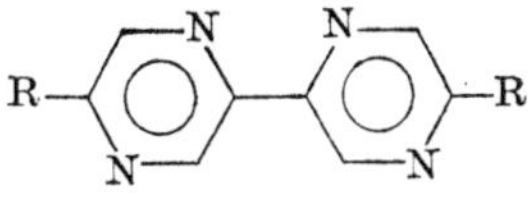

(XCVI)

and Caronna, 1968). Similar reactions of 2-aminopyrimidine, benzothiazole and acridine resulted in 4-, 2-, and 9-isomers, respectively

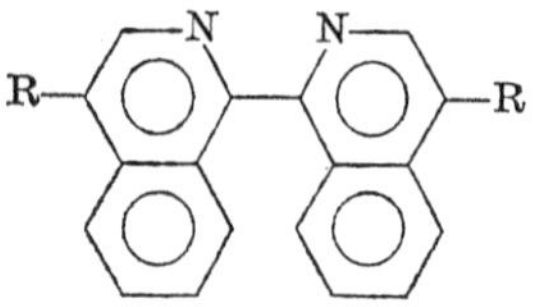

(XCVII)

(Minisci, Galli, Cecere, Malatesta and Caronna, 1968; Minisci, Galli, Malatesta and Caronna, 1970).

Mechanisms for the decomposition of oxaziridine systems have frequently been written as proceeding through an oxygen radical (XCVIII) (Emmons, 1957; Schmitz, 1963). However, Minisci and

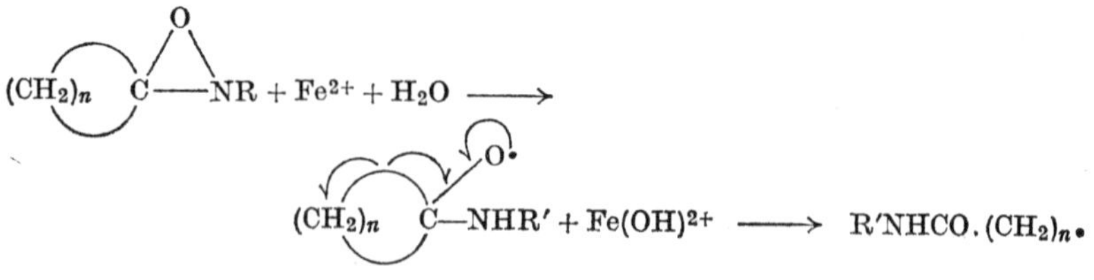

$$(CH_2)_n\text{C(O)NR} + Fe^{2+} + H_2O \longrightarrow (CH_2)_n\text{C(O}\cdot\text{)—NHR}' + Fe(OH)^{2+} \longrightarrow R'NHCO.(CH_2)_n\cdot$$

(XCVIII)

co-workers have argued that a mechanism involving nitrogen radicals (XCIV) is more plausible (Minisci, Cecere, and Galli, 1968a, 1968b; Minisci, Galli, Cecere, Malatesta and Caronna, 1968).

$$(CH_2)_n\text{C(O)NR}' + Fe^{2+} + H_2O \longrightarrow (CH_2)_n\text{C(OH)N}^{\cdot}\text{R}' + Fe(OH)^{2+} \longrightarrow$$

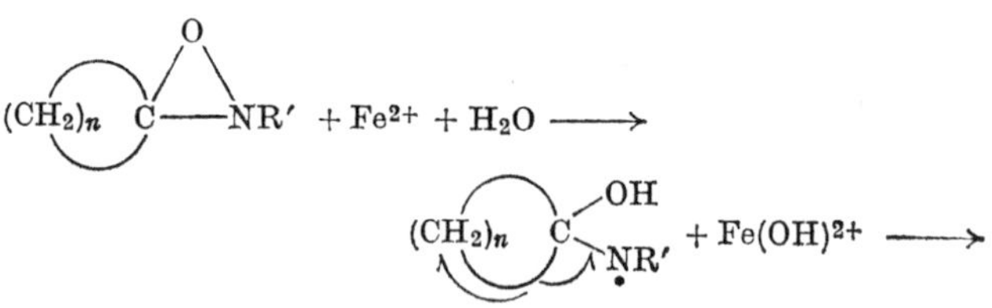

(XCIX)

$$R'N{=}C(OH)(CH_2)_n\cdot \longrightarrow R'NHCO.(CH_2)_n\cdot$$

On the basis of products of decomposition of oxaziridines used so far, no decision is possible between these two alternatives since identical products would be formed (Minisci, Galli, Malatesta and Caronna, 1970).

To prove their hypothesis, Minisci and co-workers investigated the decomposition of 2-phenylethyl-3,3-dimethyl-oxaziridine (C). In this case two redox processes can be envisaged to give either radical (CI) or (CII). By β-scission the alkoxy-radical (CI) could then give a

$$\underset{\text{(C)}}{\mathrm{Me_2C(-O-)NCH_2CH_2Ph}} \xrightarrow[H_2O]{Fe^{2+}} \underset{\text{(CI)}}{\mathrm{Me_2C(O\cdot)-NHCH_2CH_2Ph}} \quad \text{or} \quad \underset{\text{(CII)}}{\mathrm{Me_2C(OH)-\dot{N}CH_2CH_2Ph}}$$

methyl radical and the amide (CIII), whereas the amino-radical could give either a methyl radical or, more probably, because of greater stability, a benzyl radical.

$$\underset{\text{(CI)}}{\mathrm{Me_2C(O\cdot)-NHCH_2CH_2Ph}} \longrightarrow \mathrm{Me\cdot} + \underset{\text{(CIII)}}{\mathrm{MeCO.NHCH_2CH_2Ph}}$$

Because the decomposition of (C) in the presence of a mixture of ferrous sulphate and ferric chloride yielded benzyl chloride,

$$\mathrm{PhCH_2\cdot + FeCl^{2+} \longrightarrow PhCH_2Cl + Fe^{2+}}$$

it was concluded that nitrogen radicals are involved (Minisci, Galli, Malatesta and Caronna, 1970). Further support for this mechanism is based on the following considerations: (1) structural analogy of hydroxylamine-titanous ion system; (2) electron transfer on N—O bond; (3) behaviour of oxaziridines in the presence of cuprous and titanous salts (Minisci, Cecere and Galli, 1968a).

G. Dealkylation of Tertiary Amine Oxides

In 1955 an interesting dealkylation reaction of tertiary amine oxides (CIV) in the presence of ferrous ions was reported (Fish, Johnson, Lawrence and Horning, 1955). This reaction usually results

TABLE 6

Dealkylation of tertiary amine oxides

N-oxide	Initiator	Products	Yield %	Ref.
$Me_3N \rightarrow O$	$FeSO_4$	CH_2O	65	1, 2, 3, 4
	$Fe(NO_3)_3$	Me_2NH (60%), Me_3N (30%)		
	Cu(I) acetate	CH_2O	54	
$Me_2N.CH_2.CO_2H$ ↓ O	$Fe(NO_3)_3$	CH_2O, Me_2NH, CO_2, $OHC.CO_2H$, $MeNH.CH_2.CO_2H$	—	5, 6
$Et_3N \rightarrow O$	$FeSO_4$	$MeCHO$	25	3
Bu^nNMe_2 ↓ O	$FeSO_4$	CH_2O, Bu^nNMe_2, Bu^nNHMe	37	3
Bu^n_2NMe ↓ O	$FeSO_4$	Pr^nCHO CH_2O	25	3
$Bu^n_3N \rightarrow O$	$FeSO_4$	Pr^nCHO Bu^n_3N, Bu^n_2NH	25	3
Pr^iNMe_2 ↓ O	$FeSO_4$	CH_2O	50	3
Bu^tNMe_2 ↓ O	$FeSO_4$	CH_2O	42	3
$PhCH_2NMe_2$ ↓ O	$FeSO_4$	CH_2O		3, 7
	$Fe(NO_3)_3$	$PhCH_2NMe_2$, $PhCH_2NHMe$		
		$PhCHO$	85	

$p\text{-}NO_2.C_6H_4CH_2NMe_2(\rightarrow O)$	$Fe(NO_3)_3$	CH_2O, $HNMe_2$		7
		$p\text{-}NO_2.C_6H_4.CHO$	35	
4-Picoline oxide	$FeSO_4$	4-Picoline	—	3
		4-Picoline oxide	—	
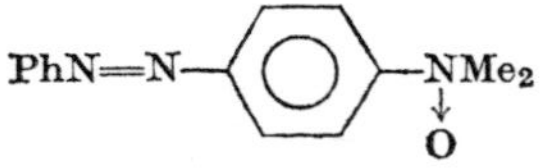 $PhN{=}N\text{-}C_6H_4\text{-}NMe_2(\rightarrow O)$	$Fe(NH_4)_2(SO_4)_2$	CH_2O	—	8, 3
		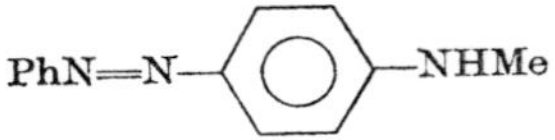 $PhN{=}N\text{-}C_6H_4\text{-}NHMe$	—	
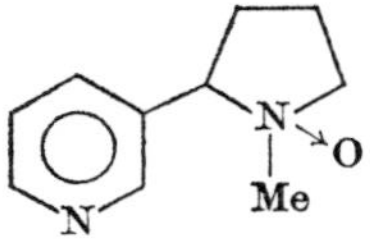 3-pyridyl-pyrrolidine, N(→O)–Me		CH_2O	72	3, 9
		3-pyridyl-pyrrolidine, N–H	77	
		3-pyridyl-pyrrolidine, N–Me	4	
		3-pyridyl-pyrroline, N–Me	2·9	

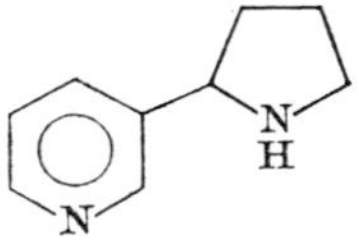

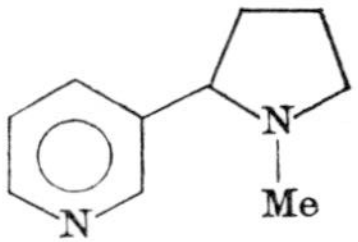

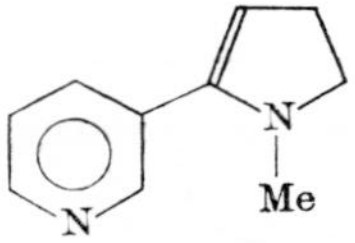

TABLE 6—*continued*

N-oxide	Initiator	Products	Yield %	Ref.
$Me_2CH.CH(NMe_2).CO_2H$ ↓ O	Fe(III) tartrate	$Me_2CH.CH(NHMe).CO_2H$	—	10
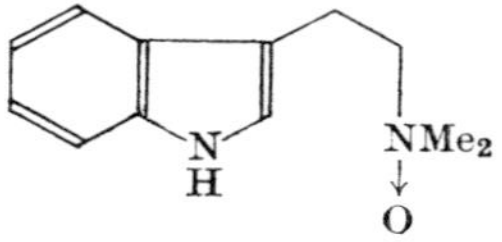 N H; NMe_2 ↓ O	$FeSO_4$ Fe(III) tartrate $Fe(NO_3)_3$	CH_2O	—	11, 10, 12
		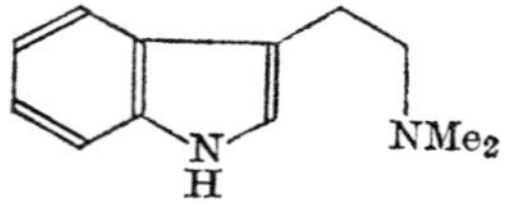 N H; NMe_2	—	
		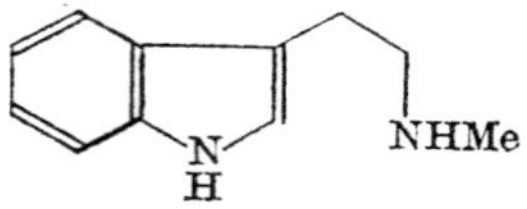 N H; $NHMe$	—	
		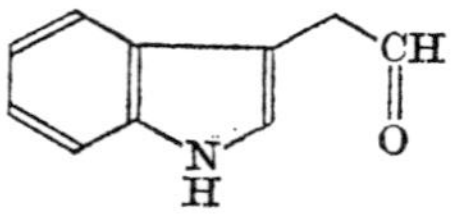 N H; CH ‖ O	—	

MeO-indole (N-H) -CH₂CH₂-$NMe_2 \rightarrow O$	$FeSO_4$	CH_2O	—	12
		MeO-indole (N-H) -CH₂CH₂-NMe_2	—	
		MeO-indole (N-H) -CH₂CH₂-HNMe	—	
		MeO-tetrahydro-β-carboline (N-H), NMe	—	

References: (1) Ferris and Gerwe (1964). (2) Ferris, Gerwe and Gapski (1967). (3) Ferris, Gerwe and Gapski (1968). (4) Craig, Dwyer, Glazer and Horning (1961). (5) Sweeley and Horning (1957). (6) Fish, Sweeley and Horning (1957). (7) Craig, Mary and Wolf (1964). (8) Ishidate and Hanaki (1961). (9) Craig, Mary, Goldman and Wolf (1964). (10) Fish, Sweeley, Johnson, Lawrence and Horning (1956). (11) Fish, Sweeley and Horning (1956). (12) Ghosal and Mukherjee (1966).

in the formation of aldehydes and secondary and tertiary amines as major products (Fish, Sweeley and Horning, 1956; Sweeley and Horning, 1957; Craig, Mary and Wolf, 1964; Ghosal and Mukherjee, 1966; Ferris and Gerwe, 1964).

$$\underset{\text{(CIV)}}{\underset{\substack{\downarrow \\ \text{O}}}{\text{RCH}_2\text{NR}_2}} + \text{Fe}^{2+} \longrightarrow \text{RCHO} + \text{R}_2\text{NH} + \text{RCH}_2\text{NR}_2 + \text{Fe}^{3+}$$

Most of the results obtained to date are summarised in Table 6.

The method can be used for preparation of aldehydes (Craig, Mary and Wolf, 1964; Ferris and Gerwe, 1964; Ferris, Gerwe and Gapski, 1968). The reaction proceeds in non-polar solvents, and is catalysed by iron ions, and to some degree by other metal ions, such as copper ions (Craig, Dwyer, Glazer and Horning, 1961; Fish, Sweeley and Horning, 1956). Although in the earlier work mixtures of ferric salts with oxalic, tartaric and other chelating agents were used, it has since been established that these systems were generating ferrous ions which are actually the initiating species (Ferris, Gerwe and Gapski, 1967; Ferris, Gerwe and Gapski, 1968). The reaction rate is dependent on the type of anion used, e.g. it decreases in the order $PO_4^{3-} > SO_4^{2-} > Cl^- > ClO_4^-$ (Ferris and Gerwe, 1964; Ferris, Gerwe and Gapski, 1968). The ease of aldehyde formation in acid media is in the following order: $PhCH_2 > Me > RCH_2$, R_2CH (Ferris, Gerwe and Gapski, 1968). The selectivity of aldehyde formation from a mixed N-oxide decreases with increasing pH values (Ferris, Gerwe and Gapski, 1968), and the formation of aldehydes is also dependent on the number of protons which are located on the carbon adjacent to the nitrogen atom. Thus, the loss of α-protons determines which carbonyl compound is formed (Ferris, Gerwe and Gapski, 1968). The reaction rate decreases as the steric hindrance of groups in the *N*-oxide increases (Ferris, Gerwe and Gapski, 1968).

The hypothesis that the reaction proceeds *via* aminium radical ions is supported by experimental evidence. Thus, the presence of radicals is demonstrated by conversion of benzyl alcohol or cyclohexanol added to the reaction mixture into benzaldehyde and cyclohexanone, respectively (Ferris and Gerwe, 1964; Ferris, Gerwe and Gapski, 1967; Ferris, Gerwe and Gapski, 1968), by polymerisation of ethyl methacrylate (Ferris and Gerwe, 1964), and by the initiation of the reaction with *N*-chloro-di-n-butylamine, *N*-chloropiperidine and benzoyl peroxide (Ferris, Gerwe and Gapski, 1967). The presence of aminium radicals can be shown by intercepting them with butadiene to give high yields of quaternary amine adducts derived from a mixture of intermediate radicals (CV) and (CVI) (Ferris, Gerwe and Gapski, 1967).

$$Me_3\overset{+\bullet}{N} + CH_2{=}CH.CH{=}CH_2 \longrightarrow Me_3\overset{+}{N}CH_2CH{=}CHCH_2\bullet + Me_3\overset{+}{N}CH_2\dot{C}HCH{=}CH_2$$

(CV) (CVI)

The following mechanism accounts for all these observations (Ferris and Gerwe, 1964; Ferris, Gerwe and Gapski, 1967; Ferris, Gerwe and Gapski, 1968).

$$R_2\underset{\downarrow O}{N}CH_2R + H^+ \longrightarrow R_2\underset{|OH}{\overset{+}{N}}CH_2R$$

$$R_2\underset{|OH}{\overset{+}{N}}CH_2R + Fe^{2+} + H^+ \longrightarrow R_2\overset{+\bullet}{N}CH_2R + H_2O + Fe^{3+}$$

$$R_2\overset{+\bullet}{N}CH_2R \longrightarrow R_2\overset{\bullet\bullet}{N}\dot{C}HR + H^+$$

$$R_2\overset{\bullet\bullet}{N}\dot{C}HR + Fe^{3+} \longrightarrow R_2\overset{+}{N}{=}CHR + Fe^{2+}$$

$$R_2\overset{+}{N}{=}CHR + H_2O \longrightarrow R_2\overset{+}{N}H_2 + RCHO$$

$$R_2\overset{+\bullet}{N}CH_2 + Fe^{2+} \longrightarrow R_2\overset{\bullet\bullet}{N}CH_2R + Fe^{3+}$$

$$R_2\overset{\bullet\bullet}{N}\dot{C}HR + Fe^{2+} + H^+ \longrightarrow R_2\overset{\bullet\bullet}{N}CH_2R + Fe^{3+}$$

Support for this type of mechanism is found in studies involving dealkylation of amines by chlorine dioxide. For example, the reaction of triethylamine gives diethylamine, acetaldehyde, hydrogen ion and chlorite ion (Rosenblatt *et al.*, 1963, 1967; Hull *et al.*, 1967).

$$Et_3N + ClO_2 + H_2O \longrightarrow MeCHO + Et_2NH + 2H^+ + 2ClO_2^-$$

Although no metal ions are involved in these reactions, results of isotope effect and pH effect studies are similar to those observed in the degradation of *N*-oxides, which suggests the formation of a common intermediate in both reactions and the α-hydrogen loss from the aminium radical ion (Rosenblatt *et al.*, 1963; 1967; Hull *et al.*, 1967).

$$R_2\ddot{N}CH_2R + ClO_2 \longrightarrow R_2\overset{+\cdot}{N}CH_2R + ClO_2^-$$

$$R_2\overset{+\cdot}{N}CH_2R \longrightarrow R_2\ddot{N}\dot{C}HR + H^+$$

$$R_2\ddot{N}\dot{C}HR + ClO_2 \longrightarrow R_2\overset{+}{N}{=}CHR + ClO_2^-$$

$$R_2\overset{+}{N}{=}CHR + H_2O \longrightarrow RCHO + R_2NH$$

Tertiary amine oxides are known as components of plant and animal metabolism, and it has been suggested that they act as intermediates in biological oxidative dealkylation processes (Fish, Johnson, Lawrence and Horning, 1955; Fish, Sweeley, Johnson, Lawrence and Horning, 1956; Fish, Sweeley and Horning, 1956). In most cases, the dealkylation reactions have been studied with the aim of comparing the dealkylation of simple model systems with biological systems (Ferris and Gerwe, 1964). Therefore, the dealkylation of a series of biologically important compounds has been investigated. Thus, the reaction of dimethylglycine oxide (CVII) was studied at various pH ranges. At pH 4–6 the reaction gave formaldehyde, dimethylamine, glyoxalic acid and sarcosine (Sweeley and Horning, 1957; Fish, Sweeley and Horning, 1956), whereas at pH 8 decarboxylation also

$$Me_2\underset{\downarrow \atop O}{N}.CH_2.CO_2H + Fe^{2+} \longrightarrow HOCH_2.NMe.CH_2.CO_2H \rightarrow CH_2O + MeNH.CH_2.CO_2H$$

(CVII)

$$Me_2N.CH_2.CO_2H + Fe^{2+} \longrightarrow Me_2N.CHOH.CO_2H \rightarrow Me_2NH + OHC.CO_2H$$

occurred, and was interpreted by mechanisms (*a*) (Sweeley and Horning, 1957) and (*b*) (Ferris, Gerwe and Gapski, 1968).

$$(a)\quad Me_2\underset{|\atop OH}{\overset{+}{N}}CH_2{-}CO.O^- \rightarrow HO^- + Me_2\overset{+}{N}{=}CH_2 + CO_2 \xrightarrow{H_2O} Me_2NH + CH_2O$$

$$(b)\quad Me_2\underset{|\atop OH}{\overset{+}{N}}CH_2CO.O^- + Fe^{2+} \rightarrow Me_2\dot{N}CH_2{-}CO.O^- \rightarrow Me_2\ddot{N}\dot{C}H_2 + CO_2$$

$$Me_2\ddot{N}\dot{C}H_2 + Fe^{3+} \rightarrow Me_2\overset{+}{N}{=}CH_2 \xrightarrow{H_2O} Me_2\overset{+}{N}H_2 + CH_2O$$

The reaction of tyrosine oxide with ferric tartrate complex at pH 5–7 at 38° yielded *N*-methyltyrosine (Fish, Sweeley, Johnson,

Lawrence and Horning, 1956). A similar reaction with *N*,*N*-dimethyltryptamine oxide (CVIII, R = H) with ferrous sulphate gave formaldehyde, formic acid, *N*-methyltryptamine (CIX) and indole-3-acetaldehyde (CX) (Fish, Johnson and Horning, 1956; Fish, Sweeley, Johnson, Lawrence and Horning, 1956; Ghosal and Mukherjee, 1966).

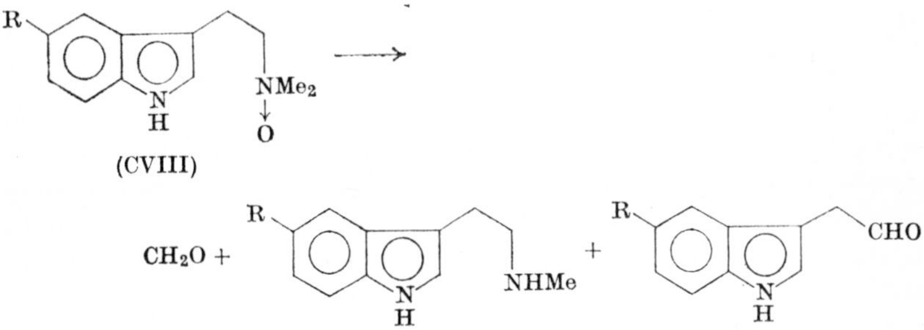

(CVIII)

CH_2O + (CIX) + (CX)

An analogous reaction of 5-methoxy-*N*,*N*-dimethyltryptamine oxide (CVIII; R = MeO) resulted in (CIX; R = MeO) and (CX; R = MeO), together with 6-methoxy-2-methyl-1,2,3,4-tetrahydro-β-carboline (CXI) (Ghosal and Mukherjee, 1966).

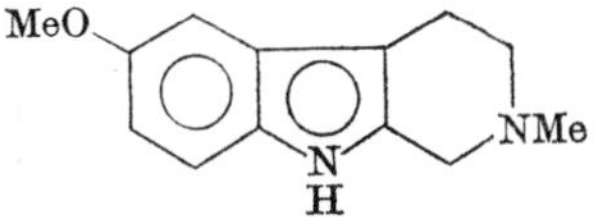

(CXI)

Similar reactions were performed with bufotenine and hordenine oxides (Fish, Sweeley and Horning, 1956). Ferrous ion catalysed conversions of pyrrolizidine alkaloid *N*-oxides to pyrrole derivatives were also achieved. Thus, the reaction of retrorsine *N*-oxide in the presence of ferrous sulphate gave 15% of a pyrrole derivative. Similar reactions were achieved with monocrotaline, heliotrine, lasiocarpine and retronecine (Mattocks, 1968).

A phosphate-buffered system of iron chelate of EDTA, ascorbic acid and oxygen at pH 7·3 and 40° demethylated 4-*N*,*N*-dimethylaminoazobenzene (DAB) (CXII) to give 4-*N*-methylaminoazobenzene (MAB) (CXIII) (Ishidate and Hanaki, 1961).

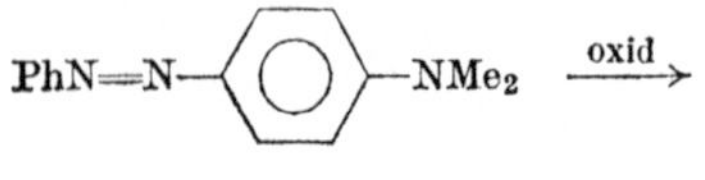

PhN=N–C₆H₄–NMe₂ $\xrightarrow{\text{oxid}}$

(CXII)

PhN=N–C₆H₄–NMe₂(→O) ⟶ PhN=N–C₆H₄–N(Me)(H) + CH_2O

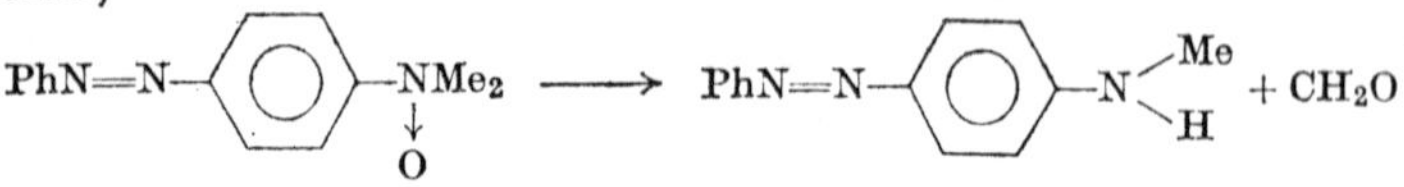

(CXIII)

The oxygen is needed to convert the amine to amine oxide. It is implied that carcinogens such as DAB are also demethylated in biological systems through an *N*-oxide intermediate. An enzyme in mouse liver homogenate will also catalyse this type of reaction (Ishidate and Hanaki, 1961). Analogous compounds such as 4′-methyl-DAB, 3′-methyl-DAB, 2′-methyl-DAB, 2-methyl-DAB and 3-methyl-DAB also underwent this type of dealkylation. Rate constants were compared and correlated with carcinogenicity (Ishidate and Hanaki, 1961).

The reaction of nicotine oxide (CXIV) gave formaldehyde and three products, namely (CXV), (CXVI) and (CXVII) which are formed directly in the rearrangement process. In addition, products (CXVIII), (CXIX) and (CXX) were formed by subsequent oxidation processes (Craig, Mary, Goldman and Wolf, 1964; Ferris, Gerwe and Gapski, 1968).

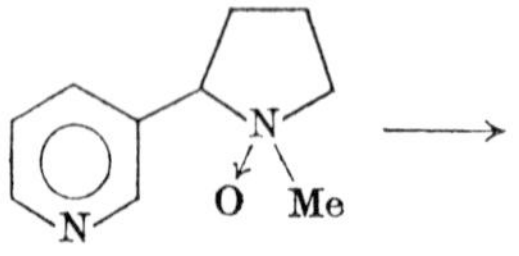

(CXIV) ⟶

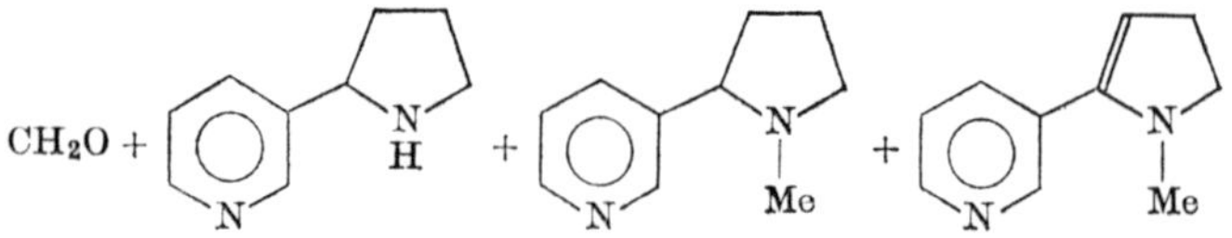

CH_2O + (CXV) + (CXVI) + (CXVII)

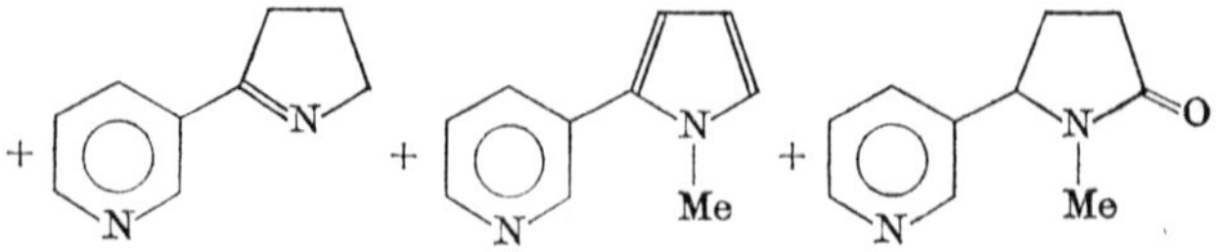

+ (CXVIII) + (CXIX) + (CXX)

We thank all of the authors who sent us their reprints. In particular, we are grateful to Dr. F. Minisci for his numerous contributions.

References

Albisetti, C. J., Coffman, D. D., Hoover, F. W., Jenner, E. L. and Mochel, W. E. (1959) *J. Amer. Chem. Soc.*, **81**, 1489.
Anderson, J. H. (1964) *Analyst*, **89**, 357.
Anderson, J. H. (1966) *Analyst*, **91**, 532.
Beckwith, A. L. J. and Goodrich, J. E. (1965) *Austral. J. Chem.*, **18**, 747.
Bernardi, R., Galli, R. and Minisci, F. (1968) *J. Chem. Soc. (B)*, 324.
Bock, H. and Kompa, K.-L. (1965) *Angew. Chem. Internat. Edn.*, **4**, 783.
Bock, H. and Kompa, K.-L. (1966a) *Chem. Ber.*, **99**, 1347.
Bock, H. and Kompa, K.-L. (1966b) *Chem. Ber.*, **99**, 1357.
Bock, H. and Kompa, K.-L. (1966c) *Chem. Ber.*, **99**, 1361.
Brikun, I. K. and Kozlovskii, M. T. (1965) *Izvest. Akad. Nauk Kazakh.* S.S.R. *Ser. Khim.*, **15**, 18; *Chem. Abstr.*, 1966, **64**, 10743d.
Bro, M. I. and Scheyer, R. C. (1962) *U.S. Pat.* 3,032,543; *Chem. Abstr.*, **57**, 1962, 2426i.
Brown, L. L. and Drury, J. S. (1967) *J. Chem. Phys.*, **46**, 2833.
Calusaru, A. (1968) *Isotopenpraxis*, **4**, 401.
Coffman, D. D. and Jenner, E. L. (1957) *Canad. Pat.* 542,162.
Corey, E. J. and Hertler, W. R. (1960) *J. Amer. Chem. Soc.*, **82**, 1657.
Corvaja, C., Fischer, H. and Giacometti, G. (1965) *Z. Phys. Chem. (Frankfurt)*, **45**, 1.
Craig, J. C., Dwyer, F. P., Glazer, A. N. and Horning, E. C. (1961) *J. Amer. Chem. Soc.*, **83**, 1871.
Craig, J. C., Mary, N. Y., Goldman, N. L. and Wolf, L. (1964) *J. Amer. Chem. Soc.*, **86**, 3866.
Craig, J. C., Mary, N. Y. and Wolf, L. (1964) *J. Org. Chem.*, **29**, 2868.
Davies, G. and Kustin, K. (1969) *Inorg. Chem.*, **8**, 484.
Davis, P., Evans, M. G. and Higginson, W. C. E. (1951) *J. Chem. Soc.*, 2563.
Dewing, J., Longster, G. F., Myatt, J. and Todd, P. T. (1965) *Chem. Comm.*, 391.
Edwards, O. E., Vocelle, C., ApSimon, J. W. and Haque, F. (1965) *J. Amer. Chem. Soc.*, **87**, 678.
Emmons, W. D. (1957) *J. Amer. Chem. Soc.*, **79**, 5739.
Ferris, J. P. and Gerwe, R. (1964) *Tetrahedron Letters*, 1613.
Ferris, J. P., Gerwe, R. D. and Gapski, G. R. (1967) *J. Amer. Chem. Soc.*, **89**, 5270.
Ferris, J. P., Gerwe, R. D. and Gapski, G. R. (1968) *J. Org. Chem.*, **33**, 3493.
Fischer, K. A. and Pilgram, K. (1961) *Germ. Pat.* 1,119,507; *Chem. Abstr.*, 1962, **57**, 2427a.
Fish, M. S., Johnson, N. M. and Horning, E. C. (1956) *J. Amer. Chem. Soc.*, **78**, 3668.
Fish, M. S., Johnson, N. M., Lawrence, E. P. and Horning, E. C. (1955) *Biochim. Biophys. Acta*, **18**, 564.
Fish, M. S., Sweeley, C. C. and Horning, E. C. (1956) *Chem. and Ind. Brit. Inds. Fair Rev., Apr.*, R24.

Fish, M. S., Sweeley, C. C., Johnson, N. M., Lawrence, E. P. and Horning, E. C. (1956) *Biochim. Biophys. Acta*, **21**, 196.

Frank, C. E., Mador, I. L., Rekers, L. J. and Tonne, C. D. (1960) *Germ. Pat.* 1,082,593; *Chem. Abstr.*, 1961, **55**, 25805d.

Gadzhiev, T. A., Mamedov, Z. F. and Gabelaya, K. A. (1969) *Azerb. khim. Zhur*, **89**; *Chem. Abstr.*, 1970, **72**, 31127n.

Ghosal, S. and Mukherjee, B. (1966) *J. Org. Chem.*, **31**, 2284.

Griffiths, W. E., Longster, G. F., Myatt, J. and Todd, P. F. (1967) *J. Chem. Soc. (B)*, 530.

Gutch, C. J. W. and Waters, W. A. (1964) *Proc. Chem. Soc.*, 230.

Gutch, C. J. W. and Waters, W. A. (1965) *J. Chem. Soc.*, 751.

Hlasivcova, N., Novak, J. and Zyka, J. (1967) *Coll. Czech. Chem. Comm.*, **32**, 4410.

Hobson, J. D. and Riddell, W. D. (1968) *Chem. Comm.*, 1178.

Hofmann, A. W. (1883) *Ber.*, **16**, 558, 586.

Hoover, F. W. (1961) *U.S. Pat.* 2,983,758; *Chem. Abstr.*, 1961, **55**, 21050i.

Howard, E. (1951) *U.S. Pat.* 2,567,109; *Chem. Abstr.*, 1952, **46**, 783c.

Howard, E. (1954) *U.S. Pat.* 2,683,140; *Chem. Abstr.*, 1954, **48**, 13273i.

Hull, L. A., Davis, G. T., Rosenblatt, D. H., Williams, H. K. R. and Weglein, R. C. (1967) *J. Amer. Chem. Soc.*, **89**, 1163.

Huyser, E. S. (1970) *Synthesis*, 7 (and refs. therein).

Ishidate, M. and Hanaki, A. (1961) *Nature*, **191**, 1198.

Jijee, K. and Santappa, M. (1969) *Proc. Indian Acad. Sci. (A)*, **69**, 117.

Kessar, S. V., Rampal, A. L. and Mahajan, K. P. (1962) *J. Chem. Soc.*, 4703.

Kovacic, P., Lowery, M. K. and Field, K. W. (1970) *Chem. Rev.*, **70**, 639.

Lingane, P. J. and Christie, J. H. (1967) *J. Electroanal. Chem.*, **13**, 227.

Löffler, K. and Freytag, C. (1909) *Ber.* **42**, 3427.

Löffler, K. and Freytag, C. (1910) *Ber.*, **43**, 2035.

Mador, I. L. and Rekers, L. J. (1962) *Germ. Pat.* 1,132,122; *Chem. Abstr.*, 1963, **58**, 2369.

Mattocks, A. R. (1968) *Nature*, **219**, 480.

Minisci, F. (1964) *Chimica e Industria*, **46**, 57.

Minisci, F. (1967) *Chimica e Industria*, **49**, 705.

Minisci, F. (1968) *Corsi Semin. Chim.*, **18**; *Chem. Abstr.*, 1970, **72**, 11710u.

Minisci, F., Bernardi, R., Grippa, L. and Trabucchi, V. (1966) *Chimica e Industria*, **48**, 264.

Minisci, F., Bernardi, R., Trabucchi, V. and Grippa, L. (1966) *Chimica e Industria*, **48**, 484.

Minisci, F. and Cecere, M. (1967) *Chimica e Industria*, **49**, 1333.

Minisci, F., Cecere, M. and Galli, R. (1968a) *Chimica e Industria*, **50**, 225.

Minisci, F., Cecere, M. and Galli, R. (1968b) *Gazzetta*, **98**, 358.

Minisci, F. and Galli, R. (1963) *Chimica e Industria*, **45**, 1400.

Minisci, F. and Galli, R. (1964a) *Tetrahedron Letters*, 167.

Minisci, F. and Galli, R. (1964b) *Chimica e Industria*, **46**, 173.

Minisci, F. and Galli, R. (1964c) *Chimica e Industria*, **46**, 546.

Minisci, F. and Galli, R. (1964d) *Tetrahedron Letters*, 3197.

Minisci, F. and Galli, R. (1965a) *Tetrahedron Letters*, 433.

Minisci, F. and Galli, R. (1965b) *Tetrahedron Letters*, 1679.

Minisci, F. and Galli, R. (1966a) *Tetrahedron Letters*, 2531.
Minisci, F. and Galli, R. (1966b) *Chimica e Industria*, **48**, 268.
Minisci, F., Galli, R. and Bernardi, R. (1966) *Tetrahedron Letters*, 699.
Minisci, F., Galli, R. and Bernardi, R. (1967a) *Chimica e Industria*, **49**, 594.
Minisci, F., Galli, R. and Bernardi, R. (1967b) *Chem. Comm.*, 903.
Minisci, F., Galli, R., Bernardi, R. and Galli, A. (1967) *Chimica e Industria*, **49**, 387.
Minisci, F., Galli, R., Bernardi, R. and Perchinunno, M. (1968) *Chimica e Industria*, **50**, 328.
Minisci, F., Galli, R., Bernardi, R. and Perchinunno, M. (1969) *Chimica e Industria*, **51**, 280.
Minisci, F., Galli, R. and Cecere, M. (1965) *Tetrahedron Letters*, 4663.
Minisci, F., Galli, R. and Cecere, M. (1966a) *Chimica e Industria*, **48**, 132.
Minisci, F., Galli, R. and Cecere, M. (1966b) *Chimica e Industria*, **48**, 347.
Minisci, F., Galli, R. and Cecere, M. (1966c) *Chimica e Industria*, **48**, 725.
Minisci, F., Galli, R. and Cecere, M. (1966d) *Tetrahedron Letters*, 3163.
Minisci, F., Galli, R. and Cecere, M. (1966e) *Chimica e Industria*, **48**, 1324.
Minisci, F., Galli, R., Cecere, M. and Bernardi, R. (1967) *Chimica e Industria*, **49**, 946.
Minisci, F., Galli, R., Cecere, M., Malatesta, V. and Caronna, T. (1968) *Tetrahedron Letters*, 5609.
Minisci, F., Galli, R., Cecere, M. and Mondelli, R. (1965) *Chimica e Industria*, **47**, 994.
Minisci, F., Galli, R., Cecere, M. and Quilico, A. (1969) *French Addn. Pat.* 93,635; *Chem. Abstr.*, 1970, **72**, 43103g.
Minisci, F., Galli, R., Cecere, M. and Trabucchi, V. (1966) *Chimica e Industria*, **48**, 1147.
Minisci, F., Galli, R., Galli, A. and Bernardi, R. (1967a) *Chimica e Industria*, **49**, 252.
Minisci, F., Galli, R., Galli, A. and Bernardi, R. (1967b) *Tetrahedron Letters*, 2207.
Minisci, F., Galli, R., Malatesta, V. and Caronna, T. (1970) *Tetrahedron*, **26**, 4083.
Minisci, F., Galli, R. and Perchinunno, M. (1969a) *Org. Prep. and Proc.*, **1**, 77.
Minisci, F., Galli, R. and Perchinunno, M. (1969b) *Org. Prep. and Proc.*, **1**, 87.
Minisci, F., Galli, R., Perchinunno, M. and Bernardi, R. (1968) *Chimica e Industria*, **50**, 453.
Minisci, F., Galli, R. and Pollina, G. (1965) *Chimica e Industria*, **47**, 736.
Minisci, F., Galli, R. and Rossetti, M. A. (1967) *Chimica e Industria*, **49**, 947.
Minisci, F., Gardini, G. P. and Bertini, F. (1970) *Canad. J. Chem.*, 48, 544.
Minisci, F., Trabucchi, V. and Galli, R. (1966) *Chimica e Industria*, **48**, 716.
Minisci, F., Trabucchi, V., Galli, R. and Bernardi, R. (1966) *Chimica e Industria*, **48**, 845.
Montecatini Edison, S.p.A. (1967) *Netherlands Pat. Appl.* 6,614,947; *Chem. Abstr.*, 1968, **68**, 68642v.
National Distillers Products Corp. (1957) *Belgian Pat.* 557,730.
Neale, R. S. (1964) *J. Amer. Chem. Soc.*, **86**, 5340.
Neale, R. S. (1967) *J. Org. Chem.*, **32**, 3263.
Neale, R. S. (1971) *Synthesis, Intern. J. Meth. Syn. Org. Chem.*, 1.

Neale, R. S. and Marcus, N. L. (1969) *J. Org. Chem.*, **34**, 1808.
Ogata, Y. and Morimoto, T. (1965) *J. Org. Chem.*, **30**, 597.
Poutsma, M. L. (1969) *Methods in Free-Radical Chemistry*, Ed. Huyser, E. S., Marcel Dekker, New York, **1**, 79.
Rosenblatt, D. H., Hayes, A. J. Jr., Harrison, B. L., Streaty, R. A. and Moore, K. A. (1963) *J. Org. Chem.*, **28**, 2790.
Rosenblatt, D. H., Hull, L. A., DeLuca, D. C., Davis, G. T., Weglein, R. C. and Williams, H. K. R. (1967) *J. Amer. Chem. Soc.*, **89**, 1158.
Schaffner, K., Arigoni, D. and Jeger, O. (1960) *Experientia*, **16**, 169.
Schmidt, W., Swinehart, J. H. and Taube, H. (1968) *Inorg. Chem.*, **7**, 1984.
Schmitz, E. (1963) *Adv. Heterocyclic Chem.*, Eds. Katritzky, A. R. *et al.*, Academic Press, New York and London, **2**, 83.
Schmitz, E. and Murawski, D. (1965) *Chem. Ber.*, **98**, 2525.
Schmitz, E. and Murawski, D. (1966) *Chem. Ber.*, **99**, 1493.
Schmitz, E., Ohme, R. and Murawski, D. (1965) *Chem. Ber.*, **98**, 2516.
Schuster, C., Gehm, R. and Völkl, E. (1958) *Germ. Pat.* 1,039,749; *Chem. Abstr.*, 1960, **54**, 20338a.
Seaman, H., Taylor, P. J. and Waters, W. A. (1954) *J. Chem. Soc.*, 4690.
Sosnovsky, G. and Rawlinson, D. J. (1970a) *Organic Peroxides*, Ed. Swern, D., Wiley-Interscience, New York, **1**, 561.
Sosnovsky, G. and Rawlinson, D. J. (1970b) *Organic Peroxides*, Ed. Swern, D., Wiley-Interscience, New York, **1**, 585.
Sosnovsky, G. and Rawlinson, D. J. (1971a) *Organic Peroxides*, Ed. Swern, D., Wiley-Interscience, New York, **2**, Ch. 2.
Sosnovsky, G. and Rawlinson, D. J. (1971b) *Organic Peroxides*, Ed. Swern, D., Wiley-Interscience, New York, **2**, Ch. 3.
Sosnovsky, G. and Zaret, E. H. (1970) *Organic Peroxides*, Ed. Swern, D., Wiley-Interscience, New York, **1**, 517.
Spanswick, J. and Ingold, K. U. (1970) *Canad. J. Chem.*, **48**, 546.
Surzur, J. M., Stella, L. and Tordo, P. (1970) *Bull. Soc. Chim. France*, 115.
Surzur, J. M., Tordo, P. and Stella, L. (1970) *Bull. Soc. Chim. France*, 111.
Sweeley, C. C. and Horning, E. C. (1957) *J. Amer. Chem. Soc.*, **79**, 2620.
Tanner, D. D. and Mosher, M. W. (1969) *Canad. J. Chem.*, **47**, 715.
Thaler, W. A. (1969) *Methods in Free-Radical Chemistry*, Ed. Huyser, E. S., Marcel Dekker, New York, **2**, 121.
Uri, N. (1952) *Chem. Rev.*, **50**, 404.
Waters, W. A. and Wilson, I. R. (1966) *J. Chem. Soc.* (*A*), 534.
Wawzonek, S. and Thelen, P. J. (1950) *J. Amer. Chem. Soc.*, **72**, 2118.
Wells, C. F. and Salam, M. A. (1967) *Chem. and Ind.*, 2079.
Wolff, M. E. (1963) *Chem. Rev.*, **63**, 55.
Yoshida, Z., Matsumoto, T. and Oda, R. (1962) *Kogyo Kagaku Zasshi*, **65**, 46; *Chem. Abstr.*, 1962, **57**, 16446i.
Yoshida, Z., Matsumoto, T. and Oda, R. (1964a) *Kogyo Kagaku Zasshi*, **67**, 64; *Chem. Abstr.*, 1964, **61**, 1789f.
Yoshida, Z., Matsumoto, T. and Oda, R. (1964b) *Kogyo Kagaku Zasshi*, **67**, 67; *Chem. Abstr.*, 1964, **61**, 1810e.
Yoshida, Z., Matsumoto, T. and Oda, R. (1964c) *Kogyo Kagaku Zasshi*, **67**, 70; *Chem. Abstr.*, 1964, **61**, 1810e.

AUTHOR INDEX

SUBJECT INDEX